Antioxidant Response in Skeletal Muscle

Antioxidant Response in Skeletal Muscle

Guest Editors

Jan Górski

Elzbieta Supruniuk

MDPI

Basel • Beijing • Wuhan • Barcelona • Belgrade • Novi Sad • Cluj • Manchester

Guest Editors
Jan Górski
Faculty of Health Sciences
University of Lomza
Lomza
Poland

Elzbieta Supruniuk
Department of Physiology
Medical University
of Bialystok
Bialystok
Poland

Editorial Office
MDPI AG
Grosspeteranlage 5
4052 Basel, Switzerland

This is a reprint of the Special Issue, published open access by the journal *Antioxidants* (ISSN 2076-3921), freely accessible at: https://www.mdpi.com/journal/antioxidants/special_issues/639M13R0TW.

For citation purposes, cite each article independently as indicated on the article page online and as indicated below:

Lastname, A.A.; Lastname, B.B. Article Title. *Journal Name* **Year**, *Volume Number*, Page Range.

ISBN 978-3-7258-7200-8 (Hbk)
ISBN 978-3-7258-7201-5 (PDF)
https://doi.org/10.3390/books978-3-7258-7201-5

Contents

About the Editors

Jan Górski

Jan Górski is a full professor at the Department of Health Sciences, University of Lomza, Poland, and at the Department of Physiology at the Medical University of Bialystok. He is a prominent Polish physiologist whose research has substantially advanced the understanding of metabolic regulation in skeletal muscle, particularly in the context of exercise and lipid metabolism. His scientific achievements include over 250 peer-reviewed publications devoted to the biochemical, cellular, and systemic mechanisms governing substrate utilization, muscle energy homeostasis, and adaptive responses to physical activity. His work has contributed to elucidating the role of lipid-derived substrates in muscle metabolism and their significance for metabolic health and exercise performance. For many years, he was the head of the Department of Physiology at the Medical University of Bialystok, and he has supervised 31 PhD dissertations. He is also the editor of the academic textbooks: *Physiological Bases of Physical Exercise, Human Physiology*, and *Physiology of Exercise and Physical Training*. He served as Rector of the Medical Academy for a total of twelve years. In recognition of his scientific and organizational achievements, he was awarded the title of Doctor Honoris Causa by the Medical University of Bialystok.

Elzbieta Supruniuk

Elzbieta Supruniuk is an assistant professor at the Department of Physiology at the Medical University of Bialystok in Poland. Her main research interests include studying the mechanisms underlying brain insulin resistance in the progression of obesity and possible strategies to counteract it in rat models. She studies the role of oxidative stress and antioxidants in exercise and various disorders, including cancer. Additionally, she conducts experimental studies in adipose-derived mesenchymal stem cells that focus on metabolic interactions between branched-chain amino acids and lipids. For her studies, she received grants from the National Science Centre in Cracow, Poland. In collaboration with New York Medical College, she works on the role of eicosanoids and 20-HETE/GPR75 in obesity.

Preface

This Reprint is dedicated to providing a comprehensive and critical presentation of current advances in the field of redox biology of skeletal muscle, with particular emphasis on antioxidant defense mechanisms operating under both physiological and pathological conditions. Its scope covers cellular and molecular aspects of redox signaling, adaptive responses to contractile activity and exercise, as well as alterations associated with metabolic diseases, muscle injury, and disuse. The principal aim of this volume is to integrate contemporary experimental and translational findings to provide a coherent framework for understanding how redox homeostasis determines muscle structure, function, and regenerative capacity.

Bringing together contributions representing diverse methodological approaches and research models creates an opportunity to highlight both the complexity of redox-dependent processes and their relevance for clinical practice, exercise physiology, and metabolic health. This Reprint is tailored towards an interdisciplinary readership, including basic scientists, physiologists, biochemists, medical researchers, and clinicians interested in muscle biology and oxidative stress.

Jan Górski and Elzbieta Supruniuk
Guest Editors

antioxidants

MDPI

Editorial

Antioxidant Response in Skeletal Muscle

Elżbieta Supruniuk [1,*] and Jan Górski [2]

[1] Department of Physiology, Medical University of Białystok, 15-222 Białystok, Poland
[2] Department of Medical Sciences, Academy of Applied Sciences, 18-400 Łomza, Poland; jan.gorski@umb.edu.pl
* Correspondence: elzbieta.supruniuk@umb.edu.pl

1. Introduction

Skeletal muscle, which in men accounts for nearly half of the body's mass, stands out as the most adaptable and energetically demanding tissue. In addition to facilitating mobility and maintaining posture, it is indispensable for regulating the body's overall metabolism and communicating with distant organs by releasing myokines. To perform these functions, muscles continuously adjust their structure and biochemical activity, which requires a precisely tuned equilibrium between energy generation, repair mechanisms, and cellular turnover. This remodeling of skeletal muscle is governed by its redox environment, which is determined by the balance between free radical production and the antioxidant systems that regulate it [1].

To control skeletal muscle function and metabolic flexibility, a well-balanced redox network is established, in which endogenous antioxidant systems and exogenous inputs interact across molecular, cellular, systemic, and temporal levels. At moderate concentrations, reactive oxygen and nitrogen species (RONS) provide the modulatory stimulus for cell adaptation, promoting exercise tolerance, mitochondrial remodeling, and stress resistance. A central hub in maintaining this equilibrium is nuclear factor erythroid 2-related factor 2 (Nrf2), which orchestrates the expression of antioxidant and detoxifying enzymes [2]. Oxidative stress occurs when RONS accumulate more rapidly than they can be neutralized. Although these molecules are intrinsic byproducts of normal metabolism, excessive levels can damage cellular constituents, including lipids, proteins, and DNA/RNA. Heightened RONS production and weakened antioxidant defenses are common features of muscle wasting, disuse, and age-related deterioration. Fast-twitch fibers, due to their high metabolic demands and relatively limited protective systems, show particular sensitivity to oxidative challenges [3].

This Special Issue focused on mechanisms by which redox processes accomplished by both endogenous antioxidant systems and exogenous antioxidant inputs govern muscle function. Because skeletal muscle repeatedly endures mechanical load and metabolic demand, redox dynamics are further challenged by factors such as intense or prolonged exercise, periods of inactivity, aging, nutritional variation, and systemic stress. Such influences modify the oxidative environment of muscle and ultimately determine whether redox signals support beneficial adaptive changes. It is essential to investigate the mechanisms by which intrinsic antioxidant defenses interact with applied stressors to maintain redox equilibrium.

 https://doi.org/10.3390/antiox14121445

2. Overview of Published Articles

2.1. Acquired Endogenous Defense Mechanisms: The Muscle's Intrinsic Redox Machinery

Skeletal muscle has developed sophisticated defense mechanisms, involving redox-mediated pathways, that allow it to adapt and even thrive under various insults, including metabolic and mechanical stressors, to protect cells against oxidative stress and resulting inflammation. Pribil Pardun et al. [Contribution 1] showed that contractile activity evoked by electrical stimulation (EPS) activates the Nrf2-dependent antioxidant pathway in C2C12 myotubes. This leads to the activation of thioredoxin and glutathione systems to the greatest extent that persisted for up to 24 h after EPS, and in turn protected the cells against hydrogen peroxide-induced cytotoxicity. Rather than being purely deleterious, moderate ROS production served as a signal that reinforced redox buffering capacity, improving mitochondrial efficiency and resilience, and controlling the degree of inflammation. This illustrates the concept of exercise-induced hormesis, wherein a transient increase in low levels of a stressor (e.g., radicals) exerts a beneficial effect on the cells [4]. However, the implementation of blood flow restriction (BFR) training protocols has added an important cue in adjusting skeletal muscle response to oxidative stress in training. Direct measurements of free radical production confirmed a systemic increase in ROS after BFR exercise associated with local reactive hyperemia, which contributed to increased muscle mass, strength, and performance [5].

The dual role of RONS in exercise was reviewed by Meng et al. [Contribution 2] based on the data from elite soccer players. They emphasized the need to monitor individuals' oxidative balance, training load, as well as physiological and psychological indicators of perceived exertion to personalize effective recovery and nutrition strategies. Antioxidant supplementation has been commonly used to overcome oxidative stress-related muscle soreness and damage. However, the timing and dosing of exogenous antioxidants require careful management to prevent blunting the adaptations that depend on repetitive exposure to redox-signaling activation, which supports force production, improves metabolic flexibility, delays fatigue development, and enhances muscle regenerative capacity. Variability in responses can be caused by genetic polymorphisms in antioxidant genes (such as SOD2, GPX1, or CAT), seasonal fluctuations in buffering capacity, dietary habits, and training history.

Sutton et al. [Contribution 3] highlighted the crosstalk between circadian rhythms and redox homeostasis, especially in the case of muscle aging. Circadian clock, governed by the transcriptional feedback loop of brain and muscle ARNT-like protein 1 (BMAL1) and circadian locomotor output cycles kaput (CLOCK), coordinates daily oscillations in metabolism, mitochondrial function, and antioxidant defenses. For instance, approximately 38% of mitochondrial proteins exhibit diurnal oscillations, while some of them were abolished by a high-fat diet [6]. With advancing age, circadian amplitude diminishes, and the rhythmic expression of redox genes becomes blunted. The authors stated that this desynchronization contributes to the deterioration of skeletal muscle function with age by impairing mitochondrial turnover and reducing resistance to oxidative stress. Moreover, they propose that interventions targeting circadian alignment, involving time-restricted feeding, light exposure, and appropriately timed exercise, may restore redox synchrony and improve muscle health [Contribution 3].

2.2. Environmental and Physiological Stressors: Redox Imbalance as a Common Factor

Spaceflight represents a unique physiological model of extreme disuse and environmental stress, characterized by muscle unloading, fluid shifts, radiation exposure, and altered circadian cues. Blottner et al. [Contribution 4] analyzed skeletal muscle biopsies obtained from astronauts who had been on the International Space Station (ISS), comparing short-duration missions (~9 days) and long-term missions (>180 days). Proteomic analyses showed extensive S-nitrosylation of metabolic, mitochondrial, and contractile proteins, including creatine kinase, glycolytic enzymes, and the tricarboxylic acid (TCA) cycle components. These molecular alterations contribute to muscle atrophy independent of mechanical unloading. Interestingly, while short-duration missions induced a pattern of over-nitrosylation associated with reduced mitochondrial activity and impaired ATP production, long-duration missions showed a partial adaptation with a distinct nitrosoproteomic profile, suggesting that human muscle can adjust redox balance under sustained microgravity conditions. This adaptation, however, appeared incomplete, pointing to the limits of physiological plasticity in the absence of gravitational loading. What is more, following about 180 days of space flight, both the soleus and gastrocnemius muscles exhibited reduced fiber size and force, with the greatest atrophy occurring in the order: soleus type I > soleus type II > gastrocnemius type I > gastrocnemius type II [7]. The above also underlines parallels between muscle disuse due to microgravity or inactivity (bed rest, aging, immobilization). It was demonstrated that antioxidant administration protects from muscle fibrosis via attenuating immobilization-induced activation of profibrogenic transforming growth factor β1 (TGF-β1) signaling [8]. Nevertheless, the direct mechanisms underlying ROS's role in the development of muscle atrophy in disuse settings in humans remain putative.

Among environmental stressors, mycotoxins are widespread contaminants to which humans and animals are exposed via food, inhalation, or contact. Certain mycotoxins can be harmful by interfering with protein synthesis, preventing particle clearance of the lungs, damaging macrophage systems, raising sensitivity to bacterial endotoxins, and also inducing oxidative stress. Due to their heat-stability, complete eradication of mycotoxins is difficult despite the variety of procedures used, including physical separation, heat treatment, cleaning, milling, washing, radiation, biological or chemical agents, and extraction with solvents [9]. One of the most common mycotoxins is zearalenone, which is present in a range of food products, including cereals, dried fruits, and spices [10]. Li et al. [Contribution 5] presented a mechanistic analysis of skeletal muscle injury in response to zearalenone administration to mice. The authors observed that this mycotoxin damages skeletal muscle through excessive ROS generation, mitochondrial dysfunction, and apoptosis. This resulted in reduced myoblast proliferation, differentiation, and slow-to-fast myofiber shift. In the long run, such muscle phenotypic changes could lead to reduced endurance and fatigue resistance, impaired muscle performance, and increased muscle atrophy [11]. Importantly, co-treatment with antioxidants such as glutathione, nicotinamide mononucleotide, or melatonin mitigated these pathological manifestations, restoring redox equilibrium and muscle integrity [Contribution 5].

2.3. *Restoring Redox Balance Through Bioactive Compounds*

Herbal medicine offers potential therapeutic strategies to prevent or reverse muscle atrophy, as explored by Kim et al. [Contribution 6]. The ethanol extract of Jakyak-Gamcho-Tang exerts potent muscle-protective effects by inhibiting C2C12 myotubes degradation when exposed to H_2O_2, dexamethasone, or palmitic acid. The superior beneficial effects of ethanol extract compared to predominantly used water-based extraction were attributed to the higher concentration of key bioactive compounds. The authors identified paeoniflorin, isoliquiritigenin, catechin, glycyrrhizic acid, and glabridin as main substances regulating mitochondrial function, ROS release, and muscle-atrophy-related signaling cascades. Transcriptomic analysis upregulation of genes associated with oxidative phosphorylation, mitochondrial biogenesis, peroxisome proliferator-activated receptor γ coactivator 1α (PGC-1α) signaling, mammalian target of rapamycin complex 1 (mTORC1) signaling, and estrogen-related receptor α (ERRα) targets, while downregulation in muscle-atrophic TGF-β signaling. The study underscores the need for standardization and optimization of herbal medicine to enhance therapeutic efficacy.

In line with those findings, Iannuzzo et al. [Contribution 7] tested the biological activity of *Rhodiola rosea* extracts obtained from wild to controlled cultivation in primary human myocytes. The authors showed that optimization of cultivation conditions resulted in a unique phytochemical profile that strengthened the antioxidant properties, being an essential step for integrating herbal agents into evidence-based antioxidant therapy. *Rhodiola rosea* preserves cellular homeostasis when challenged by oxidative load, possibly underpinning its beneficial effects on muscle performance and mitigating fatigue. The plant's role in growth and repair of muscle was also associated with the stimulation of ATP production and improved mitochondrial respiration. Moreover, the interconnection between antioxidant mechanisms and the upregulation of folate and polyamine pathways in controlled cultures may contribute to improved muscle regeneration after injury or recovery from physical exercise. By regulating redox-sensitive pathways and maintaining cellular homeostasis, *Rhodiola rosea* emerges as a promising nutraceutical for supporting muscle health and combating oxidative stress-related dysfunction [Contribution 7]. These biological properties can be especially useful in sarcopenia management, a disease characterized by accelerated loss of skeletal muscle mass and strength, usually affecting the aging population [12]. Another study by Wojszel et al. [Contribution 8] explored the relationships between redox balance and two muscle-derived myokines implicated in neuro-muscular crosstalk, such as irisin and brain-derived neurotrophic factor (BDNF), in patients over 60 years of age. These molecules participate in exercise-related adaptations by supporting substrate delivery to the contracting muscle. The authors showed that reduced circulating irisin and BDNF levels correlate with impaired antioxidant defenses and elevated oxidative stress markers, which may underlie neuromuscular decline in the elderly. Dynapenia is therefore related to a disruption in systemic communication between muscle and brain metabolism, which represents a significant clinical challenge.

3. Conclusions

The studies collected in this Special Issue highlight the central role of redox processes in maintaining skeletal muscle function and adaptability across the lifespan. Transient increases in RONS during contractions act as signals to enhance antioxidant defenses, stimulate mitochondrial remodeling, and promote adaptation. On the other hand, chronic or unmitigated oxidative and nitrosative stress, as seen during aging, unloading, or metabolic dysfunction, overwhelms defenses and drives atrophy. Notably, targeted strategies such as antioxidant supplementation could abolish training benefits resulting from intrinsic adaptive processes. For instance, redox responses to sprint interval training were blunted in type 2 fibers due to antioxidant provision, underscoring that interval training is effective without supplementation and challenging the common belief that antioxidants are universally beneficial in exercise contexts [13]. Redox dynamics are influenced by circadian alignment, training load, nutrition, genetic background, and environmental exposures, indicating that timing and personalization are crucial for effective interventions. Advances in computational modeling and omics integration provide tools to map key nodes in antioxidant networks, offering the possibility of predicting muscle responses and adjusting strategies to individual redox signatures.

Author Contributions: Conceptualization, E.S. and J.G.; writing—original draft preparation, E.S.; writing—review and editing, J.G. All authors have read and agreed to the published version of the manuscript.

Funding: This research received no external funding.

Acknowledgments: We want to thank all the authors who contributed to the research topic of this Special Issue, as well as the reviewers for their insightful suggestions and comments.

Conflicts of Interest: The authors declare no conflicts of interest.

Abbreviations

The following abbreviations are used in this manuscript:

BDNF	brain-derived neurotrophic factor
BFR	blood flow restriction
CAT	catalase
ERRα	estrogen-related receptor α
GPX1	glutathione peroxidase 1
mTORC1	mammalian target of rapamycin complex 1
Nrf2	nuclear factor erythroid 2-related factor 2
PGC-1α	peroxisome proliferator-activated receptor γ coactivator 1α
SOD2	superoxide dismutase 2, mitochondrial
TGF-β	transforming growth factor β

List of Contributions:

1. Pribil Pardun, S.; Bhat, A.; Anderson, C.P.; Allen, M.F.; Bruening, W.; Jacob, J.; Pendyala, V.V.; Yu, L.; Bruett, T.; Zimmerman, M.C.; et al. Electrical Pulse Stimulation Protects C2C12 Myotubes against Hydrogen Peroxide-Induced Cytotoxicity via Nrf2/Antioxidant Pathway. *Antioxidants* **2024**, *13*, 716. https://doi.org/10.3390/antiox13060716.
2. Meng, Q.; Su, C.-H. Antioxidant Defense and Redox Signaling in Elite Soccer Players: Insights into Muscle Function, Recovery, and Training Adaptations. *Antioxidants* **2025**, *14*, 815. https://doi.org/10.3390/antiox14070815.
3. Sutton, E.; Pekovic-Vaughan, V. Time to Reset: The Interplay Between Circadian Rhythms and Redox Homeostasis in Skeletal Muscle Ageing and Systemic Health. *Antioxidants* **2025**, *14*, 1132. https://doi.org/10.3390/antiox14091132.
4. Blottner, D.; Moriggi, M.; Trautmann, G.; Furlan, S.; Block, K.; Gutsmann, M.; Torretta, E.; Barbacini, P.; Capitanio, D.; Rittweger, J.; et al. Nitrosative Stress in Astronaut Skeletal Muscle in Spaceflight. *Antioxidants* **2024**, *13*, 432. https://doi.org/10.3390/antiox13040432.
5. Li, D.; Fu, W.; Zhang, J.; Lin, Y.; Xiong, X.; Li, J.; Xiong, Y. Zearalenone Exposure Damages Skeletal Muscle Through Oxidative Stress and Is Alleviated by Glutathione, Nicotinamide Mononucleotide, and Melatonin. *Antioxidants* **2025**, *14*, 528. https://doi.org/10.3390/antiox14050528.
6. Kim, A.; Tran, M.N.; Lee, A.Y.; Yeo, H.; Baek, S.J.; Kim, N.S.; Cha, S.; Park, S.M. Integration of Transcriptomic Analysis, Network Pharmacology, and Experimental Validation Demonstrates Enhanced Muscle-Protective Effects of Ethanol Extract of Jakyak-Gamcho-Tang. *Antioxidants* **2025**, *14*, 795. https://doi.org/10.3390/antiox14070795.
7. Iannuzzo, F.; Schiano, E.; Pastore, A.; Guerra, F.; Tenore, G.C.; Novellino, E.; Stornaiuolo, M. Controlled Cultivation Confers Rhodiola rosea Synergistic Activity on Muscle Cell Homeostasis, Metabolism and Antioxidant Defense in Primary Human Myoblasts. *Antioxidants* **2024**, *13*, 1000. https://doi.org/10.3390/antiox13081000.
8. Wojszel, A.; Śliwowski, J.; Rentflejsz, J.; Rogalska, J.; Brzóska, M.M.; Wojszel, Z.B. Irisin, Brain-Derived Neurotrophic Factor (BDNF), and Redox Balance in Geriatric Dynapenia. *Antioxidants* **2025**, *14*, 1268. https://doi.org/10.3390/antiox14101268.

References

1. Zhang, J.; Fang, W.; Chen, S.; Wang, L. Dietary total antioxidant capacity is closely associated with skeletal muscle mass: A cross-sectional study. *Br. J. Nutr.* **2024**, *132*, 1674–1683. [CrossRef] [PubMed]
2. Gao, L.; Wang, H.J.; Tian, C.; Zucker, I.H. Skeletal Muscle Nrf2 Contributes to Exercise-Evoked Systemic Antioxidant Defense Via Extracellular Vesicular Communication. *Exerc. Sport Sci. Rev.* **2021**, *49*, 213–222. [CrossRef] [PubMed]
3. Gungor-Orhan, I.; Akin, S.; Powers, S.K.; Olgaz-Bingol, S.; Demirel, H.A. Sedentary lifestyle induces oxidative stress and atrophy in rat skeletal muscle. *Exp. Physiol.* **2025**, *110*, 857–865. [CrossRef] [PubMed]
4. Powers, S.K.; Deminice, R.; Ozdemir, M.; Yoshihara, T.; Bomkamp, M.P.; Hyatt, H. Exercise-induced oxidative stress: Friend or foe? *J. Sport Health Sci.* **2020**, *9*, 415–425. [CrossRef] [PubMed]
5. Korkmaz, E.; Dönmez, G.; Uzuner, K.; Babayeva, N.; Torgutalp, Ş.Ş.; Özçakar, L. Effects of Blood Flow Restriction Training on Muscle Strength and Architecture. *J. Strength Cond. Res.* **2022**, *36*, 1396–1403. [CrossRef] [PubMed]
6. Neufeld-Cohen, A.; Robles, M.S.; Aviram, R.; Manella, G.; Adamovich, Y.; Ladeuix, B.; Nir, D.; Rousso-Noori, L.; Kuperman, Y.; Golik, M.; et al. Circadian control of oscillations in mitochondrial rate-limiting enzymes and nutrient utilization by PERIOD proteins. *Proc. Natl. Acad. Sci. USA* **2016**, *113*, E1673–E1682. [CrossRef] [PubMed]
7. Fitts, R.H.; Trappe, S.W.; Costill, D.L.; Gallagher, P.M.; Creer, A.C.; Colloton, P.A.; Peters, J.R.; Romatowski, J.G.; Bain, J.L.; Riley, D.A. Prolonged space flight-induced alterations in the structure and function of human skeletal muscle fibres. *J. Physiol.* **2010**, *588*, 3567–3592. [CrossRef] [PubMed]
8. Maezawa, T.; Tanaka, M.; Kanazashi, M.; Maeshige, N.; Kondo, H.; Ishihara, A.; Fujino, H. Astaxanthin supplementation attenuates immobilization-induced skeletal muscle fibrosis via suppression of oxidative stress. *J. Physiol. Sci.* **2017**, *67*, 603–611. [CrossRef] [PubMed]
9. Shekhar, R.; Raghavendra, V.B.; Rachitha, P. A comprehensive review of mycotoxins, their toxicity, and innovative detoxification methods. *Toxicol. Rep.* **2025**, *14*, 101952. [CrossRef] [PubMed]
10. Kowalska, K.; Habrowska-Górczyńska, D.E.; Piastowska-Ciesielska, A.W. Zearalenone as an endocrine disruptor in humans. *Environ. Toxicol. Pharmacol.* **2016**, *48*, 141–149. [CrossRef] [PubMed]
11. Lloyd, E.M.; Pinniger, G.J.; Murphy, R.M.; Grounds, M.D. Slow or fast: Implications of myofibre type and associated differences for manifestation of neuromuscular disorders. *Acta Physiol.* **2023**, *238*, e14012. [CrossRef] [PubMed]
12. Sayer, A.A.; Cooper, R.; Arai, H.; Cawthon, P.M.; Ntsama Essomba, M.J.; Fielding, R.A.; Grounds, M.D.; Witham, M.D.; Cruz-Jentoft, A.J. Sarcopenia. *Nat. Rev. Dis. Prim.* **2024**, *10*, 68. [CrossRef] [PubMed]
13. Wyckelsma, V.L.; Murgia, M.; Kamandulis, S.; Gastaldello, S.; Brazaitis, M.; Snieckus, A.; Eimantas, N.; Pääsuke, M.; Edman, S.; Apro, W.; et al. Antioxidant supplementation blunts the proteome response to 3 weeks of sprint interval training preferentially in human type 2 muscle fibres. *J. Physiol.* **2025**. [CrossRef] [PubMed]

antioxidants

MDPI

Article

Electrical Pulse Stimulation Protects C2C12 Myotubes against Hydrogen Peroxide-Induced Cytotoxicity via Nrf2/Antioxidant Pathway

Sarah Pribil Pardun [1], Anjali Bhat [1], Cody P. Anderson [2], Michael F. Allen [2], Will Bruening [1], Joel Jacob [1], Ved Vasishtha Pendyala [1], Li Yu [3], Taylor Bruett [3], Matthew C. Zimmerman [3], Song-Young Park [2], Irving H. Zucker [3] and Lie Gao [1,*]

1 Department of Anesthesiology, University of Nebraska Medical Center, Omaha, NE 68198, USA; spribil@unmc.edu (S.P.P.); anjalibhatj@gmail.com (A.B.); wbruening@unmc.edu (W.B.); jjacob@unmc.edu (J.J.); vpendyala@wisc.edu (V.V.P.)

2 School of Health and Kinesiology, University of Nebraska Omaha, Omaha, NE 68182, USA; codypanderson@unomaha.edu (C.P.A.); michaelallen@unomaha.edu (M.F.A.); song-youngpark@unomaha.edu (S.-Y.P.)

3 Department of Cellular and Integrative Physiology, University of Nebraska Medical Center, Omaha, NE 68198, USA; lyu@unmc.edu (L.Y.); tbruett@unmc.edu (T.B.); mczimmerman@unmc.edu (M.C.Z.); izucker@unmc.edu (I.H.Z.)

* Correspondence: lgao@unmc.edu; Tel.: +1-(402)-559-8491

Abstract: Skeletal muscle contraction evokes numerous biochemical alterations that underpin exercise benefits. This present study aimed to elucidate the mechanism for electrical pulse stimulation (EPS)-induced antioxidant adaptation in C2C12 myotubes. We found that EPS significantly upregulated Nrf2 and a broad array of downstream antioxidant enzymes involved in multiple antioxidant systems. These effects were completely abolished by pretreatment with a ROS scavenger, N-acetylcysteine. MitoSOX-Red, CM-H2DCFDA, and EPR spectroscopy revealed a significantly higher ROS level in mitochondria and cytosol in EPS cells compared to non-stimulated cells. Seahorse and Oroboros revealed that EPS significantly increased the maximal mitochondrial oxygen consumption rate, along with an upregulated protein expression of mitochondrial complexes I/V, mitofusin-1, and mitochondrial fission factor. A post-stimulation time-course experiment demonstrated that upregulated NQO1 and GSTA2 last at least 24 h following the cessation of EPS, whereas elevated ROS declines immediately. These findings suggest an antioxidant preconditioning effect in the EPS cells. A cell viability study suggested that the EPS cells displayed 11- and 36-fold higher survival rates compared to the control cells in response to 2 and 4 mM H_2O_2 treatment, respectively. In summary, we found that EPS upregulated a large group of antioxidant enzymes in C2C12 myotubes via a contraction-mitochondrial-ROS-Nrf2 pathway. This antioxidant adaptation protects cells against oxidative stress-associated cytotoxicity.

Keywords: electrical pulse stimulation; mitochondria; ROS; Nrf2; antioxidant preconditioning; cellular protection

1. Introduction

Skeletal muscle (SkM) is a highly plastic tissue capable of functional and structural remodeling in response to numerous physiological stimuli. Particularly during endurance training, skeletal myocytes undergo various challenges from neuronal, hormonal, mechanical, and metabolic stressors, leading to genetic and biochemical adaptations that ultimately determine the beneficial phenotypes evoked by chronic exercise [1].

A predominant stressor on contracting muscle fibers is reactive oxygen species (ROS). ROS, at high levels, can react with biomolecules to oxidize proteins, lipids, and nucleic acids. This results in cell damage and organ dysfunction, a pathological redox state referred to as "oxidative stress", which underlies the pathogenesis of numerous diseases. Indeed, in skeletal muscle, it is well-known that excessive ROS contributes to muscle fatigue during exhaustive exercise [2], doxorubicin (DOX)-induced muscle wasting [3], and an age-related progressive loss of muscle mass and strength [4]. On the other hand, ROS at low- to moderate- levels function as essential intracellular signaling molecules, playing a crucial role in many biological processes under normal conditions. Physiological ROS can induce post-translational modifications (PTMs) on specific target proteins by reversible oxidation of sulfur groups and metal centers [5], therefore altering protein activity, localization, and interactions that result in the orchestration of various biological events in cells and organs, such as cell proliferation, differentiation, migration, and angiogenesis [6]. Physiological ROS, or "normal redox signaling," is essential for maintaining normal metabolism and function of cells and organs [7]. Indeed, studies have demonstrated that selective depletion of ROS from normal skeletal muscle decreases force generation [8]. Accordingly, there is an optimal ROS concentration in skeletal myocytes, which determines the maximal force generation during contraction. Deviations of redox status away from this point, either via an increase or decrease in ROS concentration, will impair muscle contractility, forming a classic bell-shaped hormesis dose–response curve [9].

While there is no clear boundary of ROS levels to induce oxidative stress and function as physiological redox signaling, there is no doubt that these free radicals can activate intracellular antioxidant defense under both conditions via compensatory mechanisms to upregulate enzymatic and non-enzymatic antioxidants. These enhanced antioxidant mechanisms not only support the cells to overcome instant oxidative stress but also provide cells with an extended antioxidant capability to counteract subsequent similar insults, a phenomenon we term "antioxidant preconditioning". This process has been proposed as a foundation of redox adaptation of skeletal muscle to endurance exercise [9].

A well-known sensor-effector apparatus in response to a redox disturbance is the Keap1-Nrf2 system. Keap1 acts as a ROS sensor and Nrf2 repressor, while Nrf2 is responsible for activating multiple antioxidant genes, thus playing a critical role in maintaining intracellular redox homeostasis [10]. By employing muscle-specific Nrf2 deletion and overexpression transgenic mouse models, iMS-*Nrf2*$^{flox/flox}$ and iMS-*Keap1*$^{flox/flox}$, we previously identified over 200 cytoprotective and regulative proteins governed by Nrf2 in skeletal muscle, which are involved in several intracellular signaling pathways. These include antioxidant defense and mitochondrial function [11]. We further demonstrated that these Nrf2-orchestrated molecular networks contribute to chronic exercise-induced enhancement of contractility and muscle antioxidant adaptation [12]. However, it remains to be determined if short-term endurance training (STET) can evoke similar beneficial effects on skeletal myocytes to protect the cells against oxidative stress-associated injury.

In the present study, we used electrical pulse stimulation (EPS) on differentiated C2C12 myotubes, a validated in vitro and widely used acute muscle preparation [13,14], to address the hypothesis that a short-term contraction can evoke an antioxidant preconditioning in skeletal myocytes via a mitochondrial-ROS-Nrf2 pathway. Furthermore, we provide evidence to demonstrate that this antioxidant adaptation protects cells against H_2O_2-induced toxicity.

2. Materials and Methods

2.1. Cell Culture

The C2C12 cell line (CRL-1722, ATCC, Manassas, VA, USA) was used in these experiments. Cells were seeded in a 6-well plate at ~50,000 or 100,000 cells per well and grown for one day in complete media (CpM) (Dulbecco's Modification of Eagle's Medium (DMEM/HIGH GLUCOSE; HyClone, Logan, UT, USA) supplemented with 10% fetal bovine serum (FBS; Gibco/ThermoFisher Scientific New Zealand Limited, Auckland, New Zealand) and 1% penicillin/streptomycin (P/S)). After ~24 h, the media was changed from CpM to differentiation media (DM) (DMEM supplemented with 5% horse serum (HS; Gibco/ThermoFisher Scientific) and 1% P/S), where the cells differentiated from myoblasts to myotubes after four days in culture at 37 °C, 5% CO_2, 20% O_2, and 95% humidity. On the fourth day, the media were replaced by fresh DM, and the cells underwent electrical pulse stimulation (EPS).

2.2. Electrical Pulse Stimulation (EPS) and Cell Collection

The electrical current of EPS was delivered to the C2C12 myotubes via stimulation electrodes designed and fabricated in our lab. These electrodes were built on the lid of standard 6-well plates, similar to what has previously been reported [15]. On the first day of EPS, the regular 6-well plate lids were replaced by the EPS lids containing the electrodes, which were connected to a single-channel pulse generator (A310 Accupulse, WPI, Sarasota, FL, USA). The EPS lids were cleaned before and after each study using the following protocol: The EPS chamber was flooded with 70% ethanol for 10 min. The electrodes and EPS lid were washed with ~3 mL of sterile PBS. Finally, the EPS lid was exposed to UV light for 10–30 min in the cell-culture laminar flow hood. The C2C12 myotubes underwent EPS at 10 v, 50 Hz, 10 ms, 0.3 s/3 s for 1 h/day for 4 days. At the end of the EPS, the cells were rinsed with 200 μL of sterile PBS, followed by trypsinizing with 200 μL of 0.05% trypsin for 2 min at 37 °C. The cells were then collected and transferred to 1.5 mL EP tubes, which were stored at −80 °C for future analysis.

2.3. Western Blotting

2.3.1. Sample Preparation (Protein Extraction)

Cells were defrosted from −80 °C, incubated on ice in 100 μL of 100:1 RIPA:PIC for 15 min, and then homogenized utilizing a sonicator. The soluble proteins of cell lysate were then extracted by centrifugation for 20 min at 13,800× *g* and 4 °C. The protein concentration was measured via BSA assay and a Multiskan Sky Microplate Spectrometer (Thermo Scientific Multiskan FC, Singapore). Samples were then diluted to 2 μg/μL with RIPA. An equal volume of 2X SDS was added to each sample to further dilute the samples to 1 μg/μL and 1X SDS. The samples were then boiled in a 70 °C water bath for 2 min and immediately returned to ice.

2.3.2. Gel Electrophoresis

The protein samples (15 μg in 15 μL/each sample) were loaded on a precast polyacrylamide gel (NW00105BOX, Invitrogen, Carlsbad, CA, USA) along with 2 μL of Invitrogen MagicMark™ Western Standard and 4 μL of Bio Rad Precision Plus Protein Dual Color Standards in separate wells. Electrophoresis was performed using a Mini Gel Tank and Blot Module Set (NW2000, Invitrogen), initially at 70 V, for approximately 15 min, and then at 120 V until the desired separation had been reached.

2.3.3. Membrane Transfer (iBlot2)

The fractionated protein on the gel was electrically transferred onto a nitrocellulose membrane by employing Thermo Fischer Scientific iBlot2 (IB21001, Thermo Fisher Israel Ltd, Kiryat Shmona, Israel) with the preset program P0 (20 V for 1 min, then 23 V for 4 min, followed by 25 V for 2 min). The membranes were gently rinsed with Milli-Q water and then stained with Ponceau S for 5 min to visualize the total protein. Once ideal bands of the total protein were obtained, Ponceau S on the membrane was removed by rinsing it in 10 mL of 1X phosphate-buffered saline-Tween 20 solution (PBST) for 5 min.

2.3.4. Immunodetection (Antibodies)

The membranes were then blocked for 30 min in 5% milk in 1X PBST at room temperature; this was followed by two 5 min washes in 10 mL of 1X PBST at room temperature. The membranes were then incubated in primary antibody in 4% BSA overnight at 4 °C. The primary antibodies used in the present experiment were purchased from Abcam (ab-), ABclonal (A-), Santa Cruz Biotechnology (sc-), and Proteintech (-AB/-ig), including NQO1 (ab80588), GSTA2 (ab232833), GSTA4 (ab231601), GPX1 (ab22604), PRX Pathway (TRX,

TXNRD1, and PRX1; ab184868), total OXPHOS (ab110413), calmodulin 1/2/3 (A4885), MFN1 (A9880), MFF (A8700), DRP1 (A2586), FIS1 (A21527), SOD1 (sc-8637), SOD2 (sc-30080), catalase (sc-50508), MyoG (sc-576), Nrf2 (A0674 and 16396-1-AP), and mTOR (66888-1-ig). The following day, the membranes were washed three times for 5 min in 10 mL of 1X PBST; this was followed by incubation in anti-rabbit IgG horseradish peroxidase secondary antibody (1:5000) in 5% milk in 1X PBST for 30 min at room temperature. After this, the membrane was again washed three times for 5 min in 10 mL of 1X PBST. The membranes were developed in the chemiluminescence detection reagent (SuperSignal West Dura or SuperSignal Femto; Thermo Scientific, Rockford, IL, USA). The films were scanned using the G:Box Syngene and Genesys. The intensity of the bands was quantified using the ImageJ software (version number 1.52a, NIH) and normalized for loading using Ponceau S staining.

2.4. ROS Measurements and NAC Application

2.4.1. Electron Paramagnetic Resonance (EPR) Spectroscopy

C2C12 myoblasts were seeded in six-well plates (100,000 cells/well), followed by differentiation and the EPS processes, as described above. After the final EPS, the old DM was replaced with fresh DM, and the cells were processed for EPR spectroscopy analysis, as described below at their specified time points (0, 3, 12, or 24 h after EPS or NS).

The media were removed from the wells, and the cells were then washed twice with 1 mL of KDD buffer (Krebs-HEPES buffer [pH 7.4]; 99 mM NaCl, 4.69 mM KCl, 2.5 mM $CaCl_2$, 1.2 mM $MgSO_4$, 25 mM $NaHCO_3$, 1.03 mM KH_2PO_4, 5.6 mM d-glucose, 20 mM HEPES, 5 μM diethyldithiocarbamic acid sodium salt [DETC], and 25 μM deferoxamine). Then, the cells were mixed with 200 μM cell-permeable ROS-sensitive spin probe 1-hydroxy-3-methoxycarbonyl-2,2,5,5-tetramethyl pyrrolidine (CMH; Noxygen Science Transfer and Diagnostics, Elzach, Germany) and incubated for 1 h and 15 min at 37 °C. After incubation, 900 μL of the KDD buffer was removed from the wells, and the cells were gently scraped into the remaining 100 μL. A total of 50 μL of the cell suspension was pipetted into an EPR glass capillary tube and inserted into a Bruker e-scan EPR spectrometer with the following settings: field sweep width, 60.0 G; microwave frequency, 9.75 kHz; microwave power, 21.90 mW; modulation amplitude, 2.37 G; conversion time, 10.24 ms; and time constant, 40.96 ms. The remaining cell suspension was utilized to count the number of cells to normalize the EPR spectrum amplitude.

2.4.2. Confocal Imaging—MitoSOX Red (MSR, Mitochondrial ROS Indicator) and CM-H2DCFDA (Cytoplasmic ROS Indicator)

C2C12 myoblasts were seeded in six-well plates (5000 cells/well) containing polylysine-coated slides for cell growth, followed by differentiation and the EPS processes, as described above. Immediately after the final day of EPS, the well was washed with 1 mL of HBSS, and the coverslips were then transferred to a 24-well plate, where 2 mL of warm 1 μM MSR working solution or 1 mL of 5 μM CM-H2DCFDA was added, followed by incubation at 37 °C and 5% CO_2 for 30 min. The coverslips were then transferred into clean black boxes and gently washed 3 times with warm HBSS. The cells on the coverslips were fixed with cold 4% paraformaldehyde (PFA) for 10 min, washed 3 times with PBS, and then mounted onto a glass slide using mounting medium (sc-24941 with DAPI) and clear nail polish. The fluorescence was analyzed via confocal microscopy at the following wavelengths: Mitochondrial ROS: Ex 488 nm/Em 550–650 nm; Total ROS: Ex 488 nm/Em 520 nm; and Cellular Nuclear: Ex 360 nm/Ex 460 nm.

2.4.3. N-Acetylcysteine (NAC) Treatment

Three six-well plates were seeded at 100,000 cells per well and underwent differentiation as described above. On the fourth day of differentiation, 40.80 mg of NAC was dissolved in 1 mL of the CpM, which was then further diluted to a final working concentration of 5 mM NAC in 50 mL of the DM. The pH of this medium was neutralized to ~7.5 with 0.5 M NaOH and then filtered through a 0.2 μm syringe filter. After 15 h of pretreatment with this DM containing 5 mM NAC, the cells underwent EPS for 1 h at 50 Hz, 10 V, 10 ms, and 0.3 s/3 s. Eight hours after EPS, the cells underwent a second treatment of NAC following the protocol outlined above. This was repeated daily for four days. After the fourth and final day of EPS, the cells were collected as described above.

2.5. Mitochondrial Function Assays

2.5.1. Seahorse—Mito Stress Test

After undergoing 4 days of EPS, the cells were collected as described above and replated in 96-well plates at 10,000 cells per well in 100 μL of DM. The sensor cartridge was hydrated in a calibration buffer at 37 °C in a non-CO_2 incubator overnight. Sterile-filtered 10 mM glucose, 2 mM glutamate, and 1 mM pyruvate were added to the Seahorse medium, whose pH was then corrected to ~7.4 with 1 M HCl or 0.5 M NaOH. Next, 80 μL of DM was removed from each well of the XF96-well plate, followed by washing with 100 μL of Seahorse medium. A total of 160 μL of fresh Seahorse medium was added to each well, bringing the final volume to 180 μL. The cells were incubated in the Seahorse medium in a non-CO_2 incubator for 1 h. Working solutions of Oligomycin, FCCP, and Rotenone/Antimycin A were made and then added to the injection ports of the sensor cartridge in the respective order. The plate was run on a 96-well plate Seahorse instrument using Wave programming. The Seahorse XF Mito Stress Test Report Generator was used to retrieve the data for later analysis.

2.5.2. Oroboros–High-Resolution Respirometry

After undergoing 4 days of EPS or no stimulation (controls), approximately 300,000 cells suspended in DM were centrifuged at 1500× *g* for 5 min, and the cell pellet was resuspended in 2 mL of mitochondrial respiration medium [MiR05, containing 0.5 mM EGTA, 3 mM $MgCl_2$, 60 mM lactobionic acid, 20 mM taurine, 10 mM KH_2PO_4, 20 mM HEPES, 110 mM D-Sucrose, 1 g/L BSA, pH 7.1] in a high-resolution respirometer (Oxygraph-2k, Oroboros, Innsbruck, Austria) to measure the mitochondrial respiratory function. The chamber block maintained a constant temperature of 37 °C and magnetic stir bars mixed the solution continuously at 750 RPM. The chambers were hyper-oxygenated to 300 μM O_2 via a syringe with pure oxygen from an Oxia device (Oxia, Innsbruck, Austria). After the O_2 respiration rate stabilized, the cells were permeabilized with digitonin (4.05 μM) for 20 min. After permeabilization, substrates were administered in the following order: (1) malate (2 mM) and glutamate (10 mM), (2) adenosine diphosphate (ADP, 5 mM), (3) succinate (10 mM), (4) cytochrome c (10 μM), (5) rotenone (0.5 μM), (6) oligomycin (5 nM), (7) antimycin-A (2.5 μM), and (8) ascorbate (2 mM) and N,N,N′,N′-tetramethyl-p-phenylenediamine dihydrochloride (TMPD, 0.5 mM) [16]. Mitochondrial complexes were assessed as follows: (1) complex I state 2 (LEAK respiration) was assessed after malate and glutamate, (2) complex I state 3 was assessed after malate and glutamate + ADP, (3) complex I + II state 3 was assessed after malate and glutamate + ADP + succinate, (4) the membrane integrity was evaluated after the addition of cytochrome c, (5) complex II state 3 was assessed after the addition of rotenone, and (6) complex IV respiration was assessed after ascorbate and TMPD. The respiration rate for each substrate or inhibitor was recorded after the O_2 flux reached a stable value (~5 min for substrates and ~10 min for inhibitors). The O_2 respiration rates were measured as picomoles of O_2 per second per mL and were converted to attomoles of O_2 per second per cell (amol·s-1·cell).

2.6. H_2O_2 Treatment and CCK-8 Assay

After undergoing 4-day EPS, the cells were collected as described above and replated in 96-well plates at 5000 cells per well in 100 μL of CpM. The cells were allowed to adhere to the plate overnight. After the cells had adhered to the plate, the CpM was removed and replaced with 100 μL of PBS in CpM (control) or 2 or 4 mM H_2O_2 in CpM. The cells were incubated in H_2O_2 at 5% CO_2 and 37 °C for 5 h. After 5 h, 10 μL of CCK-8 was added to each well. The cells were incubated in CCK-8 at 5% CO_2 and 37 °C for 1 h. The absorbance was measured by a Multiskan Sky Microplate Spectrometer at 450 nm to evaluate cell viability.

2.7. Statistical Analyses

Data are expressed as means ± SD. A t-test was used for analyzing the differences in protein expression between the EPS and the control group using GraphPad Prism 8 software. A two-way ANOVA was used to analyze the EPS-NAC data shown in panels C and D of Figure 4. Tukey's honestly significant difference test was performed after ANOVA analysis. A *p*-value of <0.05 was taken as indicative of the statistical significance.

3. Results

3.1. EPS Evokes Voltage- and Time-Dependent Upregulation of NQO1 and GSTA2 Proteins

The basic EPS parameters (10 ms, 50 Hz, 0.3 s/3 s) were adopted from our previous experiments in mice, where EPS was employed to induce an in situ tetanic contraction for evaluating the SkM fatigue development and force generation of mice with chronic heart failure [17], Nrf2 gene manipulation [11], and chronic exercise training [12]. The frequency of EPS was set as 50 Hz because mouse motor-unit output impulses fall within the range of 30–80 Hz [18]. To optimize the EPS suitable for the present study in cultured cells, we investigated the protein expression of NQO1 and GSTA2, two primary targets of Nrf2 in SkM [11], in response to the EPS administration at different voltages (2.5–10 V; Panel A of Figure 1) and different durations (1–6 days; Panel B of Figure 1). Clearly, both NQO1 and GSTA2 were upregulated by EPS in a voltage- and duration-dependent manner, with a minor response difference between these two proteins. As shown in Panel A, 10 V-EPS evoked a maximal expression in both, whereas 2.5 V-EPS upregulated the expression only in GSTA2 but not in NQO1. As shown in Panel B, NQO1 displayed a maximal response to EPS on day 4, while GSTA2's maximal expression existed on day 5. Based on these results, we chose the EPS parameters of 10 ms, 50 Hz, 0.3 s/3 s, 10 V, and 1 h/day for 4 days for all of the following experiments.

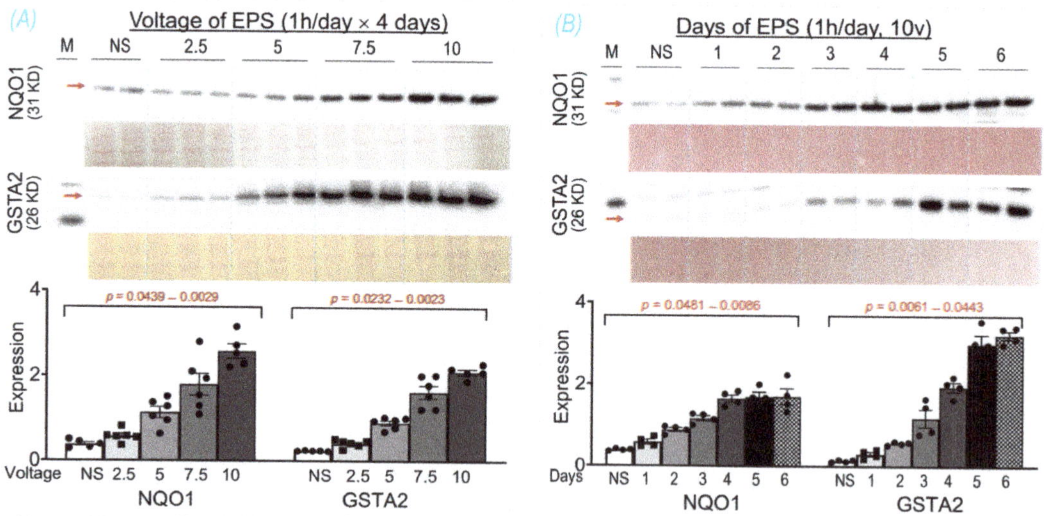

Figure 1. NQO1 and GSTA2 protein expression in C2C12 myotubes receiving EPS at different voltages (**A**) and durations (**B**). Red arrows indicate the target bands. Data are expressed as mean ± SD; n = 4–6/each group. NS, non-stimulated control. The y-axes of the bar graphs indicate protein expression normalized by Ponceau S.

3.2. EPS Increases Nrf2 Protein Content in C2C12 Myotubes

In two previous experiments from our laboratory, we employed the ab13550 antibody (AbCam) to probe Nrf2 protein in mouse SkM lysate and showed a single specific band at 110 kDa, which was significantly upregulated by administrating the Nrf2 activator curcumin [19] or by deleting the Keap1 gene [11]. Unfortunately, the current lot of ab13550 did not detect a Nrf2 band in mouse muscle tissue nor in C2C12 myotubes. We therefore tested seven other Nrf2 antibodies provided by different vendors, including AbCam (ab62352), Proteintech (16396-1-AP, 66504-1-lg), ABclonal (A0674, A1244, A11159), and Invitrogen (PA5-67520). Among these antibodies, we found that two antibodies, A0674 and 16396-1-AP, detected a specific Nrf2 band at 110 kDa, whose intensity was significantly higher in the EPS cells than in the control cells, as shown in Figure 2. Interestingly, in the blot of 16396-1-AP antibody, we found an additional Nrf2 band at 66 kDa, which was not changed by the EPS, as shown in Panel B. The different blots of these two antibodies may reflect a different immunogen used to make the antibodies where the NP_006155.2 was used for A0674 (ABclonal) and the AG9489 (Proteintech) for the 16396-1-A. Since the predicted molecular weight of Nrf2, based on its amino acid sequence, is ~66 kDa, we think that the band at ~66 kDa represents a non-modified and non-activated Nrf2 protein, whereas the band in ~110 kDa is the biologically relevant and activated Nrf2, whose molecular weight is higher due to post-translational modifications (PTMs) such as acetylation, phosphorylation, ubiquitination, and SUMOylation [20]. The observation and interpretation of the Nrf2 blots showing two bands have been discussed by two laboratories in the field [21,22]. Although the current data clearly showed an increased Nrf2 protein content in whole-cell lysates of EPS-C2C12 myotubes, the underlying molecular mechanisms remain to be elucidated. We postulated that both the activation of the Nrf2 gene (NFE2L2) and reduction of the proteasome-mediated Nrf2 protein degradation contribute to this increased Nrf2 protein content, with the latter mechanism being the predominant. This conclusion is based on the data shown in Figure 4, where we found that EPS dramatically elevated intracellular ROS, the principal mediator to dissociate Nrf2 from Keap1, and that pretreatment with the ROS scavenger, N-acetylcysteine (NAC), completely abolished the EPS-induced increase in Nrf2 protein.

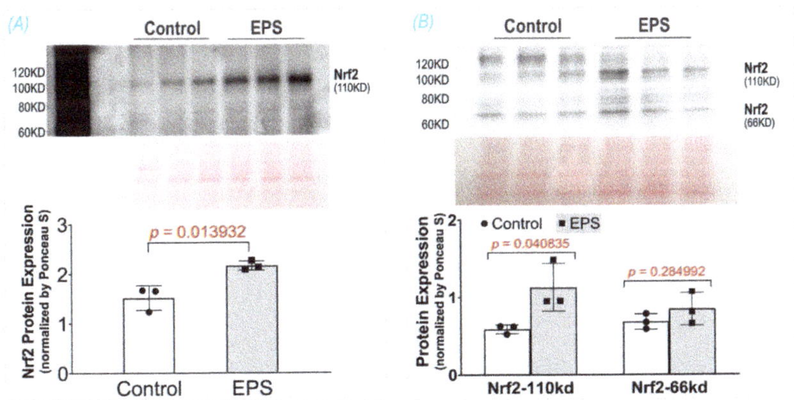

Figure 2. Protein expression of Nrf2 in the C2C12 myotubes receiving EPS, detected by using two different antibodies: (**A**) A0674 (ABclonal) and (**B**) 16396-1-AP (Proteintech). Data are expressed as mean ± SD; n = 3/each group.

3.3. EPS Upregulates Multiple Antioxidant Proteins

After demonstrating increased Nrf2 protein in the EPS cells, we then evaluated the Nrf2 downstream target proteins that constitute cellular antioxidant defenses. We found that, in the EPS cells, a large group of proteins associated with multiple antioxidant systems were significantly upregulated (Figure 3). These proteins include the components of the first-line antioxidant system (SOD1, SOD2, Cat, and GPX1; Panel A), the thioredoxin antioxidant system (Trx and Prx1; Panel B), and the glutathione antioxidant system (GSTA2, GSTA4, and NQO1; Panel C). Among these evaluated Nrf2 target proteins, NQO1, GSTA2, and GSTA4 were the top three most upregulated proteins following EPS, which were 3-, 12-, and 7-fold higher in the EPS cells than the control cells, respectively (right subpanel of Panel C). These data suggest that EPS can activate multiple antioxidant systems in C2C12 cells, where the glutathione system is the most mobilized.

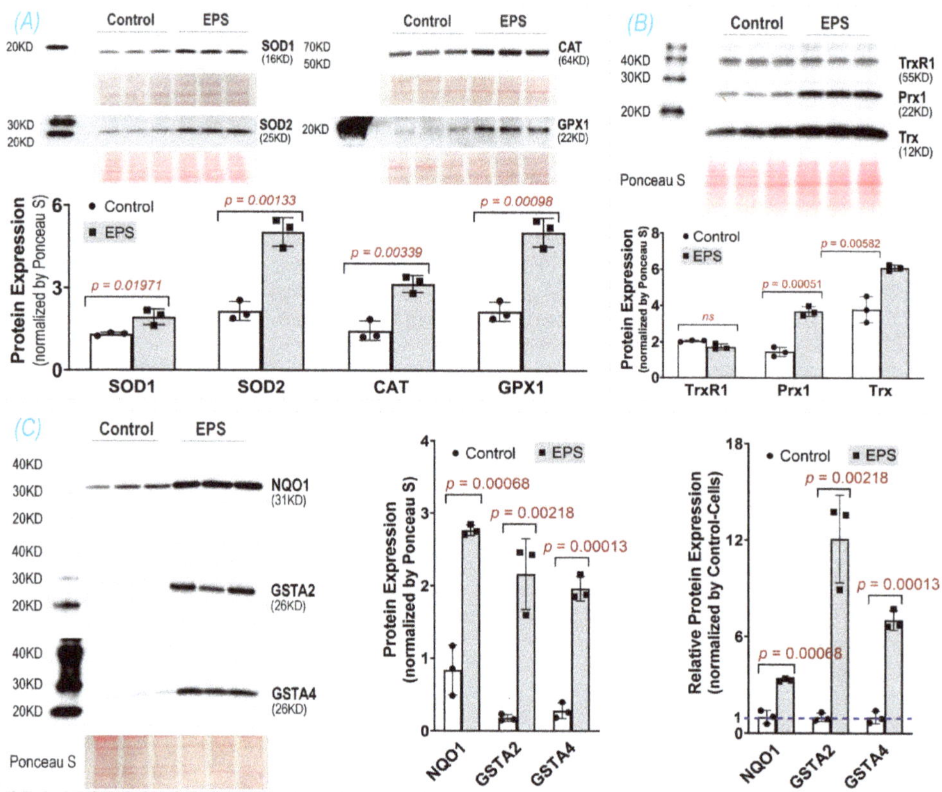

Figure 3. Protein expression of key antioxidant components in three antioxidant systems: the first-line antioxidant system (**A**), thioredoxin antioxidant system (**B**), and glutathione antioxidant system (**C**). The middle subpanel of (**C**) shows raw data; the right subpanel of (**C**) is relative data normalized by the expression in the control group. Data are expressed as mean ± SD; $n = 3$/each group.

3.4. EPS-Evoked Activation of Nrf2/Antioxidant System Relies on ROS

To investigate the role of ROS in the EPS-evoked activation of Nrf2/antioxidant systems, we employed three techniques to measure the ROS levels and used a ROS scavenger, N-acetylcysteine (NAC). As can be seen in the MitoSOX-Red, CM-H2DCFDA, and EPR spectroscopy data shown in Panel A of Figure 4, the EPS cells exhibited a significantly higher fluorescence intensity and EPR spectrum amplitude compared to the control cells. These data indicate a significantly high ROS level in the mitochondria, cytosol, and whole cells of the C2C12 myotubes that received stimulation and demonstrated an increased ROS generation from the mitochondria of the contracting muscle. Panel B of Figure 4 shows that the EPS-induced upregulation of Nrf2/antioxidant enzymes was completely abolished by the NAC pretreatment (EPS + NAC), suggesting that the increased ROS is the exclusive mediator of the Nrf2/antioxidant upregulation induced by EPS. Interestingly, we found that the NQO1 expression was significantly lower in the cells treated with NAC alone compared to the cells without any treatment, suggesting that the basal expression of this antioxidant enzyme relies on the physiological levels of ROS. Panel C of Figure 4 shows the expression of three non-antioxidant proteins, where NAC had no effects on the EPS-induced upregulation of CaM and MyoG but did abolish the mTOR response.

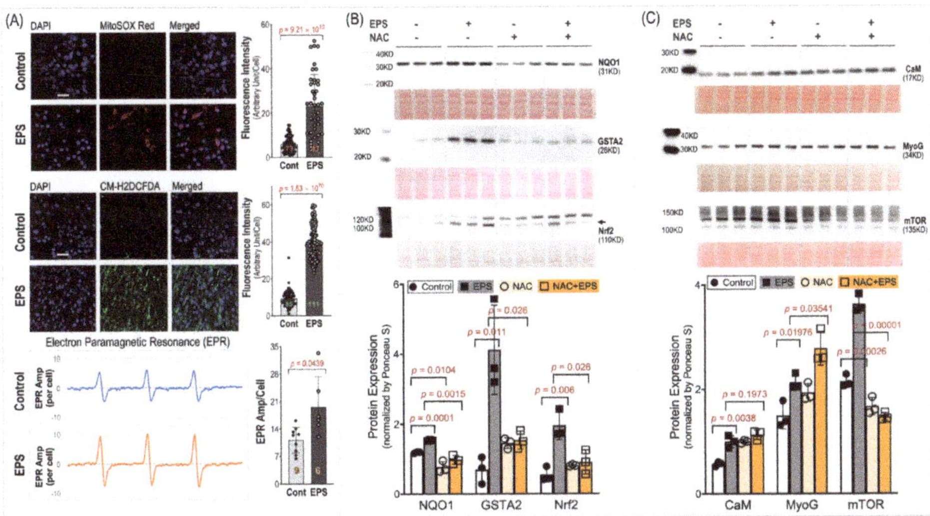

Figure 4. ROS level and its role in EPS-induced protein expression. (**A**): ROS levels in mitochondria (MitoSOX Red), cytosol (CM-H2DCFDA), and whole cells (EPR). Left subpanels: representative confocal images (scale bar: 50 μm) and EPR spectra. Right subpanels: quantitative fluorescence intensity and EPR results. The red, green, and orange numbers within the bars indicate cell and sample numbers. (**B**,**C**): Protein expression of Nrf2/antioxidant enzymes and non-antioxidant proteins. Data are expressed as mean ± SD; n = 3/each group. NAC: N-acetylcysteine; 5 mM.

3.5. EPS Enhances Mitochondrial Oxidative Phosphorylation and Dynamics

Given that mitochondria are one of the major sources of ROS in skeletal myocytes during contraction, we examined the effects of EPS on the oxidative phosphorylation (OXPHOS) complex protein expression, oxygen consumption rate, and protein expression of four factors associated with mitochondrial dynamics. We found that EPS significantly upregulated CI, the main site of ROS production [23], and CV, the ATP synthase [24], whereas the CII, CIII, and CIV expression were left unchanged (Panel A, Figure 5). Correspondingly, the FCCP-mediated uncoupled maximal oxygen consumption rate was significantly increased in the EPS cells compared to the control cells (Panel B, Figure 5). Furthermore, ADP-stimulated respiration (state 3) was increased in the EPS compared to non-stimulated cells through complex I, complex II, and complex I + II (Panel C, Figure 5). In addition, complex IV respiration was elevated, and complex I state 2 respiration was unchanged in the EPS compared to non-stimulated cells (Panel C, Figure 5). Overall, these data suggest that EPS can mediate robust increases in OXPHOS capacity in C2C12 myotubes by upregulating CI and CV expression. Mitochondria are highly dynamic cytoplasmic organelles. During contraction, the mitochondrial integrity and homeostasis in SkM have to be maintained through continual fusion and fission [25,26]. Therefore, we evaluated the expression of four proteins associated with mitochondrial dynamics and found that EPS upregulated MFN1 and DRP1 but had no effect on MFF and FIS1 (Panel D, Figure 5).

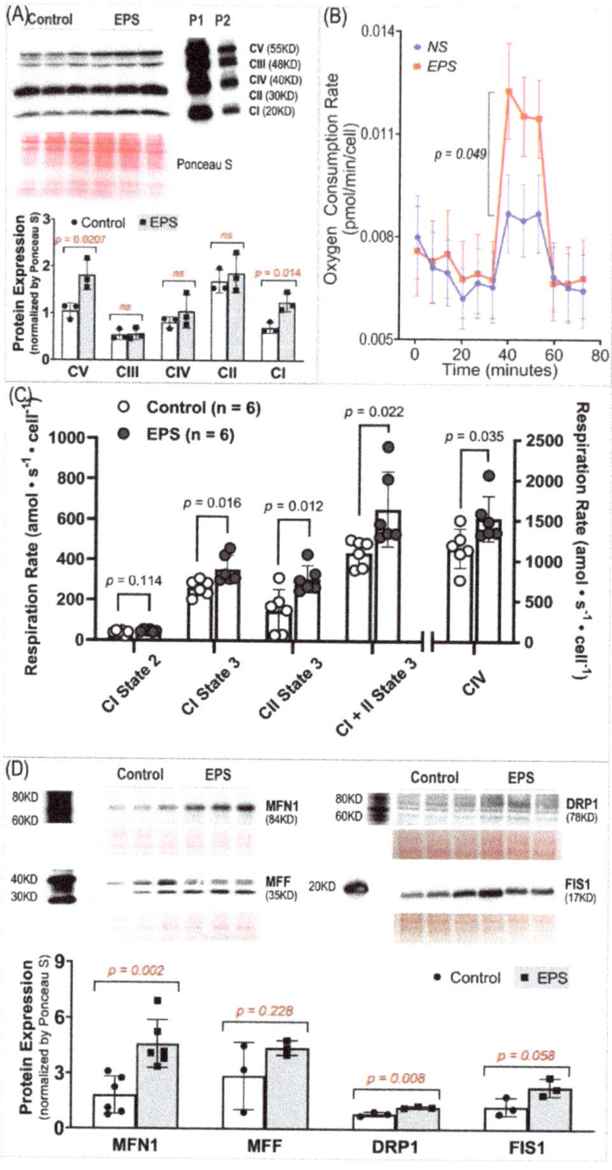

Figure 5. Mitochondrial protein expression and function. (**A**) Protein expression of respiratory complexes I–V. $n = 3$/group. (**B**) Oxygen consumption rate measured by Seahorse. $n = 46$ in the NS group and 46 in the EPS group. (**C**) Oxygen consumption rates measured by Oroboros. $n = 6$/group. (**D**) Protein expression of fission/fusion factors. $n = 3$/group. All data are expressed as mean ± SD. ns, not significant.

3.6. EPS Evokes Antioxidant Preconditioning to Protect Cells against H_2O_2-Induced Injury

To determine how long antioxidant proteins can remain upregulated after stopping stimulation, we carried out an experiment to evaluate the time course of NQO1 and GSTA2 protein expression in the samples harvested at 0, 3, 6, 12, and 24 h after EPS (Panel A of Figure 6). We did not find a significant difference in protein expression between the samples at 0, 3, 6, and 12 h. However, protein expression in the 24 h sample was significantly lower than that in other EPS samples but still significantly higher than that in the non-stimulated control sample. These data suggest that EPS-enhanced antioxidant enzyme expression can last at least 24 h. On the other hand, the EPR spectroscopy showed a significant increase in ROS immediately after EPS (i.e., 0 h), which rapidly declined in the 3–24 h samples to levels below that in the non-EPS sample (Panel B, Figure 6). These data suggest that the upregulated endogenous antioxidants by EPS provide C2C12 myotubes with excess antioxidant capacity, a status we term "antioxidant preconditioning", which can protect the cells against subsequent oxidative stressor-induced injury. As can be seen in Panel C of Figure 6, the survival rates of the EPS cells exposed to 2 mM and 4 mM H_2O_2 were 77.8% and 54.3% of the PBS-NS control group, about 11- and 36-fold higher than that of the H_2O_2-NS cells, whose survival rates were 7.2% and 1.5%, respectively.

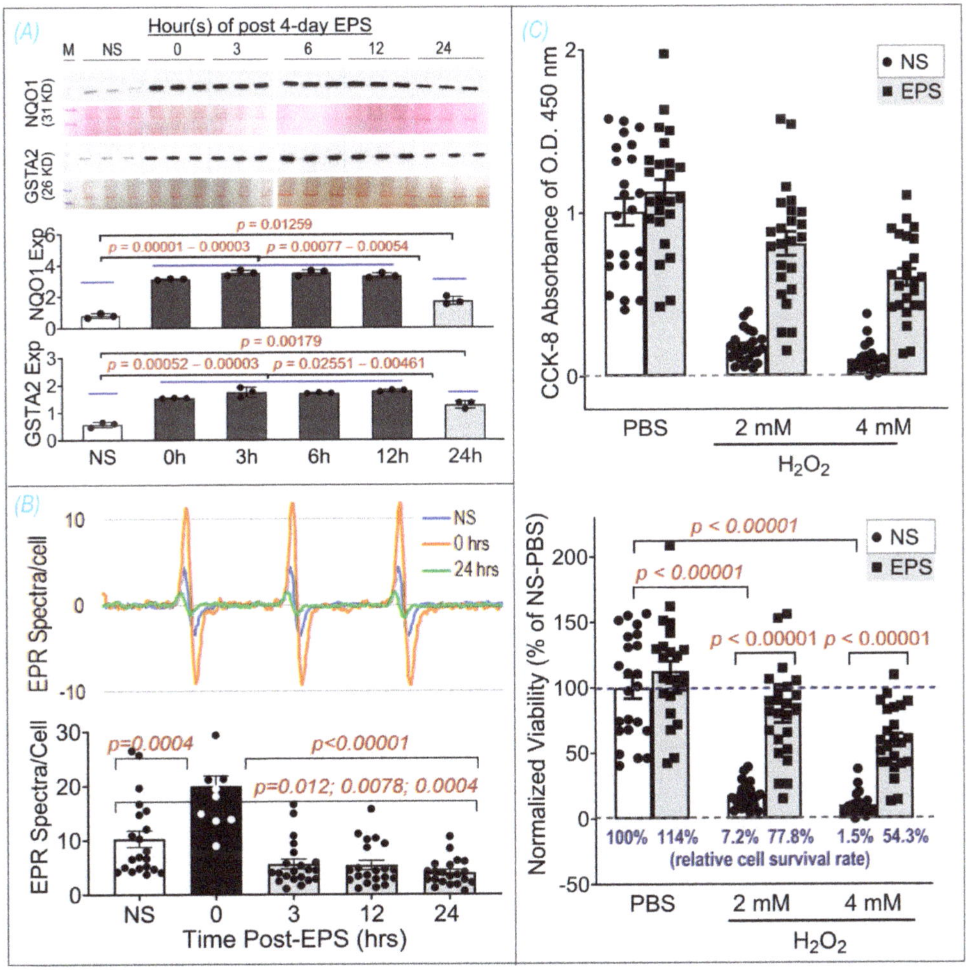

Figure 6. EPS-induced antioxidant preconditioning and protective effects. (**A**) Time course of NQO1 and GSTA2 expressions, which were normalized by Ponceau S. (**B**) Time course of ROS levels. (**C**) Cell viability assay using CCK-8 shown as raw data of OD 450 nm absorbance (upper panel) and normalized viability by NS-PBS as a percentage (lower panel). Data are expressed as mean ± SD; n = 3/group in (**A**), 12–21/group in (**B**), and 23–24/group in (**C**). NS, non-stimulation.

4. Discussion

Using EPS on cultured C2C12 myotubes to investigate the effects of contraction on the intracellular antioxidant system in SkM, we showed that EPS for up to 4 days induced an obvious time- and dose-dependent upregulation of NQO1 and GSTA2 protein expression. We found that EPS markedly activated Nrf2 and multiple antioxidant systems, including the first-line antioxidant system, the thioredoxin system, and the glutathione system, by upregulating a large group of antioxidant enzymes and antioxidant-associated proteins. Among these systems, the glutathione system displayed the largest response to EPS, with approximately a 10-fold increase in GSTA2 and GSTA4 protein expression (Panel C, Figure 3). We further demonstrated that the EPS-mediated increase in ROS played a critical role in the activation of Nrf2/antioxidant systems since EPS-mediated enhancements in antioxidant capacity were abolished when the C2C12 myotubes were pre-treated with the ROS scavenger, NAC (Panel B, Figure 4).

Importantly, our data of substantially elevated mitochondrial-derived superoxide after EPS supports the view that mitochondria may be one of the major sources of ROS and is responsible for the activation of Nrf2/antioxidant pathways (Panel A, Figure 4). Furthermore, our data suggest that EPS-induced mitochondrial adaptations may promote EPS-induced augmentation in antioxidant capacity. Of note, complex IV respiration was significantly increased after 4 days of EPS, compared to the non-stimulated cells (Panels B and C, Figure 5). Furthermore, ADP-stimulated respiration (state 3) through complex I, complex II, and complex I + II were all elevated after 4 days of EPS (Panel C, Figure 5). In addition, these adaptations were paralleled by an increased protein expression of complexes I—the main mitochondrial site for ROS production [27]—and V (Panel A, Figure 5) and upregulated fission/fusion factors (Panel D, Figure 5). Altogether, these data indicate that EPS leads to substantial adaptations to mitochondrial function and elevations in OXPHOS capacity, which is consistent with the known adaptations in human SkM mitochondria after aerobic training [28,29]. Importantly, the stepwise elevation in OXPHOS capacity due to repeated EPS exposure may also lead to elevations in mitochondrial ROS production during high-intensity EPS [30–32], which may provide an incremental stimulus necessary to induce cumulative increases in antioxidant capacity through Nrf2, as observed in the present study (Panel A, Figure 1). Furthermore, our data support the notion that ROS is transiently elevated during EPS (Panel B, Figure 6) while the antioxidant capacity remains elevated up to 24 h after EPS (Panel A, Figure 6), indicating an overcompensation of antioxidant synthesis and the development of an antioxidant reserve. Importantly, we demonstrated that "antioxidant preconditioning" induced by EPS protects the C2C12 myotubules from future oxidative injury since the EPS cells were more resistant to H_2O_2-induced cytotoxicity than the non-stimulated control cells. Overall, our data indicate that mitochondrial ROS during EPS activates the Nrf2/antioxidant pathways and may lead to an antioxidant reserve that protects cells from future oxidative injury.

C2C12 is a frequently used cell line to investigate the physiology and pathology of skeletal myocytes. After exposure to a differential medium, C2C12 are transformed from primitive myoblasts into mature myotubes, which have the primary characteristics of rodent SkM, including the ability to contract and generate force. Indeed, it has been demonstrated that EPS evokes the actual contraction of C2C12 myotubes, making this cell line a convenient in vitro model that is suitable for investigating the physiological significance and molecular responses of skeletal myocytes to stimulated exercise [14]. A single bout of EPS at 1 Hz for 1, 3, or 6 h duration has been demonstrated to upregulate the Nrf2/antioxidant system in C2C12 myotubes [33]. We employed EPS at 50 Hz and 1 h per day for 4 consecutive days in the present experiment, which may better simulate the SkM response to chronic short-term exercise. Importantly, our time-course data clearly show a significantly higher protein expression of NQO1 and GSTA2 in 4-day-EPS-cells compared to those that undergo 1 day of EPS. These findings suggest cumulative effects of repeated contraction on the activation of the muscle antioxidant system and differential responses of SkM to a single bout of exercise compared to multiple sets of short-term exercise training.

Intracellular redox homeostasis is crucial for maintaining normal muscle structure and function. However, this balance tends to be disturbed during exercise due to increased ROS production by membrane NADPH oxidases and mitochondria, where the oxidative phosphorylation process increases dramatically to meet the increased ATP demand. As shown in Panel A of Figure 4, we detected a profound increase in ROS in mitochondria, cytosol, and whole cells receiving 4 days of EPS. The exact mechanisms responsible for this increased ROS remain to be determined. However, given the data in Panels A and B of Figure 5, a significant upregulation of complexes I/V and an increased mitochondrial oxygen consumption rate were detected. We believe that mitochondria are, at least, one of the major sources of the increased ROS in these EPS cells [34].

To preserve redox homeostasis in response to oxidative stress, intracellular antioxidant defenses must be mobilized to remove the excessive ROS. Antioxidant protein expression is primarily orchestrated by the Nrf2-Keap1 system, a thiol-based sensor-effector apparatus, where ROS oxidizes three key cysteine residues (Cys151, Cys273, and Cys288) on the Keap1 molecule, resulting in dissociation of Nrf2 from Keap1 and translocation from cytosol to the nucleus. In the nucleus, Nrf2 binds to antioxidant response elements (AREs) on DNA to upregulate hundreds of proteins involved in antioxidant defense, anti-inflammation, detoxification, and metabolism [10]. This ROS-trigger and Nrf2-Keap1-dependent activation of antioxidant defense in SkM has been characterized by our laboratory by employing transgenic mouse models with muscle-specific Nrf2/Keap1 deficiency, iMS-*Nrf2*$^{flox/flox}$ and iMS-*Keap1*$^{flox/flox}$ [11], and in a chronic exercise training model in C57BL/6 mice [12]. In the present study, using 4 days of EPS on C2C12 myotubes, we describe a useful in vitro model whose Nrf2/antioxidant system can be activated to a similar level as observed in animal models. This is particularly true for the expression of NQO1 and GSTA2 proteins.

One of the most important findings from the present study, we believe, is the prolonged elevated protein expression of antioxidant enzymes for up to 24 h after ceasing 4-day-EPS (Panel A, Figure 6). This phenomenon suggests potential cumulative effects on protein expression that are induced by consecutive multiple sets of EPS when the interval of any two chronologically adjacent stimulations is less than 24 h. This cumulative effect may represent a mechanism for the gradually increased protein contents of NQO1 and GSTA2, which were observed in the time-course experiment shown in Panel B of Figure 1. The translation of this in vitro experimental data into a strategy to optimize long-term exercise thus points to the significance of intervals between two sets of exercise-induced muscle antioxidant adaptation. While the molecular mechanisms underlying this prolonged antioxidant protein overexpression remain to be elucidated, we postulate that they are involved in an extended alteration of DNA transcription, protein translation, and/or post-translational modifications.

In contrast to the expression of antioxidant enzymes, the EPS-induced increase in ROS did not last more than 3 h after ceasing stimulation (Panel B, Figure 6). This chronological dissociation between antioxidant enzymes and ROS tilts the redox balance to the antioxidant side and confers cells with excess and chronic antioxidant capacity—a state we term "antioxidant preconditioning" (Panel B, Figure 7)—thus protecting cells against subsequent oxidative injury. Indeed, as the data shows in Panel C of Figure 6, the EPS cells have 11- and 36-fold higher viability than the control cells when they are exposed to the medium containing 2 mM and 4 mM H_2O_2.

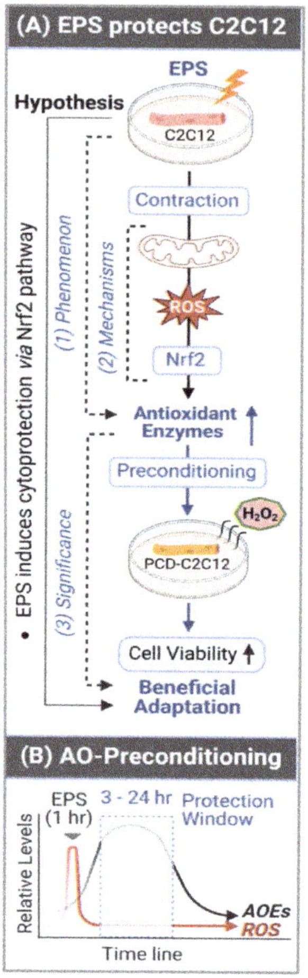

Figure 7. Graphical summary (**A**) and antioxidant preconditioning (**B**). AOEs—antioxidant (AO) enzymes (E); PCD—preconditioning.

5. Conclusions

By employing C2C12 myotubes, we demonstrated that consecutive multiple sets of EPS over 4 days profoundly elevated the protein expression of a group of antioxidant enzymes when the interval of any two chronologically adjacent stimulations was less than 24 h. The translation of this in vitro experimental data into a strategy to optimize long-term exercise thus points to the significance of intervals between two sets of exercise-induced muscle antioxidant adaptation. This study provides evidence to demonstrate that short-term contraction can protect skeletal myocytes against subsequent oxidative stress-induced cytotoxicity, likely via ROS-dependent Nrf2-evoked activation of intracellular antioxidant mechanisms (Panel A, Figure 7), which provides a novel insight into the area of SkM redox signaling responses to exercise [34].

Author Contributions: Conceptualization, L.G. and S.P.P.; Methodology, Data curation, Visualization, and Investigation: S.P.P., A.B., C.P.A., M.F.A., W.B., J.J., V.V.P., L.Y. and T.B.; Writing—original draft preparation, S.P.P. and L.G.; Writing—review and editing, I.H.Z., M.C.Z. and S.-Y.P.; Supervision, L.G. and S.P.P.; Project administration, L.G. and S.P.P.; Funding acquisition, L.G. All authors have read and agreed to the published version of the manuscript.

Funding: This research was funded by the National Institutes of Health (NIH) grant R01HL160820, awarded to L.G. I.H.Z. was supported, in part, by the Theodore F. Hubbard Foundation. The EPR spectroscopy data were collected in the UNMC EPR Spectroscopy Core, which was established with support, in part, from an NIH Centers of Biomedical Research Excellence (COBRE) grant (1P30GM103335) awarded to the University of Nebraska's Redox Biology Center.

Institutional Review Board Statement: Not applicable.

Informed Consent Statement: Not applicable.

Data Availability Statement: The data are contained within the article.

Acknowledgments: The authors thank the Seahorse Core Facility of the University of Nebraska Medical Center and Kelly Stauch and Andrew Trease for their professional assistance in the measurement of the oxygen consumption rate. The graphical summary was created with BioRender.com.

Conflicts of Interest: The authors declare no conflicts of interest.

Abbreviations

CAT, catalase; GPX1, glutathione peroxidase 1; GSTA2, glutathione S-transferase alpha 2; GSTA4, glutathione S-transferase alpha 4; NQO1, NAD(P)H dehydrogenase [quinone] 1; Prx1, paired related homeobox 1; SOD1, superoxide dismutase type 1; SOD2, superoxide dismutase type 2; Trx, thioredoxin; TrxR1, thioredoxin reductase 1.

References

1. Egan, B.; Sharples, A.P. Molecular responses to acute exercise and their relevance for adaptations in skeletal muscle to exercise training. *Physiol. Rev.* **2023**, *103*, 2057–2170. [CrossRef] [PubMed]
2. Reid, M.B. Free radicals and muscle fatigue: Of ROS, canaries, and the IOC. *Free Radic. Biol. Med.* **2008**, *44*, 169–179. [CrossRef] [PubMed]
3. Hiensch, A.E.; Bolam, K.A.; Mijwel, S.; Jeneson, J.A.; Huitema, A.D.; Kranenburg, O.; Van der Wall, E.; Rundqvist, H.; Wengstrom, Y.; May, A.M. Doxorubicin-induced skeletal muscle atrophy: Elucidating the underlying molecular pathways. *Acta Physiol.* **2020**, *229*, e13400. [CrossRef] [PubMed]
4. Fulle, S.; Protasi, F.D.; Tano, G.; Pietrangelo, T.; Beltramin, A.; Boncompagni, S.; Vecchiet, L.; Fanò, G. The contribution of reactive oxygen species to sarcopenia and muscle ageing. *Exp. Gerontol.* **2004**, *39*, 17–24. [CrossRef] [PubMed]
5. Sinenko, S.A.; Starkova, T.Y.; Kuzmin, A.A.; Tomilin, A.N. Physiological Signaling Functions of Reactive Oxygen Species in Stem Cells: From Flies to Man. *Front. Cell Dev. Biol.* **2021**, *9*, 714370. [CrossRef]
6. Sies, H.; Berndt, C.; Jones, D.P. Oxidative Stress. *Annu. Rev. Biochem.* **2017**, *86*, 715–748. [CrossRef] [PubMed]
7. Sies, H.; Jones, D.P. Reactive oxygen species (ROS) as pleiotropic physiological signalling agents. *Nat. Rev. Mol. Cell Biol.* **2020**, *21*, 363–383. [CrossRef] [PubMed]
8. Reid, M.B.; Khawli, F.A.; Moody, M.R. Reactive oxygen in skeletal muscle. III. Contractility of unfatigued muscle. *J. Appl. Physiol.* **1993**, *75*, 1081–1087. [CrossRef] [PubMed]
9. Powers, S.K.; Deminice, R.; Ozdemir, M.; Yoshihara, T.; Bomkamp, M.P.; Hyatt, H. Exercise-induced oxidative stress: Friend or foe? *J. Sport. Health Sci.* **2020**, *9*, 415–425. [CrossRef]
10. Yamamoto, M.; Kensler, T.W.; Motohashi, H. The KEAP1-NRF2 System: A Thiol-Based Sensor-Effector Apparatus for Maintaining Redox Homeostasis. *Physiol. Rev.* **2018**, *98*, 1169–1203. [CrossRef]
11. Gao, L.; Kumar, V.; Vellichirammal, N.N.; Park, S.Y.; Rudebush, T.L.; Yu, L.; Son, W.M.; Pekas, E.J.; Wafi, A.M.; Hong, J.; et al. Functional, proteomic and bioinformatic analyses of Nrf2- and Keap1- null skeletal muscle. *J. Physiol.* **2020**, *598*, 5427–5451. [CrossRef]
12. Bhat, A.; Abu, R.; Jagadesan, S.; Vellichirammal, N.N.; Pendyala, V.V.; Yu, L.; Rudebush, T.L.; Guda, C.; Zucker, I.H.; Kumar, V.; et al. Quantitative Proteomics Identifies Novel Nrf2-Mediated Adaptative Signaling Pathways in Skeletal Muscle Following Exercise Training. *Antioxidants* **2023**, *12*, 151. [CrossRef]
13. Lautaoja, J.H.; Turner, D.C.; Sharples, A.P.; Kivela, R.; Pekkala, S.; Hulmi, J.J.; Yla-Outinen, L. Mimicking exercise in vitro: Effects of myotube contractions and mechanical stretch on omics. *Am. J. Physiol. Cell Physiol.* **2023**, *324*, C886–C892. [CrossRef] [PubMed]
14. Manabe, Y.; Miyatake, S.; Takagi, M.; Nakamura, M.; Okeda, A.; Nakano, T.; Hirshman, M.F.; Goodyear, L.J.; Fujii, N.L. Characterization of an acute muscle contraction model using cultured C2C12 myotubes. *PLoS ONE* **2012**, *7*, e52592. [CrossRef]
15. Mobini, S.; Leppik, L.; Barker, J.H. Direct current electrical stimulation chamber for treating cells in vitro. *Biotechniques* **2016**, *60*, 95–98. [CrossRef] [PubMed]

16. Robinson, M.M.; Sather, B.K.; Burney, E.R.; Ehrlicher, S.E.; Stierwalt, H.D.; Franco, M.C.; Newsom, S.A. Robust intrinsic differences in mitochondrial respiration and H_2O_2 emission between L6 and C2C12 cells. *Am. J. Physiol. Cell Physiol.* **2019**, *317*, C339–C347. [CrossRef] [PubMed]
17. Wafi, A.M.; Yu, L.; Gao, L.; Zucker, I.H. Exercise training upregulates Nrf2 protein in the rostral ventrolateral medulla of mice with heart failure. *J. Appl. Physiol.* **2019**, *127*, 1349–1359. [CrossRef]
18. Manuel, M.; Chardon, M.; Tysseling, V.; Heckman, C.J.P. Scaling of motor output, from mouse to humans. *Physiology* **2019**, *34*, 5–13. [CrossRef]
19. Wafi, A.M.; Hong, J.; Rudebush, T.L.; Yu, L.; Hackfort, B.; Wang, H.; Schultz, H.D.; Zucker, I.H.; Gao, L. Curcumin improves exercise performance of mice with coronary artery ligation-induced HFrEF: Nrf2 and antioxidant mechanisms in skeletal muscle. *J. Appl. Physiol.* **2019**, *126*, 477–486. [CrossRef]
20. Walters, T.S.; McIntosh, D.J.; Ingram, S.M.; Tillery, L.; Motley, E.D.; Arinze, I.J.; Misra, S. SUMO-Modification of Human Nrf2 at K(110) and K(533) Regulates Its Nucleocytoplasmic Localization, Stability and Transcriptional Activity. *Cell Physiol. Biochem.* **2021**, *55*, 141–159.
21. Lau, A.; Tian, W.; Whitman, S.A.; Zhang, D.D. The predicted molecular weight of Nrf2: It is what it is not. *Antioxid. Redox Signal.* **2013**, *18*, 91–93. [CrossRef] [PubMed]
22. Kopacz, A.; Rojo, A.I.; Patibandla, C.; Lastra-Martinez, D.; Piechota-Polanczyk, A.; Kloska, D.; Jozkowicz, A.; Sutherland, C.; Cuadrado, A.; Grochot-Przeczek, A. Overlooked and valuable facts to know in the NRF2/KEAP1 field. *Free Radic. Biol. Med.* **2022**, *192*, 37–49. [CrossRef]
23. Zhao, R.Z.; Jiang, S.; Zhang, L.; Yu, Z.B. Mitochondrial electron transport chain, ROS generation and uncoupling (Review). *Int. J. Mol. Med.* **2019**, *44*, 3–15. [CrossRef]
24. Long, Q.; Yang, K.; Yang, Q. Regulation of mitochondrial ATP synthase in cardiac pathophysiology. *Am. J. Cardiovasc. Dis.* **2015**, *5*, 19–32. [PubMed]
25. Kitaoka, Y.; Ogasawara, R.; Tamura, Y.; Fujita, S.; Hatta, H. Effect of electrical stimulation-induced resistance exercise on mitochondrial fission and fusion proteins in rat skeletal muscle. *Appl. Physiol. Nutr. Metab.* **2015**, *40*, 1137–1142. [CrossRef] [PubMed]
26. Drake, J.C.; Wilson, R.J.; Yan, Z. Molecular mechanisms for mitochondrial adaptation to exercise training in skeletal muscle. *FASEB J.* **2016**, *30*, 13–22. [CrossRef] [PubMed]
27. Brand, M.D.; Affourtit, C.; Esteves, T.C.; Green, K.; Lambert, A.J.; Miwa, S.; Pakay, J.L.; Parker, N. Mitochondrial superoxide: Production, biological effects, and activation of uncoupling proteins. *Free Radic. Biol. Med.* **2004**, *37*, 755–767. [CrossRef] [PubMed]
28. Granata, C.; Oliveira, R.S.; Little, J.P.; Renner, K.; Bishop, D.J. Mitochondrial adaptations to high-volume exercise training are rapidly reversed after a reduction in training volume in human skeletal muscle. *FASEB J.* **2016**, *30*, 3413–3423. [CrossRef] [PubMed]
29. Greggio, C.; Jha, P.; Kulkarni, S.S.; Lagarrigue, S.; Broskey, N.T.; Boutant, M.; Wang, X.; Conde Alonso, S.; Ofori, E.; Auwerx, J.; et al. Enhanced Respiratory Chain Supercomplex Formation in Response to Exercise in Human Skeletal Muscle. *Cell Metab.* **2017**, *25*, 301–311. [CrossRef]
30. Tonkonogi, M.; Walsh, B.; Svensson, M.; Sahlin, K. Mitochondrial function and antioxidative defence in human muscle: Effects of endurance training and oxidative stress. *J. Physiol.* **2000**, *528 Pt 2*, 379–388. [CrossRef]
31. Chandwaney, R.; Leichtweis, S.; Leeuwenburgh, C.; Ji, L.L.J.A. Oxidative stress and mitochondrial function in skeletal muscle: Effects of aging and exercise training. *AGE* **1998**, *21*, 109–117. [CrossRef] [PubMed]
32. Sahlin, K.; Shabalina, I.G.; Mattsson, C.M.; Bakkman, L.; Fernstrom, M.; Rozhdestvenskaya, Z.; Enqvist, J.K.; Nedergaard, J.; Ekblom, B.; Tonkonogi, M. Ultraendurance exercise increases the production of reactive oxygen species in isolated mitochondria from human skeletal muscle. *J. Appl. Physiol.* **2010**, *108*, 780–787. [CrossRef] [PubMed]
33. Horie, M.; Warabi, E.; Komine, S.; Oh, S.; Shoda, J. Cytoprotective Role of Nrf2 in Electrical Pulse Stimulated C2C12 Myotube. *PLoS ONE* **2015**, *10*, e0144835. [CrossRef] [PubMed]
34. Powers, S.K.; Schrager, M. Redox signaling regulates skeletal muscle remodeling in response to exercise and prolonged inactivity. *Redox Biol.* **2022**, *54*, 102374. [CrossRef]

antioxidants

MDPI

Article

Integration of Transcriptomic Analysis, Network Pharmacology, and Experimental Validation Demonstrates Enhanced Muscle-Protective Effects of Ethanol Extract of Jakyak-Gamcho-Tang

Aeyung Kim [1,*,†], Minh Nhat Tran [2,3,†], A Yeong Lee [4], Heerim Yeo [2], Su-Jin Baek [4], No Soo Kim [5], Seongwon Cha [4] and Sang-Min Park [2,*]

1 KM Application Center, Korea Institute of Oriental Medicine, Daegu 41062, Republic of Korea
2 College of Pharmacy, Chungnam National University, Daejeon 34134, Republic of Korea; tnminh@huemed-univ.edu.vn (M.N.T.); yeooh1@naver.com (H.Y.)
3 Faculty of Traditional Medicine, Hue University of Medicine and Pharmacy, Hue University, Hue 530000, Vietnam
4 KM Data Division, Korea Institute of Oriental Medicine, Daejeon 34054, Republic of Korea; lay7709@kiom.re.kr (A.Y.L.); baeksj@kiom.re.kr (S.-J.B.); scha@kiom.re.kr (S.C.)
5 KM Convergence Research Division, Korea Institute of Oriental Medicine, Daejeon 34054, Republic of Korea; nosookim@kiom.re.kr
* Correspondence: aykim71@kiom.re.kr (A.K.); smpark@cnu.ac.kr (S.-M.P.); Tel.: +82-53-940-3830 (A.K.); +82-42-821-5919 (S-M.P.)
† These authors contributed equally to this work.

Abstract: Muscle atrophy, characterized by progressive loss of skeletal muscle mass and strength, remains a significant therapeutic challenge. Jakyak-gamcho-tang (JGT) is a traditional herbal formulation that has demonstrated promising muscle-protective effects; however, the key bioactive constituents and the influence of different extraction methods have not yet been fully elucidated. This study compared the muscle-protective effects of the ethanol and water extracts of JGT (JGT-E and JGT-W, respectively), while also identifying the principal bioactive compounds that contribute to the enhanced efficacy of JGT-E. An integrative methodological approach was adopted, incorporating transcriptomic profiling, network pharmacology analysis, antioxidant activity assays, and in vitro validation using C2C12 myoblasts and myotubes. This comprehensive investigation enabled a detailed assessment of the biological activities of both JGT-E and JGT-W. Transcriptomic analysis revealed that JGT-E significantly modulates key pathways involved in oxidative phosphorylation, mitochondrial biogenesis, and signaling cascades related to PGC-1α, mTORC1, and ERRα, while simultaneously inhibiting TGF-β-mediated muscle atrophic signaling. Functional assays demonstrated that under oxidative stress conditions, JGT-E preserved mitochondrial content more effectively, reduced reactive oxygen species levels, and enhanced both myoblast viability and myotube integrity. Network pharmacology analysis identified isoliquiritigenin, catechin, and glabridin as major bioactive compounds enriched in JGT-E, all of which play critical roles in mitigating oxidative stress and supporting mitochondrial function. These findings were further substantiated by antioxidant assays that confirmed the contribution of these compounds to the observed muscle-protective effects of JGT-E. Overall, JGT-E exhibited superior efficacy in preventing muscle atrophy compared to JGT-W, likely due to its enriched profile of potent bioactive constituents. These results highlight the critical role of extraction methods in herbal medicine research and support the potential of JGT-E as a promising candidate for the treatment of muscle atrophy.

Keywords: Jakyak-gamcho-tang; muscle atrophy; transcriptome analysis; network pharmacology; oxidative stress

1. Introduction

Muscle atrophy, defined as the progressive loss of skeletal muscle mass and function [1], represents a significant clinical challenge that is commonly associated with aging, chronic diseases, and prolonged periods of immobilization [2]. This decline in muscle strength not only compromises quality of life but also increases the risk of morbidity and mortality [3]. Despite ongoing advances in biomedical research, the development of effective and safe therapeutic strategies to prevent or reverse muscle atrophy remains elusive [2]. Recently, traditional herbal medicines have garnered increasing attention as potential therapeutic alternatives, owing to their multi-target mechanisms and favorable safety profiles [4,5]. One such formulation, Jakyak-gamcho-tang (JGT), is a traditional decoction composed of Paeoniae Radix and Glycyrrhizae Radix et Rhizoma that has historically been used to relieve muscle cramps and spasms [6]. Emerging in vivo evidence has further highlighted the therapeutic potential of JGT, demonstrating its efficacy in attenuating muscle atrophy. Transcriptomic analyses have provided mechanistic insights into these effects, revealing that JGT modulates key pathways related to oxidative stress, mitochondrial function, and protein degradation [7].

A significant limitation of the current literature is its predominant reliance on water-based extraction methods, reflecting the traditional preparation practices for herbal formulations such as JGT [8]. However, the choice of extraction solvent plays a critical role in shaping the chemical profile of herbal extracts and consequently their biological activity. Water primarily extracts highly polar constituents such as glycosides and polysaccharides, whereas ethanol can solubilize a wider array of phytochemicals, including moderately polar and lipophilic compounds such as flavonoids, aglycones, and terpenoids [9]. These compositional differences have been shown to influence therapeutic efficacy substantially [10–12]. For example, the ethanol extract of *Rumex acetosa* contains significantly higher levels of emodin, a lipophilic anthraquinone that confers superior gastroprotective effects compared to its water extract [13]. Similarly, transcriptomic analysis of Bupleuri Radix demonstrated that while the water extract facilitated wound healing through promotion of cell adhesion and extracellular matrix organization, the ethanol extract predominantly activated apoptotic and anti-proliferative pathways, resulting in distinct therapeutic outcomes [14]. Such findings underscore the value of transcriptome-driven evaluation of extraction protocols, as facilitated by resources like KORE-MAP v1.0, which systematically profiles gene expression changes induced by herbal prescriptions under different extraction conditions [15]. Recent studies have also revealed considerable variation in the chemical composition of JGT depending on the extraction solvent. The ethanol extract (JGT-E) has shown significantly greater efficacy than the water extract (JGT-W) in scavenging reactive oxygen species (ROS) and protecting C2C12 myoblasts from oxidative-stress-induced cell death [16]. However, the specific bioactive compounds responsible for these enhanced antioxidative and muscle-protective effects have yet to be clearly identified.

The advent of network pharmacology has provided a robust and integrative framework for elucidating the complex, multi-component interactions that characterize traditional herbal medicines. This approach combines systems biology with computational modeling to predict compound–target interactions across a range of biological pathways, thereby aligning with the multi-target and multi-effect paradigm intrinsic to herbal formulations [17,18]. Through this methodology, the functional roles of specific phytochemicals in modulating disease-associated targets and signaling cascades have been systematically characterized. In the context of JGT, network pharmacology analysis has identified naringenin as a key bioactive compound in the treatment of functional dyspepsia. Naringenin was found to regulate gastrointestinal motility by modulating factors related to gastrointestinal movement, markers of interstitial cells of Cajal, and Aquaporin-3 expression [19]. Furthermore, the potential efficacy of JGT in addressing sphincter of Oddi dysfunction has been linked to the actions of glycycoumarin, licoflavonol, echinatin, and homobutein. These compounds are predicted to influence therapeutic outcomes by targeting the B cell receptor signaling pathway, the complement cascade, and muscle-contraction pathways [20]. These predictive insights not only deepen our mechanistic understanding of JGT's therapeutic actions but also serve as a scientific basis for the rational design of targeted experimental validations.

Given that water decoction represents the traditional clinical preparation of JGT and ethanol extraction is widely adopted in modern pharmacology to capture less polar but bioactive constituents, this study aimed to systematically compare the muscle-protective effects of JGT-W and JGT-E. This comparative approach was intended to bridge traditional usage with chemical optimization. Additionally, this study sought to identify the key active constituents that may account for the differential effects observed between the two extracts. To address these objectives, we employed a comprehensive, integrative approach encompassing transcriptomic analyses, network pharmacology predictions, antioxidant activity assays, and experimental validation using C2C12 muscle cell models. This multifaceted strategy was designed not only to validate the enhanced efficacy of JGT-E but also to elucidate the specific molecular mechanisms and active compounds underlying its improved muscle-protective properties.

2. Materials and Methods

2.1. Chemicals

Ascorbic acid (A4544), 2,2′-Azino-bis(3-ethylbenzothiazoline-6-sulfonic acid) diammonium salt (ABTS, A1888), potassium persulfate (#216224), crystal violet solution (V5265), dexamethasone (D4902), dimethyl sulfoxide (DMSO, D8418), 2,2-diphenyl-1-picrylhydrazyl (DPPH, D9132), hydrogen peroxide (H_2O_2) solution (#216763), and sodium dodecyl sulfate (SDS, L3771) were obtained from Sigma-Aldrich (St. Louis, MO, USA). Palmitic acid (CFN99716) was obtained from Wuhan ChemFaces Biochemical (Wuhan, China).

2.2. Herbal Materials

Dried *Paeonia lactiflora* Pall. (Paeoniae Radix) and *Glycyrrhiza uralensis* Fisch. ex DC. (Glycyrrhizae Radix et Rhizoma) were obtained from Kwangmyungdang Co. (Ulsan, Republic of Korea). The quality of these specimens complies with the standards established by the Korean Pharmacopoeia. Reference specimens have been authenticated and are preserved under voucher numbers #2–21–0001 and #2–21–0002, respectively, at the Korean Herbarium of Standard Herbal Resources, which is affiliated with the Herbal Medicine Resources Research Center at the Korea Institute of Oriental Medicine (Naju, Republic of Korea).

2.3. Preparation of JGT-W and JGT-E

JGT-W and JGT-E were prepared by KOC Biotech Co. (Daejeon, Republic of Korea). In brief, JGT is composed of dried Paeoniae Radix and Glycyrrhizae Radix et Rhizoma in a 1:1 weight ratio. JGT-W was extracted with 10 times the volume of hot distilled water for 3 h using a reflux extraction system (MS-DM609; MTOPS, Seoul, Republic of Korea). JGT-E was produced by pulverizing the same materials, followed by the addition of 70% ethanol at a volume 4 times that of the sample weight. Ultrasonic extraction was conducted at 22–24 °C for 1 h, with this process being repeated twice. Both extracts were then filtered through a 53 μm mesh filter, concentrated using a rotary evaporator (Ev-1020, SciLab, Seoul, Republic of Korea) at 37 °C, and lyophilized in a freeze dryer (LP-20, Ilshin-Bio-Base, Dongducheon, Republic of Korea). The final extract was homogenized with a 600 μm sieve and stored in a foil Ziplock bag with desiccant silica gel. For the in vitro study, JGT-W and JGT-E (100 mg each) were vortexed for 30 min in 10 mL of phosphate-buffered saline (PBS; Thermo Fisher Scientific, Waltham, MA, USA) containing 2% DMSO, then sterilized through a 0.22 μm membrane. The resulting solution was aliquoted and stored at −80 °C until further use.

2.4. C2C12 Cell Culture and Differentiation

The C2C12 murine myoblast cell line (CRL-1772), originally derived from the thigh muscle of C3H mice after a crush injury, was obtained from the American Type Culture Collection (ATCC, Manassas, VA, USA). This cell line serves as a widely accepted in vitro model for studying skeletal muscle differentiation, oxidative stress, and muscle atrophy. C2C12 myoblasts were maintained in growth medium (GM) composed of high glucose (4.5 g/L) Dulbecco's Modified Eagle Medium (DMEM) supplemented with 10% heat-inactivated fetal bovine serum and 100 IU/mL penicillin/100 μg/mL streptomycin (P/S) at 37 °C in a humidified 5% CO_2 incubator. Upon reaching full confluence, cells were induced to differentiate into myotubes by switching to differentiation medium (DM) containing DMEM with 2% heat-inactivated horse serum and P/S. The DM was refreshed every 2 days for 5–7 days. All culture reagents were purchased from Thermo Fisher Scientific.

2.5. Assessment of Cell Viability and Myotube Degradation in C2C12 Cells

To determine the optimal non-cytotoxic dose of JGT-W and JGT-E, 5000 cells of C2C12 myoblasts were plated into each well of a 96-well culture plate. After overnight incubation, the cells were treated with JGT-W and JGT-E at concentrations of 50, 100, 250, and 500 μg/mL, or with a vehicle control (0.1% DMSO). After 24 h, the supernatant was removed, and the cells were rinsed twice with PBS. Cell viability was determined using the EZ-Cytox Enhanced Cell Viability Assay Kit (Daeil Lab Service Co., Ltd., Seoul, Republic of Korea) and SpectraMax i3 microplate reader (Molecular Devices, LLC, Sunnyvale, CA, USA) following the manufacture's guide. Based on these results and previous findings on muscle-protective activity of JGT-W [7], concentrations of 100 and 200 μg/mL were chosen for further experiments. To assess the protective role of JGT-W and JGT-E against oxidative stress-induced cellular damage, C2C12 myoblasts were pretreated with 100 and 200 μg/mL JGT-W and JGT-E for 12 h before being exposed to 0.25 mM H_2O_2. After 24 h, cell viability was assessed using EZ-Cytox Enhanced Cell Viability Assay Kit, and the results were expressed as a percentage relative to the vehicle-treated control cells. To investigate the protective effects of JGT-W and JGT-E on myotube degradation, C2C12 myotubes that had undergone differentiation for five days were pretreated with 100 and 200 μg/mL JGT-W and JGT-E for 12 h, followed by exposure to 0.25 mM H_2O_2, 0.2 mM dexamethasone, or 0.2 mM palmitic acid for an additional 24 h. After treatment, the cells were washed with PBS and stained with a crystal violet solution (0.2% crystal violet in 20% methanol) for 30 min at 25 °C. The stained cells were then rinsed with tap water, air-dried, and photographed. To quantify myotube density, the stained cells were solubilized in a 1% SDS solution at 37 °C for 30 min, and absorbance was measured at 590 nm using a SpectraMax i3 microplate reader.

2.6. Transcriptomic Analysis

The RNA-seq data of C2C12 myotubes treated with increasing concentrations of JGT-W and JGT-E (20, 100, and 500 μg/mL) for 24 h were obtained from GEO GSE227494 and GSE295069, respectively. Bioinformatics analysis of the RNA-seq data was performed using the R (v4.2.1). Differential gene expression analysis between groups was conducted using the DESeq2 package (v1.38.2), resulting in a ranked list of genes based on fold change and adjusted *p*-value corrected for multiple testing via false discovery rate. Gene set enrichment analysis determines the degree to which genes within a defined gene set that share a common biological function or pathway are disproportionately overrepresented at the top or bottom of a specific ranked list of genes, resulting in a normalized enrichment score (NES) and a false discovery rate (FDR). The gene sets of Hallmark, Gene Ontology, and canonical pathways were obtained from the Molecular Signature Database (MSigDB). Gene set enrichment analysis was performed using the fgsea package (v1.24.0) with the following parameters: 10,000 permutations, minimum size 15, and maximum size 500. GSEA results were visualized by the heatmap package (v.1.0.12).

2.7. Detection of Intracellular ROS Level

To evaluate the protective effects of JGT-W and JGT-E against oxidative stress, C2C12 myoblasts and myotubes were pretreated with 100 and 200 μg/mL of JGT-W and JGT-E for 12 h. Following this pretreatment, the cells were subjected to oxidative stress by exposure to 0.25 mM H_2O_2 for 6 h. The levels of intracellular ROS were measured using the CellROX® Green Reagent (Thermo Fisher Scientific) in accordance with the manufacturer's protocol. In brief, after exposure to H_2O_2, the cells were incubated with 5 μM CellROX® Green for 30 min at 37 °C. Subsequently, the cells were washed with Hanks' balanced salt solution. Fluorescence images were obtained using a fluorescence microscope (IX71, Olympus Co., Tokyo, Japan), and the fluorescence intensity was quantified using ImageJ software, version 1.54f (National Institute of Health, Bethesda, MD, USA).

2.8. Measurement of Mitochondrial Mass

C2C12 myotubes were pretreated with JGT-W and JGT-E for 12 h and then exposed to 0.25 mM H_2O_2 for 24 h. To evaluate mitochondrial mass, the cells were washed with PBS and then incubated with 50 nM MitoTracker™ Deep Red FM (M22426, Thermo Fisher Scientific) in serum-free DMEM for 30 min at 37 °C. After washing with PBS, red fluorescence images were captured using a fluorescence microscope, and the fluorescence intensity was quantified using ImageJ software, version 1.54f.

2.9. Free Radical Scavenging Assay

The free radical-scavenging activities of JGT-W and JGT-E were evaluated using ABTS and DPPH assays. For the ABTS assay, a stock solution was prepared by mixing equal volumes of ABTS (7.4 mM) and potassium persulfate (2.6 mM) in the dark for 2 h. The working solution was diluted to an optical density of ~0.7 at 734 nm (OD734). Serially diluted JGT-W and JGT-E were combined with the ABTS working solution in a 96-well plate and incubated in the dark for 7 min at 22–24 °C, after which OD734 was measured using a SpectraMax i3 microplate reader. In the DPPH assay, a working solution of 0.2 mM DPPH was prepared in absolute methanol in the dark. Equal volumes of serially diluted JGT-W and JGT-E were mixed with the DPPH solution and incubated for 30 min at 22–24 °C, followed by measurement at OD514. Ascorbic acid was used as a positive control. Free radical-scavenging activities were calculated using the formula: inhibition (%) = (1 − OD of JGT-W, JGT-E, or ascorbic acid/OD of distilled water) × 100.

2.10. Quantitative Analysis of JGT Using UHPLC-MS/MS

Quantitative analyses of 16 components of JGT were carried out using an Agilent 1290 Infinity II ultrahigh-performance liquid chromatography (UHPLC) system and 6495 C triple quadrupole mass spectrometer (Agilent Technologies, Santa Clara, CA, USA). The sample was detected as positive and negative modes of electrospray ionization (ESI) source. A UHPLC column was used to Kinetex 1.7 µm C18 100A (150 × 2.10 mm; Phenomenex, Torrance, CA, USA), set to 40 °C. The temperature of the auto-sampler was kept at 4 °C, and injection volume was 2 µL. Water with 0.1% formic acid (A) and acetonitrile (B) were used as mobile phase applying a gradient program; 5–20% B for 0–1 min, 20–25% B for 1–6.5 min, 25–50% B for 6.5–17 min, 50–100% B for 17–20 min, isocratic 100% B for 20–25 min, 100–5% B for 25–26 min, and isocratic 5% B for 26–30 min. The capillary voltage was set to positive of 3500 V and negative of 3000 V. Nozzle voltage of positive and negative was 500 V and 1500 V, respectively. Both gas temperature and nebulizer gas was set to 130 °C and 25 psi. Drying gas flow was 11.0 L/min, and the MS/MS parameters of 16 standard components were optimized using dynamic multiple reaction monitoring (MRM), shown in Table 1. The software used for quantitative analysis was MassHunter version 10.1. The 16 components for quantitative analysis were 1,2,3,4,6-O-Pentagalloylglucose (CFN90192), benzoylpaeoniflorin (CFN99536), glabridin (CFN99731), glycyrrhizin (CFN99151), isoliquiritin apioside (CFN90800), liquiritin apioside (CFN90707), ononin (CFN99136, ≥98.0%; ChemFaces, Hubei, China), albiflorin (HY-N0037, ≥98.67%); 18beta-glycyrrhetinic acid (HY-N0180, ≥99.88%); isoliquiritigenin (HY-N0102, ≥ 98.07%), Oxypaeoniflorin (HY-N0748, ≥98.37%; MedChemExpress, Princeton, NJ, USA), isoliquiritin (A0041), liquiritigenin (A0042), liquiritin (A0040, ≥99.0%; Chengdu Must Bio-Technology Co., Ltd., Chengdu, China), (+)-catechin (PHR1963, USP reference standard), and paeoniflorin (P0038, ≥98.0%; Merck KGaA, Darmstadt, Germany). All standards were mixed and quantitatively analyzed using a simultaneous multi-compound analysis method. To verify the analytical accuracy, recovery tests were conducted, and data are summarized in Supplementary Table S1.

Table 1. The condition of MRM for UHPLC-TQ-MS/MS.

Peak No.	Components	Rt (min)	Precursor Ion (*m/z*)	Product Ion (*m/z*)	Collision Energy (V)	Ion Polarity
1	Oxypaeoniflorin	2.985	495.1	137.1	26.0	Negative
2	(+)-Catechin	3.115	291.1	139.0	14.0	Positive
3	Albiflorin	3.566	481.2	105.0	22.0	Positive
4	Paeoniflorin	3.759	525.2	449.1	10.0	Negative
5	Liquiritin apioside	4.341	549.2	255.1	34.0	Negative
6	Liquiritin	4.535	417.2	255.0	20.0	Negative
7	1,2,3,4,6-Pentagalloyl glucose	4.680	958.1	153.0	40.0	Positive
8	Isoliquiritin apioside	7.169	549.1	255.1	30.0	Negative
9	Ononin	7.634	431.1	269.1	18.0	Positive
10	Isoliquiritin	7.789	417.0	255.1	18.0	Negative
11	Liquiritigenin	9.025	257.0	137.0	26.0	Positive
12	Benzoylpaeoniflorin	11.531	629.1	553.2	10.0	Negative
13	Isoliquiritigenin	13.811	257.0	137.0	22.0	Positive
14	Glycyrrhizin	15.030	821.4	351.0	40.0	Negative
15	Glabridin	19.529	325.1	189.1	14.0	Positive
16	Glycyrrhetinic acid	21.004	471.2	189.1	38.0	Positive

2.11. Network Pharmacology Analysis

A network pharmacology approach was employed to investigate the potential compounds and mechanisms of action of JGT in relation to muscle atrophy. Information on 16 compounds identified from HPLC analysis was retrieved from PubChem, including compound names, IDs, and SMILES structures. The targets of these compounds were collected and predicted using TCMSP v2.3, HIT 2.0, BATMAN-TCM 2.0, and SwissTargetPrediction v2019. Experimentally validated known targets were extracted from HIT, TCMSP, and BATMAN-TCM, ensuring comprehensive coverage of bioactive interactions. Additionally, predicted targets from TCMSP were also collected, while in BATMAN-TCM, those with a score cutoff ≥ 0.84 were included. SwissTargetPrediction was utilized to predict targets based on SMILES structures, and only those with a probability > 0.5 were selected for further analysis. To identify potential therapeutic targets related to muscle atrophy, data were extracted from DisGeNET v24.4, GeneCards v5.23, TTD v2024, and CTD (2025 updated). Targets were selected based on specific cutoff criteria, including a gene-disease associations score > 0.1 in DisGeNET, a Relevance score > 10 in GeneCards, and an Inference Score > 60 in CTD, while all available targets from TTD were included. For pathway enrichment analysis, Gene Ontology Biological Process (GOBP) enrichment was performed using enrichGO in R, with significantly enriched pathways selected based on an adjusted *p*-value ($p < 0.05$, Benjamini–Hochberg correction). The top 10 terms were prioritized based on adjusted *p*-values, providing insight into the biological processes influenced by the identified targets. Finally, herb-compound–target networks were constructed and analyzed using Cytoscape v3.10.3 to visualize and explore the interactions among compounds and target proteins, enabling the identification of potential compounds involved in muscle atrophy-related mechanisms.

2.12. Statistical Analysis

Data analysis was conducted with GraphPad Prism (v 9.5.1, GraphPad Software, San Diego, CA, USA). The results are expressed as the mean ± standard deviation (SD). Group differences were assessed using one-way analysis of variance, followed by Dunnett's multiple comparison test, with statistical significance established at $p < 0.05$.

3. Results

3.1. Transcriptomic Analysis Shows That JGT-E Regulated Muscle-Protective Pathways Under Oxidative Stress More Effectively than Jgt-W

To compare the muscle-protective effects of JGT-W and JGT-E, we performed transcriptomic profiling of differentiated C2C12 myotubes. Differential gene expression analysis revealed that treatment with JGT-E resulted in the regulation of a significantly larger number of genes compared to JGT-W, indicating a stronger transcriptional response (Figure 1A). While both extracts shared several transcriptional signatures, JGT-E uniquely modulated a broader set of genes, implying a more extensive impact on gene expression (Figure 1B). Pathway enrichment analysis revealed clear dose-dependent enrichment of key molecular pathways implicated in muscle protection (Figure 1C). Notably, cells treated with JGT-E exhibited robust upregulation of pathways related to oxidative phosphorylation, mitochondrial function, PGC-1α signaling, mTORC1 signaling, and ERRα targets, along with marked downregulation of muscle-atrophic TGF-β signaling (Figure 1D). Furthermore, JGT-E significantly enhanced mitochondrial metabolic processes, cellular respiration, and ATP biosynthesis across all tested concentrations (Supplementary Figure S1A,B). Genes associated with mitochondrial biogenesis and myogenic differentiation were also more significantly changed by JGT-E (Supplementary Figure S1C). Collectively, these findings imply that the enriched bioactive components in JGT-E may be responsible for its superior muscle-protective efficacy compared to JGT-W.

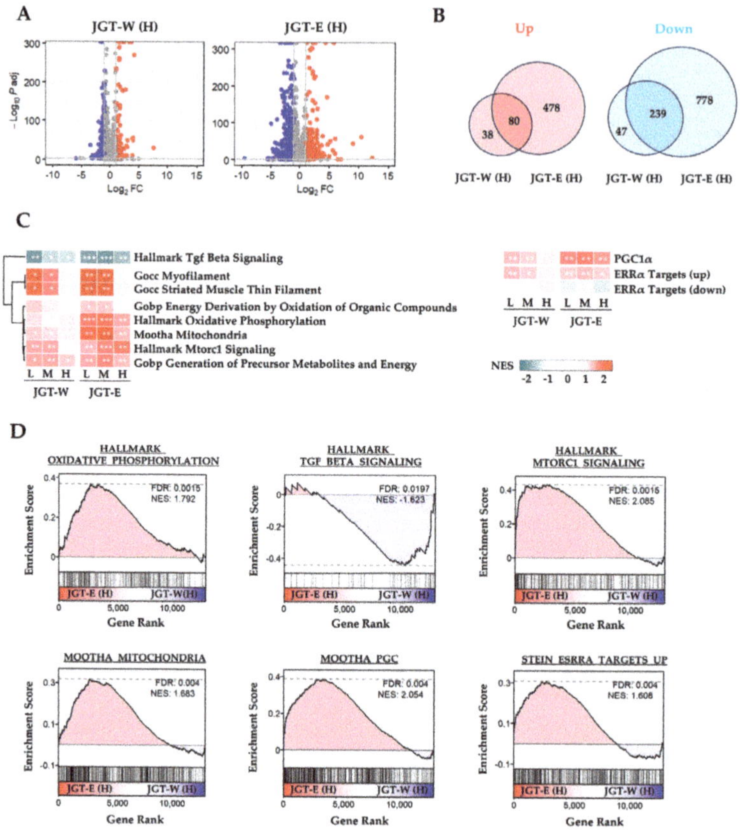

Figure 1. Transcriptomic analysis of C2C12 muscle cells treated with JGT-W or JGT-E. (**A**) Volcano plots depicting differentially expressed genes (DEGs) in C2C12 myotubes treated with JGT-W or JGT-E at a high dose (H). Genes significantly upregulated or downregulated ($|\log_2$ fold-change$| > 1$, adjusted $p < 0.05$) are marked in red or blue, respectively. (**B**) Venn diagrams illustrating the overlap and differences between significantly upregulated (left) and downregulated (right) DEGs from JGT-W and JGT-E treatments. (**C**) Gene set enrichment analysis (GSEA) results comparing enriched molecular pathways between JGT-W and JGT-E-treated cells at low (L, 20 μg/mL), medium (M, 100 μg/mL), and high (H, 500 μg/mL) doses. Normalized enrichment score (NES) is indicated by color intensity. Statistical significance of the false discovery rate (FDR) is indicated by asterisks: * FDR < 0.05, ** FDR < 0.01, *** FDR < 0.001. (**D**) Representative GSEA plots highlighting key pathways, including oxidative phosphorylation, TGF-β signaling, mitochondrial function, PGC-1α signaling, mTORC1 signaling, and ERRα targets. The dotted horizontal line in each GSEA plot represents the maximum or minimum enrichment score for the gene set across the ranked gene list.

3.2. JGT-E Demonstrated Greater Efficacy than JGT-W in Mitigating Cell Damage and ROS Production in C2C12 Myoblasts Under Oxidative Stress

Building on transcriptomic evidence demonstrating superior muscle-protective effects of JGT-E over JGT-W, we further validated these findings using a C2C12 muscle-cell model. To determine the appropriate treatment concentration, C2C12 myoblasts were incubated with various concentrations of JGT-W and JGT-E, up to 500 μg/mL, for 24 h. Cell viability assays indicated no cytotoxicity at concentrations up to 250 μg/mL (Figure 2A), and thus, 200 μg/mL was selected as the maximum concentration for subsequent in vitro experiments. To evaluate the protective effects of JGT-W and JGT-E under oxidative stress, C2C12 myoblasts were pretreated with either extract or a vehicle control (0.1% DMSO) for 12 h, followed by exposure to 0.25 mM H_2O_2 for 24 h. As shown in Figure 2B, H_2O_2 treatment significantly reduced cell viability to approximately 56.3% of control levels. In contrast, pretreatment with JGT-W and JGT-E markedly mitigated cytotoxic damage. At concentrations of 100 μg/mL and 200 μg/mL, JGT-W preserved cell viability at approximately 89.4% and 102.3% of control levels, while JGT-E maintained it at 102.6% and 115.4%, respectively. Morphological analysis further confirmed these protective effects: cells pretreated with either JGT-W or JGT-E exhibited minimal morphological changes following H_2O_2 exposure, closely resembling the morphology of untreated control cells (Figure 2C). Intracellular ROS levels were assessed using CellRox™ Green dye. As expected, H_2O_2 exposure induced marked green fluorescence, indicative of elevated ROS levels. Pretreatment with both extracts significantly attenuated ROS accumulation in a dose-dependent manner, with JGT-E demonstrating superior efficacy (Figure 2D). Specifically, at 100 μg/mL, JGT-W reduced ROS levels by approximately 29%, while JGT-E achieved a 79% reduction. At 200 μg/mL, the reductions were approximately 57% for JGT-W and 89% for JGT-E (Figure 2E). Collectively, these findings indicate that both JGT-W and JGT-E confer protective effects against oxidative stress-induced cytotoxicity in myoblasts, with JGT-E exhibiting significantly greater efficacy.

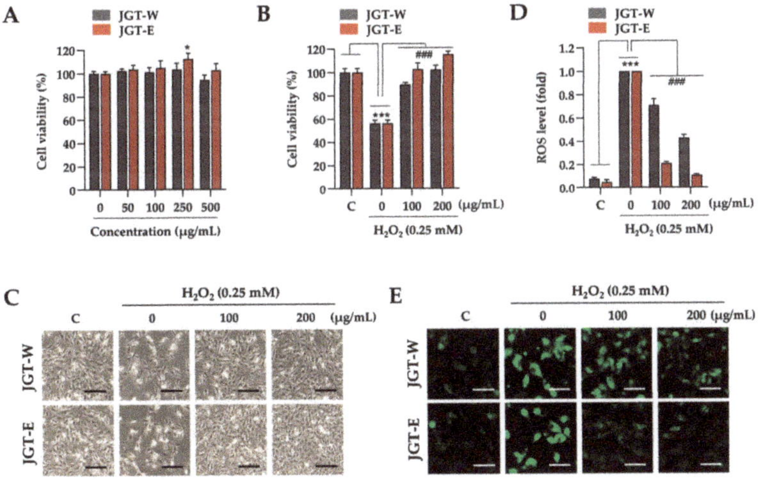

Figure 2. Effects of JGT-W and JGT-E under oxidative stress condition in C2C12 myoblasts. (**A**) C2C12 myoblasts were treated with increasing concentrations of JGT-W and JGT-E, ranging from 0 to 500 μg/mL. After 24 h, relative cell viability, compared to vehicle-treated control cells, was determined. (**B**,**C**) Myoblasts were pretreated with 100 or 200 μg/mL of JGT-W, JGT-E, or vehicle for 12 h, followed by exposure to 0.25 mM H_2O_2. After 24 h, relative cell viability was assessed. (**B**) Additionally, cellular morphology was captured using an inverted (**D**,**E**) C2C12 myoblasts were microscope (**C**). pretreated with 100 or 200 μg/mL of JGT-W, JGT-E, or vehicle for 12 h, and then exposed to 0.25 mM H_2O_2. After

6 h, intracellular levels of ROS were assessed using the ROS-sensitive CellROX™ Green dye. The fold increase in ROS levels relative to the vehicle-treated controls was quantified using ImageJ, version 1.54f. (**D**) The production of intracellular ROS was visualized through green fluorescence under a fluorescence microscope (**E**). All data are presented as the mean ± SD (n = 3). Statistical significance was evaluated by one-way ANOVA followed by Dunnett's multiple comparison test. * $p < 0.05$, *** $p < 0.001$ vs. vehicle-treated control cells; ### $p < 0.001$ vs. H_2O_2 + vehicle-treated control cells. Scale bar = 100 μm.

3.3. JGT-E Was More Effective than JGE-W in Promoting Mitochondrial Biogenesis, Reducing ROS Levels, and Inhibiting the Degradation of C2C12 Myotubes Exposed to H_2O_2

Next, C2C12 myotubes were treated with JGT-W and JGT-E at concentrations of up to 500 μg/mL for 24 h, during which time no cytotoxic effects were observed (Figure 3A). Accordingly, 200 μg/mL was selected as the maximum concentration for subsequent experiments, consistent with the protocol used in C2C12 myoblast studies. To assess the protective effects of JGT-W and JGT-E against oxidative-stress-induced muscle atrophy, myotubes were pretreated with 100 μg/mL or 200 μg/mL of either extract for 12 h, followed by exposure to 0.25 mM H_2O_2 for 24 h. As illustrated in Figure 3B, H_2O_2 treatment resulted in pronounced myotube degradation, characterized by morphological collapse and shortening. In contrast, pretreatment with JGT-W or JGT-E significantly ameliorated these effects and preserved the characteristic morphology of the myotubes, as confirmed by crystal violet staining (Figure 3C). Quantitative analysis showed that myotube density decreased by approximately 79.1% following H_2O_2 exposure compared to vehicle-treated controls. Pretreatment with 100 μg/mL or 200 μg/mL of JGT-W restored myotube density to 67.8% and 43.2%, respectively. Notably, pretreatment with JGT-E at the same concentrations preserved myotube density at 48.9% and 78.9%, respectively, indicating a more robust protective effect. To investigate further the impact on mitochondrial integrity, mitochondrial mass was assessed using MitoTracker™. As shown in Figure 3D, H_2O_2 markedly diminished mitochondrial mass in C2C12 myotubes; however, pretreatment with either JGT-W or JGT-E substantially restored mitochondrial content even after oxidative insult. Additionally, pretreatment with JGT-W or JGT-E significantly reduced H_2O_2-induced intracellular ROS accumulation in a dose-dependent manner, with JGT-E demonstrating superior efficacy (Figure 3E). Under conditions of chemically induced muscle atrophy using dexamethasone and palmitic acid, both JGT extracts significantly attenuated myotube degradation and increased myotube density. In these models, JGT-E consistently exhibited slightly greater efficacy than JGT-W (Supplementary Figure S2A). Collectively, these results indicate that both JGT-W and JGT-E effectively mitigate oxidative stress-induced myotube damage, with JGT-E showing marginally superior muscle-protective effects. In addition, we validated our injury models by evaluating two established positive control compounds, resveratrol and arachidonic acid [21–23], under identical conditions. Pretreatment with either compound significantly inhibited myotube degradation caused by H_2O_2, dexamethasone, and palmitic acid. These results, shown in Supplementary Figure S2B, confirm the robustness of our in vitro atrophy models and establish a pharmacological benchmark for the muscle-protective potential of JGT extracts.

3.4. Quantitative Analysis of JGT-W and JGT-E Using UHPLC-MS/MS

To characterize the primary constituents of JGT-W and JGT-E, LC-MS/MS was conducted at a concentration of 1.00 mg/mL. Samples exceeding the calibration curve range were subjected to a 10-fold dilution. Quantitative analysis of 16 target compounds was performed using a dynamic MRM method coupled with a UHPLC-triple quadrupole mass spectrometer. Representative chromatograms of JGT-W, JGT-E, and the standard mixture are presented in Figure 4 and Figure S3, while the quantitative data for all 16 compounds are summarized in Table 2. In JGT-W, isoliquiritin, liquiritigenin, and glycyrrhetinic acid were detected at concentrations below the limit of quantification and were, therefore, reported as trace amounts. Overall, most compounds were present at higher concentrations in JGT-E than in JGT-W. Notably, albiflorin and oxypaeoniflorin were detected at slightly higher levels in JGT-W. Paeoniflorin was the most abundant compound in JGT-W, whereas glycyrrhizin was predominant in JGT-E.

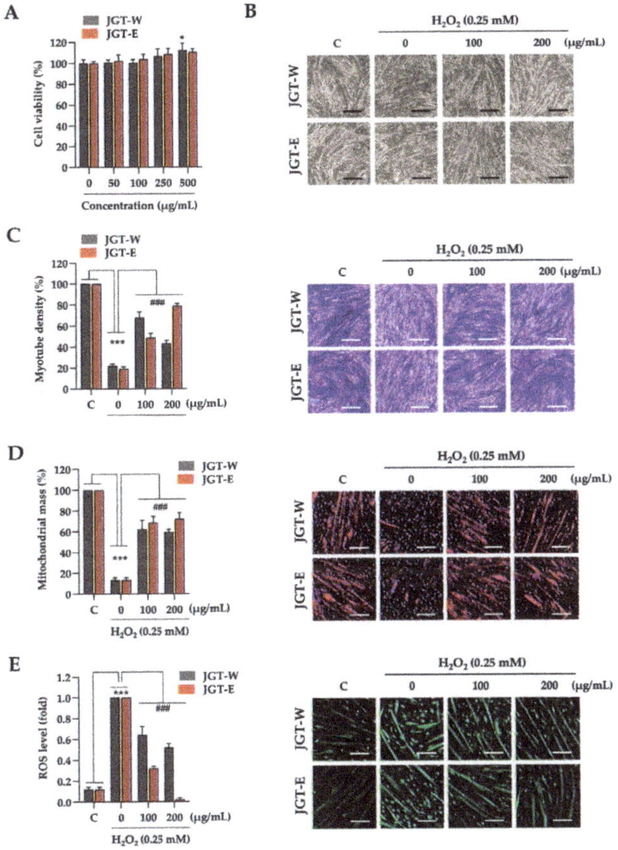

Figure 3. Effects of JGT-W and JGT-E under oxidative stress condition in C2C12 myotubes. (**A**) C2C12 myotubes were treated with specified concentrations of JGT-W and JGT-E. After 24 h, the cell viability, relative to the vehicle-treated control cells, was determined. (**B–D**) C2C12 myotubes were pretreated with 100 and 200 μg/mL of JGT-W and JGT-E for 12 h and subsequently exposed to 0.25 mM H_2O_2 for an additional 40 h. Morphological changes were observed using an inverted microscope (**B**). Myotube density was quantified through crystal violet staining. (**C**) After staining with MitoTracker^{TM} Deep Red, mitochondrial mass was visualized under a fluorescence microscope and quantified using ImageJ, version 1,54f. (**D**,**E**) C2C12 myotubes pretreated with 100 and 200 μg/mL of JGT-W and JGT-E for 12 h were subsequently exposed to 0.25 mM H_2O_2. After 6 h, intracellular ROS levels were detected using CellROX^{TM} Green dye under a fluorescence microscope. The fold increase in ROS levels was quantified using Image J, version 1,54f. All data are presented as the mean ± SD (n = 3). Statistical significance was evaluated by one-way ANOVA followed by Dunnett's multiple comparison test. * $p < 0.05$, *** $p < 0.001$ vs. vehicle-treated controls; ### $p < 0.001$ vs. H_2O_2 + vehicle-treated controls. Scale bar = 100 μm.

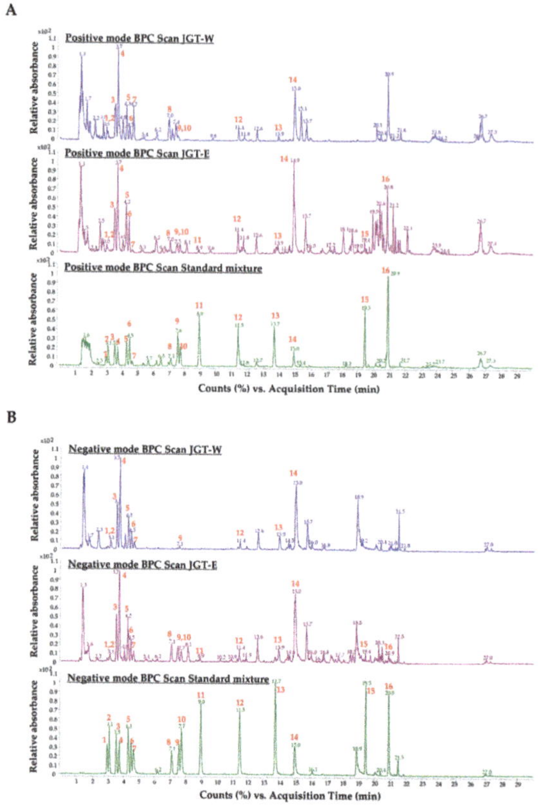

Figure 4. Analysis of JGT-W and JGT-E using UHPLC-MS/MS. Positive-ion mode of base peak ion chromatograms (BPC) (**A**) and negative-ion mode of BPC (**B**) of JGT-W, JGT-E, and standard mixture using a UHPLC-TQ-MS/MS. Red numbers indicate standard compounds; 1: oxypaeoniflorin, 2: (+)-catechin, 3: albiflorin, 4: paeoniflorin, 5: liquiritin apioside, 6: liquiritin, 7: 1,2,3,4,6-pentagalloyl glucose, 8: isoliquiritin apioside, 9: ononin, 10: isoliquiritin, 11: liquiritigenin, 12: benzoylpaeoniflorin, 13: isoliquiritigenin, 14: glycyrrhizin, 15: glabridin, 16: glycyrrhetinic acid.

Table 2. The linearity, range, and contents of 16 components in JGT-W and JGT-E using UHPLC-MS/MS analysis.

No.	Components	Regression Equation	R^2	Range (μg/mL)	Concentration (mg/g, *n* = 6)	
					JGT-W	JGT-E
1	Oxypaeoniflorin	y = 86,522x + 6367	0.9990	0.16–2.50	0.959 (±0.024)	0.875 (±0.012)
2	(+)-Catechin	y = 484,425x + 23,507	0.9990	0.08–2.5	0.089 (±0.003)	1.101 (±0.016)
3	Albiflorin	y = 590,324x + 8051	0.9992	0.04–5.00	15.141 (±1.636)	14.881 (±0.675)
4	Paeoniflorin	y = 205,795x + 93,092	0.9990	0.63–10.00	66.180 (±6.085)	70.331 (±3.824)
5	Liquiritin apioside	y = 379,464x + 22,860	0.9991	0.04–2.50	9.013 (±0.931)	10.271 (±0.748)
6	Liquiritin	y = 459,755x + 37,366	0.9990	0.08–1.25	5.158 (±0.461)	6.956 (±0.584)
7	1,2,3,4,6-Pentagalloyl glucose	y = 220,570x + 3267	0.9991	0.04–5.00	0.295 (±0.022)	2.240 (±0.047)
8	Isoliquiritin apioside	y = 162,498x + 80,392	0.9991	0.63–10.00	0.165 (±0.013)	6.092 (±0.066)
9	Ononin	y = 2,566,720x + 52,6143	0.9991	0.31–5.00	3.896 (±0.039)	4.234 (±0.069)
10	Isoliquiritin	y = 406,963x + 15,5124	0.9991	0.31–5.00	tr *	1.849 (±0.029)
11	Liquiritigenin	y = 1,525,781x + 42,803	0.9992	0.04–1.25	tr	1.038 (±0.013)
12	Benzoylpaeoniflorin	y = 306,585x + 50,791	0.9990	0.16–5.00	0.643 (±0.015)	0.975 (±0.013)

Table 2. *Cont.*

No.	Components	Regression Equation	R^2	Range (μg/mL)	Concentration (mg/g, n = 6)	
					JGT-W	JGT-E
13	Isoliquiritigenin	y = 1,244,869x + 29,884	0.9992	0.04–2.50	0.006 (±0.002)	1.184 (±0.017)
14	Glycyrrhizin	y = 80,641x + 18,808	0.9990	0.16–10.00	50.324 (±4.754)	75.053 (±4.004)
15	Glabridin	y = 1,288,421x + 10,894	0.9996	0.04–0.63	0.022 (±0.001)	0.838 (±0.077)
16	Glycyrrhetinic acid	y = 558,961x + 28,138	0.9990	0.08–1.25	tr	0.241 (±0.004)

* tr; trace.

3.5. Prediction of Potential Compounds in JGT Involved in Muscle-Protective Mechanisms

To identify the bioactive compounds responsible for the enhanced muscle-protective efficacy of JGT-E, we performed a network pharmacology analysis. In total, 206 unique molecular targets were retrieved from 16 compounds in JGT, after eliminating duplicates across multiple databases (Figure 5A). Pathway enrichment analysis of these targets revealed significant associations with oxidative stress responses (GO:0006979, adjusted p-value = 3.04×10^{-19}, gene count = 35), implying strong antioxidant potential for the constituents of JGT (Figure 5B, top). To identify the compounds involved in modulating this pathway, we constructed a compound–target interaction network focused on the 35 oxidative-stress-associated targets of JGT (Figure 5B, bottom). Among the top-ranking compounds (degree > 7) identified were paeoniflorin, isoliquiritigenin, catechin, glycyrrhizic acid, and glabridin. To investigate further whether these key compounds also regulate mitochondrial function and muscle-atrophy-related pathways, we expanded our analysis by constructing additional compound–target networks based on mitochondrial- and muscle-atrophy-related targets (Figure 5C,D). These analyses consistently highlighted the importance of the top bioactive compounds across multiple relevant pathways. To assess the significance of each compound comprehensively across the three biological contexts (oxidative stress, mitochondrial function, and muscle atrophy), we aggregated the degree centrality of each compound (Figure 5E). This analysis revealed that isoliquiritigenin, catechin, glabridin, and liquiritigenin—compounds that are more abundant in JGT-E—ranked among the most influential bioactive constituents, with isoliquiritigenin emerging as the most pivotal overall. Collectively, these network pharmacology findings imply that the superior muscle-protective effects of JGT-E, compared to JGT-W, are likely attributable to its higher concentrations of these key bioactive compounds, which notably enhance the modulation of oxidative-stress-, mitochondrial-function-, and muscle-atrophy-related pathways.

3.6. Identification of Compounds That Contribute to Protection Against Oxidative Stress and Muscle Damage

The muscle-protective properties of the bioactive compounds identified in JGT were further evaluated based on the results from the network pharmacology analysis. Initially, cell-free biochemical assays were conducted to assess the antioxidant capacity of the 16 compounds, with a focus on their free-radical scavenging activity, using established methods such as the ABTS and DPPH assays. The results indicate that catechin, glabridin, isoliquiritigenin, and 1,2,3,4,6-pentagalloyl glucose exhibited significant efficacy in ABTS radical scavenging (Figure 6A), while catechin, glabridin, and 1,2,3,4,6-pentagalloyl glucose demonstrated notable effectiveness in DPPH free-radical scavenging (Figure 6B). Ascorbic acid, used as a positive control, showed dose-dependent free-radical scavenging activity in both assays. Subsequently, we assessed the ability of these four selected compounds (catechin, glabridin, isoliquiritigenin, and 1,2,3,4,6-pentagalloyl glucose) to modulate intracellular ROS production under oxidative stress induced by H_2O_2. The results show a significant reduction in intracellular ROS levels following pretreatment with these compounds (Figure 6C). To evaluate further the protective effects of the 16 compounds in JGT against H_2O_2-induced myotube degradation, C2C12 myotubes were pretreated with the compounds for 12 h, followed by exposure to 0.25 mM H_2O_2. After 24 h, the cells were stained with crystal violet, and myotube density was quantified (Supplementary Figure S4). As shown in Figure 6D, myotube density was maintained at approximately 88.4%, 101.6%, 81%, and 93% of control levels for catechin, glabridin, isoliquiritigenin, and 1,2,3,4,6-pentagalloyl glucose, respectively. These findings demonstrate that these compounds effectively preserve muscle tissue, which may contribute to the superior muscle-protective efficacy of JGT-E compared to JGT-W.

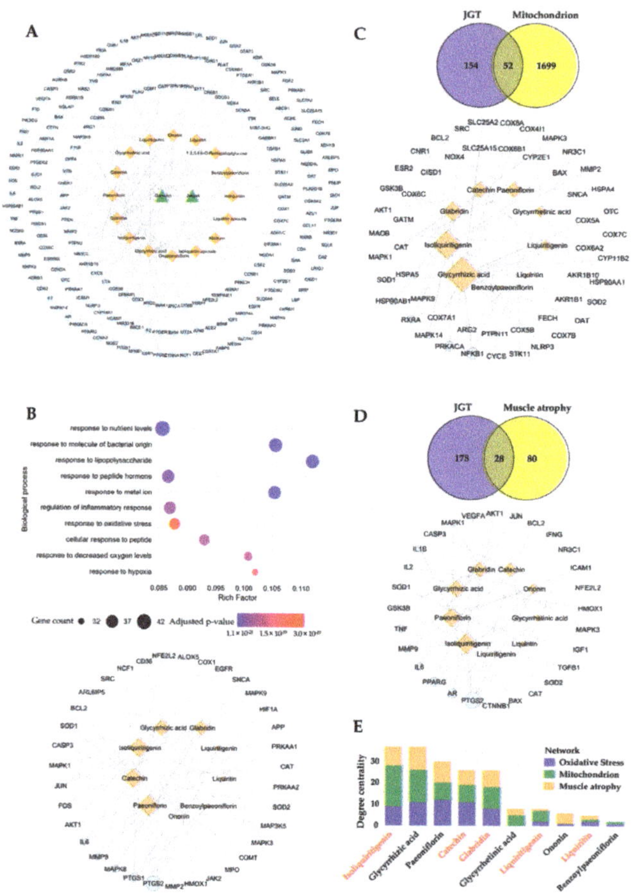

Figure 5. Network pharmacology-based prediction of compounds in JGT effective for muscle atrophy. (**A**) The herb-compound–target network of JGT, where green triangle nodes represent herbs (2 nodes), orange diamond nodes represent compounds (16 nodes), and light blue circle nodes represent targets (206 nodes). (**B**) Gene Ontology Biological Process (GOBP) enrichment analysis of the 206 JGT-associated targets (top) and the compound–target network (bottom), focusing on 35 targets linked to the "Response to Oxidative Stress" (GO:0006979) pathway. (**C**) Overlap between JGT targets and mitochondria-related genes (52 overlapping targets) (top), along with the compound–target network

of JGT compounds interacting with these overlapping mitochondrial targets (bottom). (**D**) Overlap between JGT targets and muscle atrophy-related genes (28 overlapping targets) (top), with the corresponding compound–target network of JGT compounds interacting with these overlapping muscle atrophy targets (bottom). The node sizes represent the degree centrality of each node in networks (**B–D**). (E) Total degree centrality of JGT compounds across the three networks (**B–D**): oxidative stress, mitochondrion, and muscle atrophy. Compounds highlighted in red are present at higher concentrations in JGT-E compared to JGT-W.

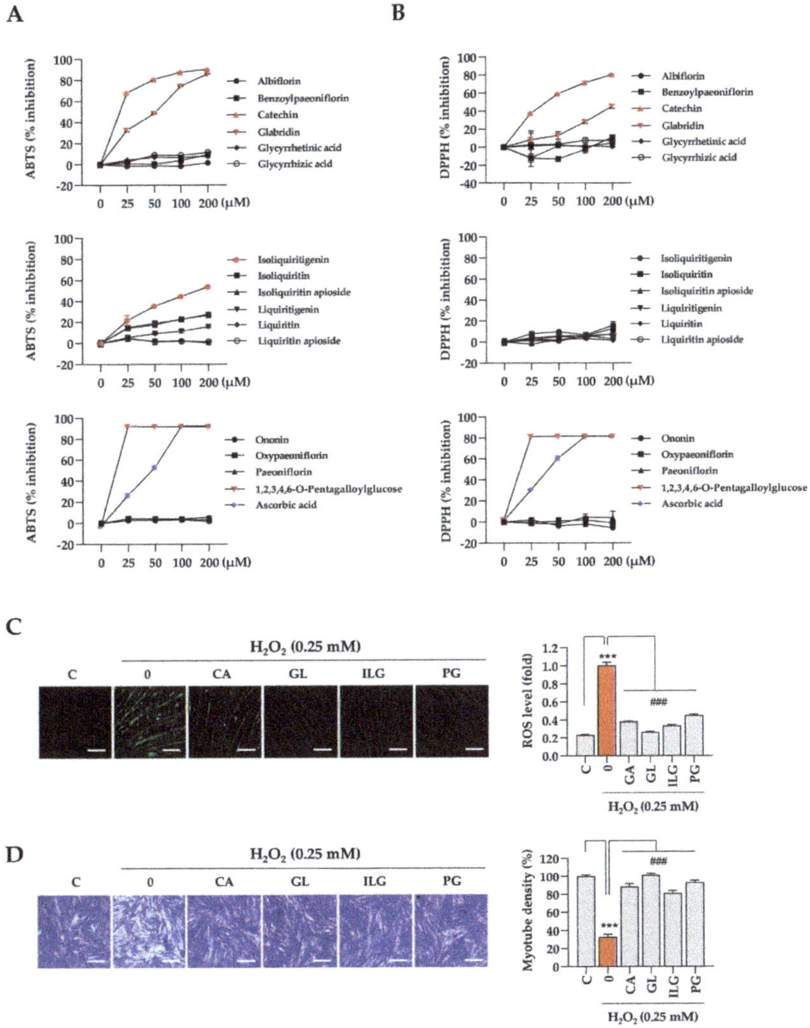

Figure 6. Antioxidant and muscle-protective activity of compounds in JGT. (**A**,**B**) Free radical-scavenging activities of 16 compounds were evaluated using the ABTS (**A**) and DPPH (**B**) assays. Ascorbic acid was used as a positive control in both assays. (**C**) C2C12 myotubes were pretreated with non-toxic concentrations of catechin (CA, 50 μM), glabridin (GL, 10 μM), isoliquiritigenin (ILG, 20 μM), or 1,2,3,4,6-pentagalloyl glucose (PG, 20 μM) for 12 h and subsequently exposed to 0.25 mM H_2O_2 for 6 h. Intracellular ROS were detected following labeling with CellROX^{TM} Green dye. (**D**) C2C12 myotubes pretreated with CA, GL, ILG, or PG were incubated for an additional 40 h in the presence of 0.25 mM H_2O_2. After staining with crystal violet solution, myotube density was quantified. Data are presented as the mean ± SD (*n* = 3 except for free radical scavenging assay, *n* = 2).

Statistical significance was evaluated by one-way ANOVA followed by Dunnett's multiple comparison test. *** $p < 0.001$ vs. vehicle-treated controls; ### $p < 0.001$ vs. H_2O_2 + vehicle-treated control cells. Scale bar = 100 μm.

4. Discussion

JGT shows promise as a therapeutic option for muscle atrophy, as demonstrated by the efficacy of JGT-W in our previous research [7]. However, the specific roles of the individual active compounds within JGT remain insufficiently defined, and the impact of different extraction methods on its therapeutic properties has not been comprehensively explored. In this study, both transcriptomic and network pharmacology analyses consistently revealed that JGT-E exhibited superior muscle-protective effects compared to JGT-W. Transcriptomic analysis demonstrated that JGT-E regulated a greater number of genes than JGT-W, significantly enhancing pathways related to oxidative phosphorylation, mitochondrial function, PGC-1α, mTORC1, and ERRα signaling, while concurrently suppressing muscle-atrophic TGF-β pathways (Figure 1).

Our previous research demonstrated that JGT-W provides protection against dexamethasone-induced muscle atrophy by promoting muscle protein synthesis, inhibiting degradation, and preserving mitochondrial integrity and bioenergetic function [7]. In the current study, transcriptomic data indicate that JGT-E exerted even more pronounced effects on these protective mechanisms. Specifically, JGT-E upregulated genes associated with oxidative phosphorylation, mitochondrial biogenesis, PGC-1α signaling, mTORC1 signaling, and ERRα targets, while simultaneously downregulating muscle-atrophic TGF-β signaling (Figure 1D). Mitochondrial biogenesis, which is crucial in maintaining mitochondrial mass and function, directly impacts skeletal muscle health. Any impairment of this process can lead to muscle dysfunction. Mitochondrial dysfunction is regulated by complex signaling networks, including the AMPK-SIRT1-PGC-1α, IGF-1-PI3K-Akt-mTOR, FoxOs, JAK-STAT3, TGF-β-Smad2/3, and NF-κB pathways [24]. Among these, PGC-1α activation, a key regulator of mitochondrial biogenesis [25], supported the hypothesis that JGT-E would enhance mitochondrial maintenance and function. This finding aligns with previous studies showing that PGC-1α protects skeletal muscle from atrophy by inhibiting FoxO3 signaling. Moreover, PGC-1α activation has been shown to prevent denervation-induced muscle atrophy [26] and age-related sarcopenia [27]. Additionally, the activation of mTORC1 signaling in cells treated with JGT-E further implies its role in promoting muscle protein synthesis and anabolic growth. mTORC1 regulates protein and lipid synthesis, supporting cellular growth, and its decline during muscle disuse or immobilization results in reduced protein synthesis and subsequent atrophy [28,29]. TGF-β, a key regulator of fibrosis and inflammation, induces skeletal muscle atrophy and fibrosis by activating Atrogin-1 and Scleraxis [30]. Therefore, the suppression of TGF-β signaling by JGT-E may contribute to its superior muscle-protective effects. Collectively, these findings highlight JGT-E as a more potent therapeutic candidate than JGT-W, likely due to its enhanced regulation of mitochondrial function and anabolic signaling, and its inhibition of muscle-atrophic pathways. To validate the findings from the transcriptome analysis, we performed a comparative evaluation of the muscle-protective effects of JGT-W and JGT-E under conditions of oxidative stress-induced muscle damage, using a C2C12 mouse muscle-cell line model. Our results demonstrate that both JGT-W and JGT-E significantly reduced intracellular ROS production in C2C12 myoblasts and myotubes. This resulted in preserved cell viability in myoblasts, increased myotube density, and enhanced mitochondrial content in the myotubes. Importantly, JGT-E demonstrated superior muscle-protective effects compared to JGT-W (Figures 2 and 3).

These transcriptomic changes were further supported by the differential expression of key genes associated with mitochondrial biogenesis and myogenic differentiation, as shown in Figure S1C. JGT-E treatment markedly upregulated Sirt1, Esrra, Mfn1, and Dnm1l, which play essential roles in mitochondrial function and remodeling. Notably, Sirt1 enhances the activity of PGC-1α through deacetylation, while Esrra (encoding ERRα) acts as a downstream effector of PGC-1α to promote oxidative metabolism [31,32]. These gene-level changes are consistent with the increased mitochondrial content observed in JGT-E-treated myotubes (Figure 3D). In addition, the elevated expression of Myog and Myod1, two master regulators of myogenic differentiation, corresponds to the preserved myotube morphology and increased density under oxidative stress (Figure 3C). The suppression of Foxo3, a pro-atrophic transcription factor [33], may further contribute to reduced expression of muscle-specific E3 ubiquitin ligases and lower ROS levels, as observed in both myoblasts (Figure 2E) and myotubes (Figure 3E). Taken together, these findings provide mechanistic insight into how the transcriptional reprogramming induced by JGT-E leads to improved mitochondrial integrity, reduced oxidative damage, and preservation of muscle fiber structure.

Network pharmacology analysis has identified several key compounds in JGT, including isoliquiritigenin, glycyrrhizic acid, paeoniflorin, catechin, and glabridin, which play significant roles in oxidative-stress responses, mitochondrial function, and regulation of muscle atrophy (Figure 5). Notably, isoliquiritigenin, catechin, and glabridin were found in significantly higher concentrations in JGT-E compared to JGT-W (Table 2). These compounds have potent anti-inflammatory and antioxidant properties, and their protective effects on muscle have been substantiated in both in vitro and in vivo studies [34–38]. Isoliquiritigenin exhibits potent antioxidant effects and improves mitochondrial function through activation of the Nrf2 signaling pathway in aging-related neurodegenerative diseases [39,40]. For musculoskeletal diseases, isoliquiritigenin has been shown to mitigate dexamethasone-induced muscle atrophy by increasing myotube diameters and decreasing expression of Atrogin-1 and MuRF-1 through the Akt/mTOR signaling pathway [41]. Catechin enhances skeletal muscle mass and strength by promoting mitochondrial biogenesis, which provides adequate energy for physiological functions and reduces the oxidative stress associated with aging, disease, and acute exercise [42]. Catechin can also enhance muscle regeneration by activating and promoting the differentiation of muscle stem cells [43], while also mitigating muscle atrophy by upregulating anabolic factors and increasing muscle fiber cross-sectional area [44]. Additionally, glabridin has been reported to inhibit dexamethasone-induced protein degradation in C2C12 myotubes and muscle atrophy in murine models. It does so by attenuating the expression of MuRF-1 and Cbl-b through suppression of glucocorticoid receptor activation. As an antagonist, glabridin binds directly to the glucocorticoid receptor, preventing its activation and subsequent downstream signaling [45]. These findings imply that the enhanced efficacy of JGT-E in the treatment of muscle atrophy may be primarily attributed to the elevated concentrations of these bioactive compounds, which collectively contribute to mitochondrial protection, reduction in oxidative stress, and preservation of muscle integrity.

The investigation into the free-radical scavenging activity of 16 compounds in JGT revealed that catechin, glabridin, and isoliquiritigenin exhibited significant antioxidant properties, which aligned with the primary compounds identified through network pharmacology analysis (Figure 6A,B). These compounds were shown to reduce ROS production and mitigate myotube degradation induced by H_2O_2 (Figure 6C,D). Additionally, 1,2,3,4,6-pentagalloyl glucose was selected for its notable antioxidant activity and its ability to protect against myotube degradation. Importantly, the contents of these four compounds were found to be higher in JGT-E than in JGT-W, which may contribute to the enhanced efficacy observed in JGT-E.

One limitation of network pharmacology analysis is its reliance on available data, which may result in underrepresentation of certain bioactive compounds with fewer known or predicted targets. As a consequence, some compounds that demonstrate promising effects in experimental studies may not be adequately represented in network analyses; 1,2,3,4,6-O-pentagalloylglucose is an example of such a compound. Therefore, additional research is needed to better understand its mechanisms and ensure a more comprehensive evaluation of its therapeutic potential. Another limitation of this study is the absence of in vivo validation. While the current study focused on in vitro analyses, our previous work has already confirmed the in vivo efficacy of JGT-W in a dexamethasone-induced muscle atrophy model [7]. Given the enhanced effects of JGT-E and its key bioactive constituents, such as isoliquiritigenin, catechin, and glabridin, demonstrated in this study, follow-up evaluations in diverse preclinical models of muscle injury are required to further verify their therapeutic potential and safety profiles.

5. Conclusions

This study demonstrated that JGT-E provides superior protection against muscle atrophy compared to JGT-W, largely due to its higher contents of key bioactive compounds such as isoliquiritigenin, catechin, and glabridin. Through an integrative approach combining transcriptome analysis with network pharmacology, we showed that JGT-E enhances mitochondrial biogenesis, oxidative phosphorylation, and anabolic signaling more effectively, while simultaneously suppressing muscle-atrophic TGF-β pathways. Functional assays using C2C12 mouse muscle cells confirmed that JGT-E significantly reduces oxidative stress and preserves myotube integrity under stress conditions, outperforming JGT-W (Figure 7). These findings highlight the importance of integrating systems-level analysis with experimental validation and emphasize the critical role of extraction methods in optimizing the therapeutic efficacy of herbal medicines for treating muscle atrophy.

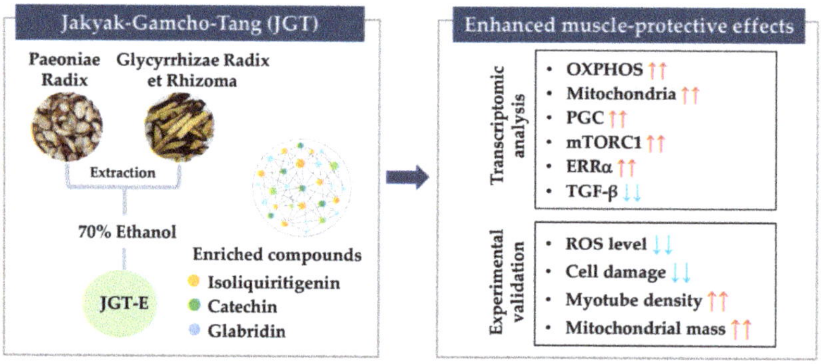

Figure 7. Schematic summary of the muscle-protective mechanism of JGT-E. Jakyak-Gamcho-Tang (JGT), a traditional herbal formulation consisting of Paeoniae Radix and Glycyrrhizae Radix et Rhizoma, was extracted using 70% ethanol to yield JGT-E, which is enriched in key bioactive compounds including isoliquiritigenin, catechin, and glabridin. Transcriptomic analysis revealed that JGT-E significantly upregulated genes associated with oxidative phosphorylation (OXPHOS), mitochondrial biogenesis, PGC-1α, mTORC1, and ERRα signaling, while downregulating TGF-β signaling. These molecular changes were validated experimentally and were accompanied by decreased intracellular ROS levels and cell damage, as well as increased mitochondrial content and myotube integrity. Collectively, these findings support the enhanced muscle-protective efficacy of JGT-E. Red arrows indicate upregulation, and blue arrows indicate downregulation.

Supplementary Materials: The following supporting information can be downloaded at: https://www.mdpi.com/article/10.3390/antiox14070795/s1. Figure S1: Transcriptomic effects regulated by JGT-E and JGT-W, Figure S2: Effects of JGT-W, JGT-E, resveratrol, and arachidonic acid under conditions of muscle atrophy in C2C12 myotubes, Figure S3: Dynamic multiple reaction monitoring (MRM) mode of JGT-W (A), JGT-E (B), and standard mixture (C) using a UHPLC-TQ-MS/MS, Figure S4: Muscle-protective activity of 16 compounds in JGT, Table S1: Recovery data of the 16 compounds in JGT.

Author Contributions: Conceptualization, A.K. and S.-M.P.; investigation, A.K. and N.S.K.; formal analysis, A.Y.L., M.N.T., and S.-M.P.; data curation, S.-J.B. and H.Y.; writing—original draft preparation, A.K., M.N.T., A.Y.L., and S.-M.P.; writing—review and editing, A.K., M.N.T., S.-M.P., N.S.K., and S.C.; funding acquisition, S.C. and S.-M.P. All authors have read and agreed to the published version of the manuscript.

Funding: This work was supported by grants from the Korea Institute of Oriental Medicine (KSN1739122, Republic of Korea), the National Research Foundation of Korea (RS-2023-00210923), and the BK21 FOUR Program by Chungnam National University Research Grant, 2024.

Institutional Review Board Statement: Not applicable.

Informed Consent Statement: Not applicable.

Data Availability Statement: The raw sequencing data (FASTQ files) and processed count data were uploaded to the Gene Expression Omnibus under the accession number GSE295069 (reviewer's access link: https://www.ncbi.nlm.nih.gov/geo/query/acc.cgi?acc=GSE295069 (accessed on 25 June 2025).

Conflicts of Interest: The authors declare no conflicts of interest.

Abbreviations

The following abbreviations are used in this manuscript:

JGT	Jakyak-gamcho-tang
ABTS	2,2′-Azino-bis(3-ethylbenzothiazoline-6-sulfonic acid) diammonium salt
BPC	Base peak ion chromatogram
DEG	Differentially expressed genes
DM	Differentiation medium
DMEM	Dulbecco's Modified Eagle Medium
DMSO	Dimethyl sulfoxide
DPPH	2,2-diphenyl-1-picrylhydrazyl
FDR	False discovery rate
GM	Growth medium
GO	Gene Ontology
GOBP	Gene Ontology Biological Process
GSEA	Gene Set Enrichment Analysis
H_2O_2	Hydrogen peroxide
JGT	Jakyak-gamcho-tang
JGT-E	Ethanol extract of Jakyak-gamcho-tang
JGT-W	Water extract of Jakyak-gamcho-tang
MRM	Multiple reaction monitoring
NES	Normalized enrichment score
OD	Optical density
PBS	Phosphate-buffered saline
P/S	Penicillin and streptomycin
ROS	Reactive oxygen species
SD	Standard deviation
SDS	Sodium dodecyl sulfate
UHPLC	Ultrahigh-performance liquid chromatography

References

1. Cohen, S.; Nathan, J.A.; Goldberg, A.L. Muscle wasting in disease: Molecular mechanisms and promising therapies. *Nat. Rev. Drug. Discov.* **2015**, *14*, 58–74. [CrossRef] [PubMed]
2. Yin, L.; Li, N.; Jia, W.; Wang, N.; Liang, M.; Yang, X.; Du, G. Skeletal muscle atrophy: From mechanisms to treatments. *Pharmacol. Res.* **2021**, *172*, 105807. [CrossRef] [PubMed]
3. Gadelha, A.B.; Neri, S.G.R.; Oliveira, R.J.; Bottaro, M.; David, A.C.; Vainshelboim, B.; Lima, R.M. Severity of sarcopenia is associated with postural balance and risk of falls in community-dwelling older women. *Exp. Aging Res.* **2018**, *44*, 258–269. [CrossRef]
4. Yadav, A.; Yadav, S.S.; Singh, S.; Dabur, R. Natural products: Potential therapeutic agents to prevent skeletal muscle atrophy. *Eur. J. Pharmacol.* **2022**, *925*, 174995. [CrossRef]
5. Jung, H.W.; Kang, A.N.; Kang, S.Y.; Park, Y.K.; Song, M.Y. The root extract of *Pueraria lobata* and its main compound, puerarin, prevent obesity by increasing the energy metabolism in skeletal muscle. *Nutrients* **2017**, *9*, 33. [CrossRef]
6. Ota, K.; Fukui, K.; Nakamura, E.; Oka, M.; Ota, K.; Sakaue, M.; Sano, Y.; Takasu, A. Effect of Shakuyaku-kanzo-to in patients with muscle cramps: A systematic literature review. *J. Gen. Fam. Med.* **2020**, *21*, 56–62. [CrossRef]
7. Kim, A.; Kim, Y.R.; Park, S.M.; Lee, H.; Park, M.; Yi, J.M.; Cha, S.; Kim, N.S. Jakyak-gamcho-tang, a decoction of Paeoniae Radix and Glycyrrhizae Radix et Rhizoma, ameliorates dexamethasone-induced muscle atrophy and muscle dysfunction. *Phytomedicine* **2024**, *123*, 155057. [CrossRef]
8. Abubakar, A.R.; Haque, M. Preparation of medicinal plants: Basic extraction and fractionation procedures for experimental purposes. *J. Pharm. Bioallied Sci.* **2020**, *12*, 1–10. [CrossRef]
9. Plaskova, A.; Mlcek, J. New insights of the application of water or ethanol-water plant extract rich in active compounds in food. *Front. Nutr.* **2023**, *10*, 1118761. [CrossRef]
10. Stanciauskaite, M.; Marksa, M.; Babickaite, L.; Majiene, D.; Ramanauskiene, K. Comparison of ethanolic and aqueous *Populus balsamifera* L. Bud Extracts by different extraction methods: Chemical composition, antioxidant and antibacterial activities. *Pharmaceuticals* **2021**, *14*, 1018. [CrossRef]
11. Ju, J.B.; Kim, J.S.; Choi, C.W.; Lee, H.K.; Oh, T.K.; Kim, S.C. Comparison between ethanolic and aqueous extracts from Chinese juniper berries for hypoglycaemic and hypolipidemic effects in alloxan-induced diabetic rats. *J. Ethnopharmacol.* **2008**, *115*, 110–115. [CrossRef] [PubMed]
12. Borges, A.; Jose, H.; Homem, V.; Simoes, M. Comparison of techniques and solvents on the antimicrobial and antioxidant potential of extracts from *Acacia dealbata* and *Olea europaea*. *Antibiotics* **2020**, *9*, 48. [CrossRef] [PubMed]
13. Bae, J.Y.; Lee, Y.S.; Han, S.Y.; Jeong, E.J.; Lee, M.K.; Kong, J.Y.; Lee, D.H.; Cho, K.J.; Lee, H.S.; Ahn, M.J. A Comparison between water and ethanol extracts of *Rumex acetosa* for protective effects on gastric ulcers in mice. *Biomol. Ther.* **2012**, *20*, 425–430. [CrossRef] [PubMed]
14. Park, S.M.; Kim, A.; Lee, H.; Baek, S.J.; Kim, N.S.; Park, M.; Yi, J.M.; Cha, S. Systematic transcriptome analysis reveals molecular mechanisms and indications of bupleuri radix. *Front. Pharmacol.* **2022**, *13*, 1010520. [CrossRef]
15. Park, M.; Park, S.M.; Lee, H.; Kim, A.; Kim, N.S.; Kim, Y.R.; Yi, J.M.; Cha, S. KORE-Map 1.0: Korean Medicine Omics Resource Extension Map on transcriptome data of tonifying herbal medicine. *Sci. Data* **2024**, *11*, 974. [CrossRef]
16. Kim, Y.S.; Yuk, H.J.; Kim, D.S. Effect of Jakyakgamcho-Tang extracts on H_2O_2-induced C2C12 myoblasts. *Molecules* **2021**, *26*, 215. [CrossRef]
17. Zhang, G.B.; Li, Q.Y.; Chen, Q.L.; Su, S.B. Network pharmacology: A new approach for chinese herbal medicine research. *Evid. Based Complement. Alternat. Med.* **2013**, *2013*, 621423. [CrossRef]
18. Noor, F.; Tahir Ul Qamar, M.; Ashfaq, U.A.; Albutti, A.; Alwashmi, A.S.S.; Aljasir, M.A. Network pharmacology approach for medicinal plants: Review and assessment. *Pharmaceuticals* **2022**, *15*, 572. [CrossRef]
19. Choi, W.G.; Choi, N.R.; Park, E.J.; Kim, B.J. A study of the therapeutic mechanism of Jakyakgamcho-Tang about functional dyspepsia through network pharmacology research. *Int. J. Med. Sci.* **2022**, *19*, 1824–1834. [CrossRef]
20. Hu, Y.H.; Wang, X.Y.; Zhang, X.W.; Chen, J.; Li, F. Investigation of the mechanisms and experimental verification of Shao yao gan cao decoction against Sphincter of Oddi Dysfunction *via* systems pharmacology. *Math. Biosci. Eng.* **2022**, *19*, 13374–13398. [CrossRef]
21. Alamdari, N.; Aversa, Z.; Castillero, E.; Gurav, A.; Petkova, V.; Tizio, S.; Hasselgren, P.O. Resveratrol prevents dexamethasone-induced expression of the muscle atrophy-related ubiquitin ligases atrogin-1 and MuRF1 in cultured myotubes through a SIRT1-dependent mechanism. *Biochem. Biophys. Res. Commun.* **2012**, *417*, 528–533. [CrossRef] [PubMed]
22. Bagherniya, M.; Mahdavi, A.; Shokri-Mashhadi, N.; Banach, M.; Von Haehling, S.; Johnston, T.P.; Sahebkar, A. The beneficial therapeutic effects of plant-derived natural products for the treatment of sarcopenia. *J. Cachexia Sarcopenia Muscle* **2022**, *13*, 2772–2790. [CrossRef] [PubMed]

23. Mitra, A.; Shanavas, S.; Chaudhury, D.; Bose, B.; Das, U.N.; Shenoy, P.S. Mitigation of chronic glucotoxicity-mediated skeletal muscle atrophy by arachidonic acid. *Life Sci.* **2023**, *333*, 122141. [CrossRef] [PubMed]
24. Chen, X.; Ji, Y.; Liu, R.; Zhu, X.; Wang, K.; Yang, X.; Liu, B.; Gao, Z.; Huang, Y.; Shen, Y.; et al. Mitochondrial dysfunction: Roles in skeletal muscle atrophy. *J. Transl. Med.* **2023**, *21*, 503. [CrossRef]
25. Popov, L.D. Mitochondrial biogenesis: An update. *J. Cell. Mol. Med.* **2020**, *24*, 4892–4899. [CrossRef]
26. Sandri, M.; Lin, J.; Handschin, C.; Yang, W.; Arany, Z.P.; Lecker, S.H.; Goldberg, A.L.; Spiegelman, B.M. PGC-1alpha protects skeletal muscle from atrophy by suppressing FoxO3 action and atrophy-specific gene transcription. *Proc. Natl. Acad. Sci. USA* **2006**, *103*, 16260–16265. [CrossRef]
27. Handschin, C.; Spiegelman, B.M. The role of exercise and PGC-1α in inflammation and chronic disease. *Nature* **2008**, *454*, 463–469. [CrossRef]
28. Gatica, D.; Klionsky, D.J. New insights into MTORC1 amino acid sensing and activation. *Biotarget* **2017**, *1*, 2. [CrossRef]
29. Rudrappa, S.S.; Wilkinson, D.J.; Greenhaff, P.L.; Smith, K.; Idris, I.; Atherton, P.J. Human skeletal muscle disuse atrophy: Effects on muscle protein synthesis, breakdown, and insulin resistance-A qualitative review. *Front. Physiol.* **2016**, *7*, 361. [CrossRef]
30. Mendias, C.L.; Gumucio, J.P.; Davis, M.E.; Bromley, C.W.; Davis, C.S.; Brooks, S.V. Transforming growth factor-beta induces skeletal muscle atrophy and fibrosis through the induction of atrogin-1 and scleraxis. *Muscle Nerve* **2012**, *45*, 55–59. [CrossRef]
31. Gurd, B.J. Deacetylation of PGC-1alpha by SIRT1: Importance for skeletal muscle function and exercise-induced mitochondrial biogenesis. *Appl. Physiol. Nutr. Metab.* **2011**, *36*, 589–597. [CrossRef] [PubMed]
32. Schreiber, S.N.; Emter, R.; Hock, M.B.; Knutti, D.; Cardenas, J.; Podvinec, M.; Oakeley, E.J.; Kralli, A. The estrogen-related receptor alpha (ERRalpha) functions in PPARgamma coactivator 1alpha (PGC-1alpha)-induced mitochondrial biogenesis. *Proc. Natl. Acad. Sci. USA* **2004**, *101*, 6472–6477. [CrossRef] [PubMed]
33. Sandri, M.; Sandri, C.; Gilbert, A.; Skurk, C.; Calabria, E.; Picard, A.; Walsh, K.; Schiaffino, S.; Lecker, S.H.; Goldberg, A.L. Foxo transcription factors induce the atrophy-related ubiquitin ligase atrogin-1 and cause skeletal muscle atrophy. *Cell* **2004**, *117*, 399–412. [CrossRef] [PubMed]
34. Shi, D.; Yang, J.; Jiang, Y.; Wen, L.; Wang, Z.; Yang, B. The antioxidant activity and neuroprotective mechanism of isoliquiritigenin. *Free Radic. Biol. Med.* **2020**, *152*, 207–215. [CrossRef]
35. Wang, T.T.; Chen, Z.Z.; Xie, P.; Zhang, W.J.; Du, M.Y.; Liu, Y.T.; Zhu, H.Y.; Guo, Y.S. Isoliquiritigenin suppresses the proliferation and induced apoptosis *via* miR-32/LATS2/Wnt in nasopharyngeal carcinoma. *Eur. J. Pharmacol.* **2019**, *856*, 172352. [CrossRef]
36. Al-Qahtani, W.H.; Alshammari, G.M.; Ajarem, J.S.; Al-Zahrani, A.Y.; Alzuwaydi, A.; Eid, R.; Yahya, M.A. Isoliquiritigenin prevents doxorubicin-induced hepatic damage in rats by upregulating and activating SIRT1. *Biomed. Pharmacother.* **2022**, *146*, 112594. [CrossRef]
37. Higdon, J.V.; Frei, B. Tea catechins and polyphenols: Health effects, metabolism, and antioxidant functions. *Crit. Rev. Food Sci. Nutr.* **2003**, *43*, 89–143. [CrossRef]
38. Zhang, J.; Wu, X.; Zhong, B.; Liao, Q.; Wang, X.; Xie, Y.; He, X. Review on the diverse biological effects of glabridin. *Drug Des. Devel. Ther.* **2023**, *17*, 15–37. [CrossRef]
39. Denzer, I.; Munch, G.; Pischetsrieder, M.; Friedland, K. S-allyl-L-cysteine and isoliquiritigenin improve mitochondrial function in cellular models of oxidative and nitrosative stress. *Food Chem.* **2016**, *194*, 843–848. [CrossRef]
40. Yang, E.J.; Min, J.S.; Ku, H.Y.; Choi, H.S.; Park, M.K.; Kim, M.K.; Song, K.S.; Lee, D.S. Isoliquiritigenin isolated from *Glycyrrhiza uralensis* protects neuronal cells against glutamate-induced mitochondrial dysfunction. *Biochem. Biophys. Res. Commun.* **2012**, *421*, 658–664. [CrossRef]
41. Jiao, X.; Zhang, Y.; Wang, Z.; Chang, H.; Li, Y.; Wang, B.; Gan, Y.; Gu, D. Isoliquiritigenin, an extract from Licorice, attenuates dexamethasone-induced muscle atrophy *via* Akt/mTOR pathway. *Mol. Nutr. Food Res.* **2025**, *69*, e70000. [CrossRef] [PubMed]
42. Li, P.; Liu, A.; Xiong, W.; Lin, H.; Xiao, W.; Huang, J.; Zhang, S.; Liu, Z. Catechins enhance skeletal muscle performance. *Crit. Rev. Food Sci. Nutr.* **2020**, *60*, 515–528. [CrossRef] [PubMed]
43. Kim, A.R.; Kim, K.M.; Byun, M.R.; Hwang, J.H.; Park, J.I.; Oh, H.T.; Kim, H.K.; Jeong, M.G.; Hwang, E.S.; Hong, J.H. Catechins activate muscle stem cells by Myf5 induction and stimulate muscle regeneration. *Biochem. Biophys. Res. Commun.* **2017**, *489*, 142–148. [CrossRef] [PubMed]
44. Li, P.; Liu, A.; Liu, C.; Qu, Z.; Xiao, W.; Huang, J.; Liu, Z.; Zhang, S. Role and mechanism of catechin in skeletal muscle cell differentiation. *J. Nutr. Biochem.* **2019**, *74*, 108225. [CrossRef]
45. Yoshioka, Y.; Kubota, Y.; Samukawa, Y.; Yamashita, Y.; Ashida, H. Glabridin inhibits dexamethasone-induced muscle atrophy. *Arch. Biochem. Biophys.* **2019**, *664*, 157–166. [CrossRef]

antioxidants

MDPI

Review

Time to Reset: The Interplay Between Circadian Rhythms and Redox Homeostasis in Skeletal Muscle Ageing and Systemic Health

Elizabeth Sutton and Vanja Pekovic-Vaughan *

Institute of Life Course and Medical Sciences, Department of Musculoskeletal & Ageing Science, University of Liverpool, Liverpool L7 8TX, UK; ems33@liverpool.ac.uk
* Correspondence: vpv35@liverpool.ac.uk

Abstract: Skeletal muscle plays vital roles in locomotion, metabolic regulation and endocrine signalling. Critically, it undergoes structural and functional decline with age, leading to a progressive loss of muscle mass and strength (sarcopenia) and contributing to a systemic loss of tissue resilience to stressors of multiple tissue systems (frailty). Emerging evidence implicates misalignments in both the circadian molecular clock and redox homeostasis as major drivers of age-related skeletal muscle deterioration. The circadian molecular clock, through core clock components such as BMAL1 and CLOCK, orchestrates rhythmic gene, protein and myokine expression impacting diurnal regulation of skeletal muscle structure and metabolism, mitochondrial function, antioxidant defence, extracellular matrix organisation and systemic inter-tissue communication. In parallel, the master redox regulator, NRF2, maintains cellular antioxidant defence, tissue stress resistance and mitochondrial health. Disruption of either system impairs skeletal muscle contractility, metabolism, and regenerative capacity as well as systemic homeostasis. Notably, NRF2-mediated redox signalling is clock-regulated and, in turn, affects circadian clock regulation. Both systems are responsive to external cues such as exercise and hormones, yet studies do not consistently include circadian timing or biological sex as key methodological variables. Given that circadian regulation shifts with age and differs between sexes, aligning exercise interventions with one's own chronotype may enhance health benefits, reduce adverse side effects, and overcome anabolic resistance with ageing. This review highlights the essential interplay between circadian and redox systems in skeletal muscle homeostasis and systemic health and argues for incorporating personalised chrono-redox approaches and sex-specific considerations into future experimental research and clinical studies, aiming to improve functional outcomes in age-related sarcopenia and broader age-related metabolic and musculoskeletal conditions.

Keywords: skeletal muscle; ageing; circadian rhythms; NRF2; redox homeostasis; oxidative stress; chronotherapy; chronotype; exercise; sarcopenia; frailty; myokines

1. Introduction to Skeletal Muscle: From Structural to Metabolic and Endocrine Functions

The human body comprises >600 different skeletal muscles, accounting for ~40–60% of total body mass. Skeletal muscles contain ~50–75% of all proteins within the body and are responsible for ~30–50% of all protein turnover [1]. Certain properties of skeletal muscles are determined from an early developmental stage, such as anatomical structure, shape, spatial organisation and orientation, along with connections to correct tendons and bones [2]. Yet, skeletal muscle is one of the most dynamic and plastic tissues within the body, with skeletal muscle mass, force and fatigue resistance being adaptive to physiological or pathological conditions. For example, exercise training regimens can lead to increased skeletal muscle mass in young adults, alongside improved force and resistance to fatigue, whilst in disease conditions or with ageing, skeletal muscles lose mass and/or strength leading to sarcopenia and associated frailty [2].

The highly organised architecture of skeletal muscle is well-characterised. Briefly, myofibrils form the basis of the skeletal muscle structure, composed of thick myosin filaments and thin filaments of actin, troponin and tropomyosin. The myofibrils are bundled to form myofibres, the myofibres themselves are bundled to create the fascicles, and the bundles of the fascicles form the skeletal muscle [3,4] (see Figure 1). A complex network of supportive extracellular matrix (ECM) consisting of collagens, glycoproteins, proteoglycans and elastin [5] surrounds the myofibres and the fascicles, providing mechanical support [6]. Skeletal muscle-specific stem cells known as satellite cells reside within the ECM, where they exert key functions in myogenesis and muscle morphogenesis, thereby playing an important role in the growth and repair of skeletal muscle in response to injury, disease or sedentary behaviour [7].

Extracellular matrix (ECM) is the non-cellular structural scaffold which surrounds all tissues and organs, with >1000 characterised proteins, called the matrisome. The ECM is essential for tissue growth and remodelling as well as stabilising tissue structures and providing elasticity [8]. The ECM is particularly vital in supporting the function of skeletal muscle, providing important cues for transmission of force and regulating skeletal muscle homeostasis through myogenesis, repair and degradation [9]. The thickest and strongest of the structures of ECM, the epimysium, covers the whole skeletal muscle (see Figure 1). The tendon that joins the skeletal muscles to the bones continues as the epimysium, which contains the main blood vessels and nerves that nourish the skeletal muscles. There are many myofibres that surround the ECM layer called the perimysium to form fascicles. The perimysium contains larger blood vessels and nerves and transmits the force to tendons. Each muscle fibre is encircled by a thin and delicate component of the ECM, called the endomysium. Small arteries and neurons are found in the endomysium, which plays a significant role in myogenesis, regeneration and preserves the integrity of the skeletal muscle [10].

Skeletal muscle is involved in a wide array of physiological functions, which can be broadly split into three groups: mechanical, metabolic and endocrine. The mechanical functions of skeletal muscle enable the maintenance of posture, generation of movement and maintenance/improvement of health through the conversion of chemical into mechanical energy [1]. In terms of metabolism, skeletal muscle is responsible for the uptake of ~80% glucose following a meal [11], and is involved in the regulation of energy metabolism and heat production for thermoregulation through the mitochondrial consumption of oxygen and fuel during daily activity. Skeletal muscle also acts as a metabolic reserve of amino acids following nutritional intake. In addition, during times of poor nutritional intake, it helps provide those amino acids to tissues that need maintenance of proteostasis for survival, such as the brain and heart, though this can only occur if skeletal muscle mass is adequate [12].

Skeletal muscle has endocrine functions that have crucial roles in inter-tissue communication and systemic health effects. This was first proposed by Goldstein ~60 years ago, who viewed skeletal muscle as an endocrine-like organ, releasing an unidentified humoral factor involved in regulating blood glucose in response to exercise [13]. Later in the early 2000s, Pedersen and colleagues identified and demonstrated in a series of studies that interleukin-6 (IL-6), irisin and brain-derived neurotrophic factor (BDNF), produced by skeletal muscles during contraction, play an important role in the beneficial health effects of exercise by mobilising extracellular substrates and/or supporting substrate delivery during exercise [14–16]. Further research established that skeletal muscle produces numerous cytokines, peptides and metabolites, termed myokines [17]. Hartwig and colleagues catalogued the secretome of human primary muscle cells, identifying 548 non-redundant proteins in conditioned media, with 305 proteins classified as potential myokines [18]. These myokines exhibit autocrine, paracrine and endocrine effects on various cells and tissues, placing skeletal muscle as an endocrine organ [19].

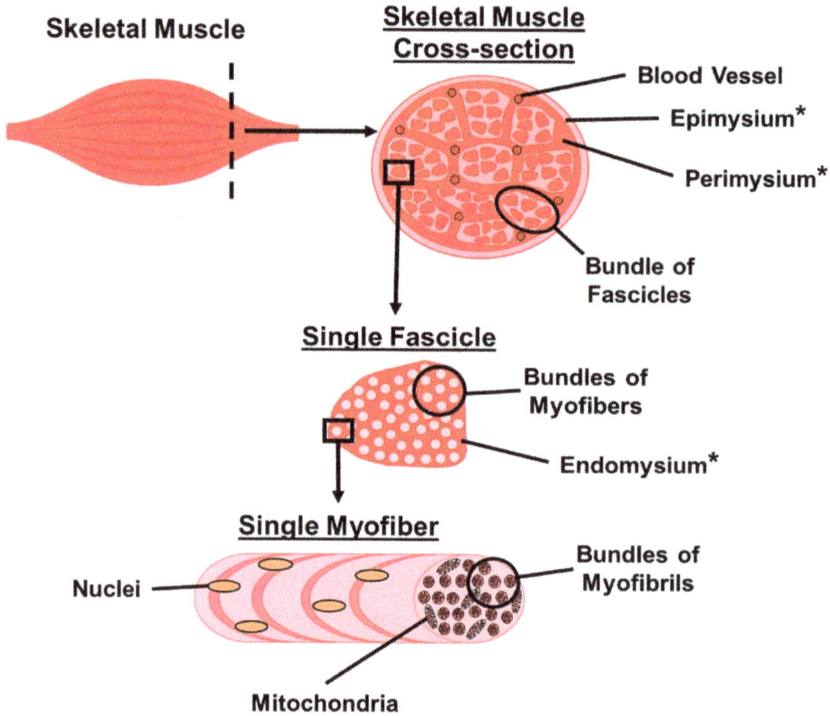

Figure 1. Schematic depicting a highly organised structure of skeletal muscle. Skeletal muscle is one of the most abundant tissues in the body, possessing structural, metabolic and endocrine functions. Skeletal muscle is a highly organised tissue composed of individual myofibrils. The myofibrils bundle into a single myofibre, which contains many nuclei and mitochondria. The myofibres are bundled to create a fascicle. The bundles of myofibres are supported by a layer of extracellular matrix (ECM) called the endomysium. The fascicles bundle to form the skeletal muscle. The fascicles are supported by an ECM layer called the perimysium, and they are highly oxygenated through a blood supply. An additional layer of ECM called the epimysium surrounds the whole skeletal muscle. The epimysium mainly contains types I, III collagens and fibronectin; the perimysium contains collagen types I, III, V and VI, proteoglycans and fibronectin, while the endomysium contains types I, III, and V collagens, laminins and fibronectin. * Supportive ECM.

Exercise-induced myokine release, termed exerkines, contribute to the broad beneficial effects of physical activity on overall health. They control adaptive processes in skeletal muscle by acting as regulators of hypertrophy, angiogenesis, inflammation and ECM remodelling, and in doing so, mediate communication between skeletal muscle and other organs, including the brain, adipose tissue (AT), pancreas, liver, intestines and other musculoskeletal tissues [20]. The interaction between skeletal muscle and adipose tissue, both white (WAT) and brown (BAT), has been of great metabolic interest, with WAT being a key energy storage depot, releasing hormones and cytokines to regulate whole body metabolism, and BAT being necessary for thermogenesis. Both skeletal muscle and AT are affected by exercise and help the body to adapt to metabolic challenges. The tissue crosstalk, through the release of myokines, mediates inter-tissue communication to improve total metabolic health [21]. Some examples of myokines include IL-6, which leads to increases in glucose uptake and β-oxidation within the skeletal muscle itself, but also promotes lipolysis in WAT and glucagon-like peptide-1 (GLP1) secretion in the pancreas and gut that help insulin release to regulate blood sugar levels. Other examples include IL-15, which induces hypertrophy in skeletal muscle, as well as increases thermogenesis and β-oxidation in BAT, whilst BDNF increases β-oxidation in skeletal muscle and promotes learning and memory in the brain [20].

Skeletal muscle extracellular vesicles (EVs) have emerged as another crosstalk mechanism through which the beneficial effects of exercise are transmitted to other tissues. EVs are membrane-bound vesicles with the capacity to transfer functional biomolecules such as lipids, proteins, nucleic acids and sugars. EVs are currently divided into three subtypes: exosomes (50–150 nm in diameter), microvesicles (100–1000 nm) and apoptotic bodies (50–5000 nm) [22]. Mitochondrial DNA was found to be taken up into exosomes and released into extracellular medium of C2C12 cell cultures [23]. Several miRNAs are highly abundant in skeletal muscle including *miR-1, miR-133a, miR-133b* and *miR-206*. These account for ~25% of miRNA expression in human skeletal muscle, contributing to myoblast differentiation via an autocrine route, and become excreted from skeletal muscle cells via exosomes [24]. A further study by Guescini and colleagues reported that circulation of *miR-133b* and *miR181a-5p* in exosomes was increased following acute exercise, with a positive correlation between aerobic fitness and circulating skeletal muscle-abundant miRNAs [25]. It has been suggested that miRNA-containing exosomes excreted from the skeletal muscle mediate the stress response of cells by regulating gene expression to restore homeostasis. Resistance training has been shown to reduce levels of *miR-1* and *miR-133a*, targeting the insulin-like growth factor 1 (IGF1) pathways and inducing hypertrophy, whilst endurance training increased *miR-1* and *miR-181* to induce mitochondrial biogenesis via the peroxisome proliferator-activated receptor gamma coactivator 1-α (PGC-1α) pathway [26].

2. Age-Associated Sarcopenia and the Role of Redox Homeostasis

Sarcopenia is an age-related condition characterised by the loss of skeletal muscle mass, quality and strength, the rate of which is dependent on activity levels, co-morbidities, nutritional status and circadian regulatory factors. Given its structural, metabolic and endocrine roles, skeletal muscle loss can have serious functional and metabolic consequences on the body, including increased risk of frailty that is characterised by reduced resilience to stressors, physical immobility and a functional decline in several organ systems such as cardiovascular, cognitive and metabolic [27]. As skeletal muscle undergoes pathophysiological changes with ageing, the functions of its major components such as myofibres, ECM, satellite cells and mitochondria become adversely affected.

Type II myofibres undergo a significant atrophy with ageing and an irreversible decline in number due to loss of innervation [28]. Several studies have determined that in the elderly, type II myofibres become more deformed from their typical tessellations of pentagons or hexagons seen in younger adults, though type I myofibres appear to be less affected by ageing [29,30]. The reduction in type II myofibre number directly results in a loss of myosin heavy chain (MHC) isoforms IIa and IIx, which are associated with strength and speed of muscle contraction. Contractile properties of older adults are therefore weaker, slower and less powerful than that of younger adults [31]. The loss of myosin impedes the formation of actin–myosin cross-bridges required for contractility, further impairing skeletal muscle activity [32]. Loss of muscle fibres is associated with the loss of motor units, leading to myofibre denervation, and increased susceptibility to atrophy [33].

Protein homeostasis of the ECM is severely impacted with age resulting in loss of new ECM protein synthesis, reduced turnover of old ECM and increased matrix metalloproteinase degradation activity [34]. During ageing, collagens, mainly type I and III, undergo glycation and form harder-to-cleave crosslinks. Furthermore, collagens can no longer align in linear conformations, and deteriorate into erratically grouped fibres, thus stiffening the matrix and inducing the build-up of fibrotic connective tissue within the skeletal muscle [34–36]. Another consequence of skeletal muscle ageing is the diminishing function of the satellite cell (SC) niche within the ECM. Upon skeletal muscle injury in young adults, the satellite cells proliferate and differentiate into new muscle fibres. However, with age, satellite cells are driven towards a fibro-adipogenic fate leading to fibrosis and impaired recovery from injury [37]. In addition, the increased deposition of basal lamina proteins with skeletal muscle ageing can drive SCs out of their niches, affecting the regulation of their divisions [10].

Furthermore, in line with the mitochondrial theory of ageing, mitochondrial mass, oxygen consumption and ATP synthesis become impaired with ageing with a ~40–50% decrease in oxidative phosphorylation, resulting in a major decline in energy homeostasis [38]. Reactive oxygen and nitrogen species (RO/NS) are produced in cells predominately as a by-product of mitochondrial oxidative phosphorylation. Build-up of RO/NS can be detrimental to nucleic acids, proteins and organelles, thus cells and tissues have evolved a protective network of detoxifying enzymes and non-enzymatic antioxidants to prevent excessive RO/NS accumulation [39]. This cellular antioxidant defence system is controlled by the master antioxidant transcription factor NRF2 [40].

NRF2 belongs to the Cap 'n' Collar (CNC) basic region-leucine zipper (bZIP) family [41] and its main endogenous regulator is KEAP1 (Kelch-like erythroid cell-derived protein with CNC homology-associated protein 1), an E3 ubiquitin ligase substrate adaptor. Under low oxidative stress, KEAP1 sequesters NRF2 in the cytoplasm, enabling the binding of E3 ligase complex CULLIN3, which polyubiquitinylates NRF2 leading to its proteasomal degradation [42]. Upon cellular oxidative stress, cysteines on KEAP1 become oxidised, so that NRF2 can no longer be sequestered for degradation, thus allowing de novo synthesised NRF2 to accumulate in the cytoplasm and translocate to the nucleus. In the nucleus, NRF2 heterodimerises with small musculoaponeurotic fibrosarcoma (sMAF) proteins, binds to antioxidant response elements (AREs) within the promoter regions of cytoprotective genes encoding proteins such as phase 2 and antioxidant enzymes, and stimulates their timely transcription. Some of the NRF2-dependent antioxidant enzymes include Quinone oxidoreductase 1 (*Nqo1*), Heme oxygenase-1 (*Hmox-1*), Peroxiredoxin-1 (*Prdx-1*) and glutathione enzymes such as Glutamate–cysteine ligase catalytic subunit (*Gclc*), Glutamate–cysteine ligase modifier subunit (*Gclm*) and Glutathione S-transferases (*Gsts*) [43] (see Figure 2). Other antioxidant proteins work independently of NRF2, such as mitochondrial coenzyme Q10, a key cofactor of the electron transport chain which scavenges RO/NS [44] and vitamin E, which down-regulates the generation of RO/NS and superoxide within mitochondria, thereby attenuating oxidative damage [45].

With age, there is a shift in redox homeostasis to a more oxidative environment, with the accumulation of RO/NS via impaired mitochondrial function and/or defective RO/NS elimination processes, and this build-up of oxidative stress with age has been implicated in the development of sarcopenia [46]. In addition, nuclear accumulation of NRF2 is reduced with ageing, leading to impaired cytoprotective responses to oxidative stress and reduced elimination of RO/NS [47]. In older mice lacking *Nrf2*, markers of oxidative stress burden were robustly increased, alongside decreased expression of antioxidant genes, rate and efficiency of respiration, leading to the higher production of RO/NS per O_2 consumed. Loss of *Nrf2* with age exacerbated age-associated mitochondrial oxidative stress [48], impaired skeletal muscle mitochondrial biogenesis and dynamics and worsened both sarcopenia and frailty [49]. Loss of *Nrf2* in mice also triggered ubiquitination and pro-apoptotic signals in the skeletal muscle, leading to protein degradation, cell death and skeletal muscle deterioration with age [50,51]. Thus, *Nrf2* plays a protective role against early development of sarcopenia, as older *Nrf2* KO mice had increased RO/NS production and reduced O_2 consumption by the mitochondria, along with reduced skeletal muscle mass and diminished generation of contractile force [52] . In line with this, sedentary older wild type mice also showed an impaired *Nrf2-Keap1* signalling axis, which promoted oxidative damage to skeletal muscle and impaired redox homeostasis [53].

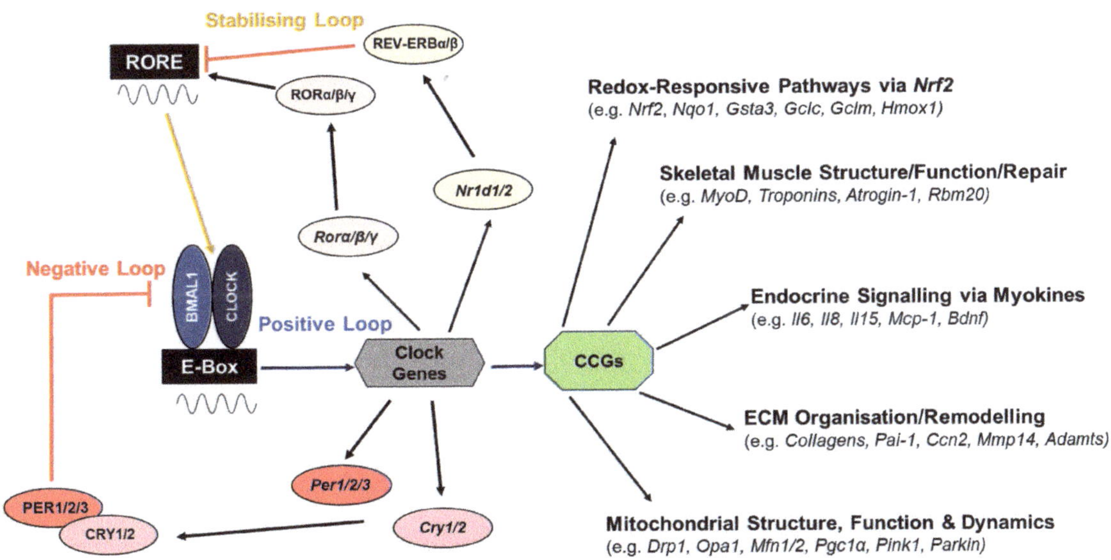

Figure 2. Schematic depicting the molecular circadian clock TTFL mechanism and downstream CCGs and molecular pathways under circadian control. BMAL1 and CLOCK within the positive arm of the molecular clock heterodimerise to activate rhythmic transcription of the negative and stabilising loop target genes as well as numerous clock-controlled genes (CCGs) by binding to E-box elements in the gene promoters. PER1/2/3 and CRY1/2, the negative loop components, inhibit BMAL/CLOCK activity to repress their own expression and that of CCGs. The stabilising loop components activated by the BMAL/CLOCK heterodimer include RORα/β/γ and REV-REBα/β, which regulate rhythmic expression of core clock genes such as *Bmal1* and *Clock* in a positive or negative manner, respectively, via binding to ROREs in the gene promoters. Examples of downstream CCGs within various molecular pathways under circadian control include the following: redox-responsive pathways via *Nrf2*, pathways regulating muscle structure, function and repair, endocrine signalling via myokines and pathways regulating ECM homeostasis and mitochondrial structure, function and dynamics.

3. Introduction to the Circadian Clock Timing System

Organisms from across the three domains of life, bacteria, archaea and eukaryotes, all display time-of-day dependent changes to their physiology, behaviour and metabolism, which ensures their ability to adapt to daily environmental changes [54]. This evolutionarily conserved, endogenous ~24 h time-keeping system is termed the circadian clock (from the Latin 'circa diem' meaning 'around a day'), which is responsible for the generation of biological oscillations in physiological and metabolic processes called the 'circadian rhythms'. Some of the daily physiological processes exhibiting circadian rhythms include the sleep/wake cycles, thermoregulation, energy homeostasis, metabolic homeostasis, hormone secretion and locomotor activity [55]. In mammals, according to the traditional model, circadian clock regulation is hierarchical with the master or central clock located in the suprachiasmatic nuclei (SCN) of the brain's anterior hypothalamus receiving the main entrainment signal from light, and coordinating the synchronicity of peripheral clocks found in various tissues and cells around the body, such as skeletal muscle, lungs and liver. The light-dependent time cues from the SCN are relayed via the retina through the neuronal and endocrine pathways to the peripheral clocks (SCN-dependent synchronisation) [56]. Recent research suggests a bidirectional circadian network system, whereby peripheral clocks can also maintain their rhythms independently of the SCN through responding to external non-photic cues such as cycles of feeding/fasting, activity/rest and drug administration (SCN-independent synchronisation), with feedback signals from these tissues acting upon the master clock [57]. Additional features of the circadian timing system include inter-tissue communication via hormones, myokines and metabolites, which integrate the whole circadian system and regulate the body's response to daily cues; see Figure 3 [58–60]. These external cues produce minor but biologically predictable circadian shifts, many of which optimise the body's response to a challenge [61]. Circadian misalignment, a mismatch between the internal circadian clock and the daily external cues (e.g., light/dark cycles, activity/rest or feeding/fasting cycles), can be caused by factors such as irregular or disrupted sleep, light pollution (e.g., light at night (LAN) exposure), shift work or frequent transmeridian travel, and has been associated with many diseases including metabolic disorders, cardiovascular diseases, musculoskeletal conditions, immune system impairments, neurodegenerative diseases and cancer.

The SCN, present in the anterior region of the brain's hypothalamus, is a bilateral structure, composed of two nuclei comprising 10,000 neurons each [62]. The SCN can be sub-divided into the dorsal 'shell' and ventral 'core' regions. The 'shell' contains cells expressing arginine vasopressin (AVP), whilst the 'core' contains gastrin-releasing peptide (GRP) and vasoactive intestinal peptide (VIP) [63]. Entrainment of mammals to the environmental ~24 h light/dark cycle is regulated by the retino-hypothalamic tract (RHT) via direct input from photoreceptive retinal ganglion cells, which depolarises the retinorecipient neurons in the ventral core of the SCN [64]. γ-aminobutyric acid (GABA) acts as the primary inhibitory neurotransmitter in the brain [65]. GABA propagates the signal received by the SCN core to the dorsal SCN shell [64]. The SCN clock projects to other regions of the brain, which are responsible for regulating brain-specific behavioural, autonomic and neuroendocrine circadian rhythms, such as cognition, mood and brain homeostasis. The SCN clock utilises the autonomic, neuroendocrine and behavioural pathways, via hormones and temperature, as systemic cues to synchronise clocks of peripheral tissues. The peripheral clocks regulate tissue-specific physiological rhythms, whose correct functioning is required for optimal health, including blood pressure, glucose metabolism and skeletal muscle force and function [64,66].

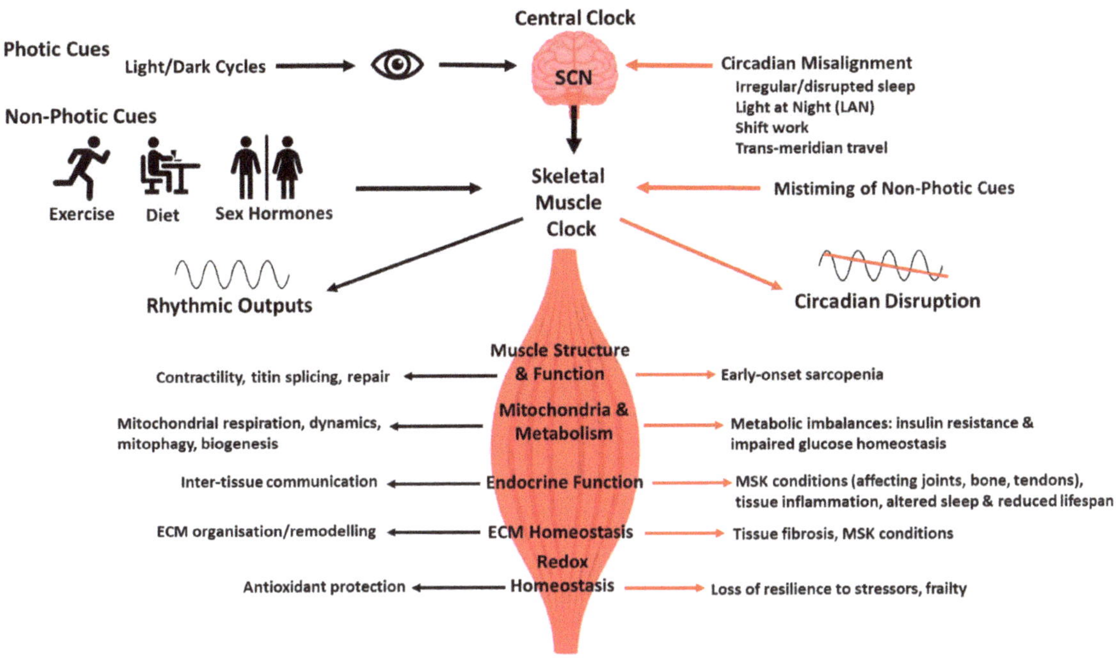

Figure 3. Schematic depicting the roles of mammalian skeletal muscle clock in regulating tissue-specific rhythmic outputs and systemic health and the consequences of circadian disruption. Photic cues (e.g., light/dark cycles) entrain the mammalian SCN clock, a central clock within the brain's anterior hypothalamus. SCN relays photic information through the retina to the peripheral clocks in tissues and cells around the body, including skeletal muscle, to coordinate their synchronisation to the external environment. Non-photic cues can also synchronise the peripheral clocks independently of SCN. Skeletal muscle clock leads to several key rhythmic outputs encompassing its structural, metabolic and endocrine roles. Circadian misalignments in light/dark cycles or mistiming of non-photic cues lead to functional consequences to skeletal muscle homeostasis (e.g., early-onset sarcopenia and frailty) as well as systemic health effects, such as metabolic imbalances, MSK conditions affecting joints, bone and tendons, tissue fibrosis and inflammation, altered sleep and reduced lifespan.

4. The Molecular Network Regulating the Clock Circuitry

The mammalian circadian rhythms are governed by the molecular clock mechanism that relies on the interlocked network of auto-regulatory transcriptional/translational feedback loops (TTFLs), which leads to rhythmic expression of genes, proteins and metabolites. Temporal homeostasis is maintained as the molecular clock responds to metabolic and environmental stimuli [67]. The molecular clock comprises positive, negative and stabilising feedback loops; see Figure 2. The positive arm is formed by the basic helix-loop-helix-PER-ARNT-SIM (bHLH-PAS) transcription factors CLOCK (Circadian locomotor output control kaput) and BMAL1 (Brain muscle arnt-like 1) [68]. CLOCK and BMAL1 heterodimerise to activate the transcription of core clock genes, as such the negative loop genes *Per1/2/3 and Cry1/2*, by binding to E-box elements in their gene promotor regions. The PER (Period) and CRY (Cryptochrome) proteins accumulate in the cytoplasm before forming multimers, translocate back to the nucleus and inhibit CLOCK:BMAL1 transcriptional activity, thereby leading to their own repression [54]. The PER and CRY proteins are regulated by phosphorylation through casein kinase I δ/ε and AMP-regulated protein kinase (AMPK), respectively, which leads to their ubiquitination and subsequent degradation, which plays a crucial role in regulating the period length of the circadian cycle [69]. As PER and CRY levels decrease, this enables the CLOCK:BMAL1 heterodimer to re-initiate transcriptional activity. The feedback mechanism between the positive and negative arms of the molecular clock takes ~24 hours to complete and represents the driving force behind circadian rhythm generation [68].

The CLOCK:BMAL1 heterodimer also initiates the E-box mediated transcription of the stabilising loop components which act to fine-tune the molecular clock. The stabilising arm consists of the retinoic acid-related orphan receptors (RORs) encoded by *Rorα/β/γ* and nuclear receptor subfamily 1, group D (REV-ERBs) encoded by *Nr1d1/2*. RORs activate *Bmal1* and *Clock* transcription via ROR-response elements (RORE) in their gene promotors, whilst REV-ERBs repress their transcription [67,68]. In addition, numerous clock-controlledgenes (CCGs) are transcriptionally activated by the CLOCK:BMAL1 heterodimer. A total of ~43% of the mammalian genome is expressed in a circadian manner, and across any two tissues <10% of genes are shared, suggesting a high level of tissue specificity [70]. The core molecular clock regulates thousands of CCGs in a cell type- and tissue-specific manner that are outside of the molecular core clock mechanism, and the protein products of some of these targets can often feedback onto the core clock mechanism. CCGs rhythmically oscillate in their expression, and many encode for transcription factors or proteins involved in rate-limiting steps in metabolic pathways, thereby producing rhythmic changes in tissue-specific physiology [71]. Figure 2 details the molecular clock mechanism underlying the circadian rhythm generation and examples of downstream CCGs within various molecular pathways under circadian control discussed in sections below.

5. Skeletal Muscle Clock: From Experimental Models to Human Studies

Skeletal muscle has been known to be under circadian regulation for well over two decades. Functions regulated by the skeletal muscle clock range from mitochondrial biogenesis and oxidative metabolism, skeletal muscle force and function [66], titin splicing [72], proteostasis [73] to satellite cell-mediated muscle regeneration and repair [74]. More recent work has shown that the skeletal muscle clock also regulates systemic health effects ranging from musculoskeletal homeostasis of bones, tendons and joints [75] to glucose homeostasis [76] and total sleep amount [77]; see Figure 3.

One of the first reports showing skeletal muscle as a peripheral tissue clock was in 1998, when Zylka and colleagues discovered the circadian expression of the equine *Per* genes [78]. Miller and colleagues were the first to report the presence of the circadian transcriptome in mouse skeletal muscle [79]. Further investigations through global transcriptomic profiling identified >800 genes which oscillate in *Gastrocnemius* mouse skeletal muscle [76,80]. A large number of discovered skeletal muscle CCGs are involved in physiological processes specific to the skeletal muscle, such as Myoblast determination protein (*MyoD*), Troponins and *Atrogin-1* [81]; see Figure 2. With the continuing development of bioinformatic analysis tools, Harfmann and colleagues were able to expand the number of genes expressed in a circadian manner within skeletal muscle to >2300 [82]. Further transcriptomic data for mouse *Tibialis anterior* (TA) and *Soleus* muscles identified a clock-controlled output of ~1000–1500 genes, and these two muscles only shared ~110 CCGs [83]. Overlapping genes included all core clock genes as well as *MyoD* and genes encoding proteins for muscle structure, signalling, metabolism and transcription factors. Of the muscle CCGs identified, a large proportion was specific to each skeletal muscle. For example, antioxidant genes *Hmox1* and *Gsta3* were rhythmically expressed in the TA along with angiogenesis gene Angiopoietin-1 (*Angpt1*), whereas enzymatic metalloproteinase family members, such as A disintegrin and metalloproteinase with thrombospondin motifs (*Adamts*), were rhythmically expressed in *Soleus* [83]; see Figure 2. Interestingly, mouse *Gastrocnemius* and human *Vastus lateralis* skeletal muscles shared 430 clock output genes, thereby suggesting conserved features of the skeletal muscle clock in both mouse and human muscle [70,84].

The circadian regulation of skeletal muscle has not only been demonstrated in murine models, but also in human studies. Human skeletal muscle plays an important metabolic role in the body, regulating proteostasis as well as glucose and lipid homeostasis. Desynchrony of human circadian rhythms, through irregular shift work patterns, leads to increased development of metabolic disorders such as obesity and type 2 diabetes [85]. Irregular shift work also increases the risk of skeletal muscle atrophy [86], whilst sleep disorders developed as a result of circadian desynchrony increase the degradation rate of skeletal muscle proteins [87]. Participants on a 14-day calorie restriction diet. who had undergone restricted sleep, lost 60% more muscle mass than those having regular sleep hours [88]. Furthermore, circadian desynchrony caused by insufficient sleep perturbed peripheral tissue clocks and led to impaired metabolism of skeletal muscle and liver as well as a whole-body imbalance in energy homeostasis. During such circadian desynchrony, hormesis is also affected, shifting the body to a catabolic state, thereby increasing the breakdown of skeletal muscle proteins and reducing the rate of protein synthesis [89].

Furthermore, in a study by van Moorsel and colleagues, a diurnal rhythm in skeletal muscle oxidative capacity was demonstrated in lean and healthy male participants [90]. Mitochondrial oxidative capacity peaked in the late evening (~23 p.m.) and troughed in the early afternoon (~13 p.m.). Interestingly, this aligned with the peak of resting energy expenditure (~23 p.m.-midnight) and coincided with the peak levels of core clock protein BMAL1 (~23 p.m.) and the trough of negative loop protein PER2, indicating a robust human skeletal muscle clock. In a follow-up study involving overweight male participants, mitochondrial respiration rates were lower when compared to the leaner participants, and the rhythmicity seen in the mitochondrial oxidative capacity was completely lost [91]. This suggests that restoration of the human muscle skeletal muscle clock may be an important step to improve metabolic health in those who suffer from metabolic disorders.

6. Circadian Regulation of Structural, Metabolic and Endocrine Functions of Skeletal Muscle with Ageing

Many studies have linked the disrupted genetic molecular clock to impaired skeletal muscle function, particularly with ageing. In 2006, Kondratov et al. demonstrated that a whole body knock-out of *Bmal1* induced early onset of sarcopenia, loss of skeletal muscle mass and strength with age, in addition to the loss of activity/rest rhythms and a reduction in locomotor activity [92]. *Bmal1* knock-out (KO) and *ClockΔ19* mice (with an ablation of exon 19 in the *Clock* gene) both exhibit ~30% reductions in normalised maximal skeletal muscle force, which is reciprocated at the single-fibre level, with changes in the myofilament architecture impairing contractile ability [66]. When a stabilising loop gene *Nr1d1* was knocked out in mice (encoding *Rev-erbα*), this resulted in a similar early ageing phenotype found by Andrews et al., (2010), with disrupted myofibre structure and an impaired ability to exercise [93]. Furthermore, using genetic approaches targeting *Bmal1* in the Duchenne muscular dystrophy (DMD^{mdx}) mouse model, it was shown that a loss of *Bmal1* significantly accelerated dystrophic disease progression due to impaired myogenic progenitor proliferation [94]. In contrast, a loss-of-function of circadian repressor *Rev-erbα* in vivo augmented satellite cell proliferative expansion and regenerative progression during regeneration [95] through modulating Wnt/Notch signalling, and pharmacological inhibition of Rev-erbα (which stimulates *Bmal1* expression) ameliorates the severity of muscular dystrophy in mdx mice [96,97]. Similarly to *Rev-erbα* loss, satellite cell-specific knock-out of *Cry2* enhanced muscle regeneration by stimulating the proliferation of myoblasts and enhancing their differentiation potential [98]. Pre-clinical mouse models of DMD and muscle biopsies from DMD patients show deregulation of *Bmal1*, *Cry1* and *Cry2* in dystrophic muscles [99], which may result from disrupted dystrophin- RhoA-actin-SRF (serum response factor) signalling required for the molecular clock resetting [100]. Furthermore, deep RNA profiling of another muscular dystrophy caused by mutations in collagen VI A1-3 genes, using both the *Col6a1* knockout mouse model and patients with Collagen VI pathology [101], discovered alterations of the molecular clock genes, and specifically of the circadian gene CLOCK, as a key pathophysiological signature. These findings linking the altered molecular clock to muscular dystrophies could provide a rationale for exploring clock-augmenting interventions in conjunction with current or in-development therapies to ameliorate the dystrophic muscle phenotype by reinforcing sarcolemma integrity [102].

New research by Riley and colleagues (2022) has identified a mechanistic link between the molecular clock and the organisation of skeletal muscle sarcomere filaments. Using an inducible skeletal muscle-specific *Bmal1* KO, the group determined that the loss of *Bmal1* led to heterogeneity in sarcomere length due to altered titin isoform expression, caused by impaired transcriptional regulation of RNA-binding motif protein 20 (*Rbm20*), a splicing regulator of titin [72]. Titin provides passive stiffness to the sarcomeres and prevents overstretching during contraction, and dysfunction in titin has been linked to a range of skeletal muscle diseases including Late-onset autosomal dominant tibial muscular dystrophy, Limb–girdle muscular dystrophy type 2J and Adult-onset recessive proximal muscular dystrophy [103].

Further to impairments in skeletal muscle structure and contractility, a muscle-specific knock-out of *Bmal1* resulted in impaired insulin-dependent glucose uptake and reduced glucose oxidation in skeletal muscle as well as impaired systemic glucose homeostasis [76,104]. In line with skeletal muscle's role in endocrine functions, it was found that basal secretion of cytokines such as IL-6, IL-8 and monocyte chemoattractant protein 1 (MCP-1) is under cell-autonomous circadian regulation in synchronised primary human skeletal myotubes, which was abolished when the molecular clock was disrupted through the knock-down of *Clock* [105] (see Figure 2).

Moreover, circadian regulation of skeletal muscle has revealed key roles in maintaining systemic health. Skeletal muscle plays a critical role in promoting systemic synthesis and accumulation of lipids in fat storage tissues such as adipose tissue and intestine through the cytokine unpaired 2 (*Upd2*), which maintains daily rhythms in circulating lipids in Drosophila [106]. More recent work has shown that scheduled exercise training led to marked modification of skeletal muscle clock gene expression and myokine production, in a manner that is dependent on the schedule of exercise training, which also impacted tumour progression [107]. Interestingly, muscle-specific *Bmal1* KO mice demonstrated a blunted transcriptional response to acute exercise, with gene expression signatures of inflammation across many tissues, suggesting that the skeletal muscle clock regulates systemic exercise effects [108]. Furthermore, rescuing *Bmal1* expression specifically in the skeletal muscle of *Bmal1* KO mice led to extended lifespan with improvements in skeletal muscle strength (but not muscle mass, fibre type or size), glucose tolerance, metabolic flexibility and reduction in inflammatory signatures across several peripheral tissues, including liver, lung and white adipose fat [109]. Moreover, exercise timing influenced the intra-tissue and inter-tissue metabolic signatures, revealing time- and tissue-dependent exerkines in liver and skeletal muscle, such as 2-hydroxybutyrate [110]. Therefore, disruption of the core clock genes in skeletal muscle leads to impaired structural, metabolic and endocrine functions of the skeletal muscle at the intrinsic cellular and systemic levels; see Figure 3. Given that daily voluntary exercise [111] and ageing [112] significantly remodel the murine muscle circadian transcriptomes, future work is warranted to decipher circadian muscle-secreted factors induced by exercise or ageing and their role in inter-tissue communication and systemic homeostasis.

7. Redox Homeostasis, NRF2 and Circadian Rhythms: A Bi-Directional Relationship

Cells and tissues maintain redox homeostasis and prevent RO/NS accumulation and associated oxidative damage through the activation of powerful antioxidant stress defence systems which sense variations in redox potential and trigger the required homeostatic processes to return to a steady state [113]. At low concentrations, RO/NS is considered to be a 'eustress' which is a physiologically beneficial oxidative stress [113]. Under these low oxidative stress conditions, RO/NS act as secondary messenger molecules and are known to activate several intracellular signalling pathways. RO/NS can enable adaptation to changes in environmental and internal redox environments via hydrogen peroxide signalling [114]. As skeletal muscle is a highly contractile and metabolic tissue, it requires large numbers of mitochondria to produce ATP. Due to aerobic respiration, skeletal muscle fibres are a major source of superoxide and nitric oxide, which can be converted into RO/NS [115]. When RO/NS levels rise above a physiological threshold, the oxidative stress ensues. Oxidative stress is thus a result of an imbalance between the production and the elimination of RO/NS, and represent a major risk factor for the onset and/or progression of vast array of age-related diseases including musculoskeletal conditions, cancer, cardiovascular and respiratory diseases, metabolic disorders (e.g., type 2 diabetes) and neurodegenerative conditions [116].

It is well established that there is coupling between the circadian clocks and redox systems, where the molecular clock directly regulates the levels of redox metabolites and antioxidant enzymes [117]. Self-sustained rhythms in the cellular redox state [92,118–120] as well as ROS-responsive antioxidant genes [121,122] have been observed in the brain and/or peripheral tissues. Moreover, diurnal rhythms in organismal survival to ROS-generating agents have been reported [123]. One of these antioxidant enzymes, Peroxiredoxin 1, is an evolutionarily conserved antioxidant protein that exhibits circadian redox oscillations in mouse tissues and in human red blood cells lacking nuclei [119,124], through a non-transcriptional/translational feedback loop.

Furthermore, there is a distinct interaction between the circadian clock and redox homeostasis via the NRF2-mediated redox signalling. The traditional paradigm of NRF2 regulation has mainly focused on oxidative stress-induced post-translational mechanisms, including regulation by its negative regulators (KEAP1, BACH1 and β-TrCP), positive regulators (DJ1, p62 and p21), or protein modifications [125–129]. However, the circadian paradigm of NRF2 regulation suggests that the endogenous levels of *Nrf2* are much more dynamic than previously thought. Indeed, *Nrf2* and its target genes have been identified as clock-controlled genes under the regulation of the core molecular clock in several peripheral tissues such as liver, lung, pancreas, macrophages and the brain, leading to rhythmic balance of redox homeostasis [130–134]. In this manner, the regulation of NRF2 nuclear activity appears to be restricted to a particular circadian window, gated by both the clock-controlled mRNA synthesis and the KEAP1-based protein degradation.

In mouse liver, diurnal expression of mRNAs of *Nrf2* and *Nqo1* reached their highest expression in early evening (~6 p.m.), whilst troughing in early morning (~2–6 a.m.).

Interestingly, the expression of NRF2 target genes was higher in females than males, especially for *Nqo1* [132]. Similarly, in mouse pancreas, *Nrf2* peaked in early evening (ZT14), and this regulation was abolished in *Bmal1* KO-specific pancreatic islets (Zeitgeber ZT0= lights on, ZT12= lights off). In contrast, in mouse peritoneal cells, both *Bmal1* and *Nrf2* troughed during mid-afternoon (ZT8) and peaked in the early morning (ZT20), whilst a rapid induction of *Nrf2* was seen upon stimulation by BMAL1 [131]. Similarly to macrophages, in the mouse lung, *Nrf2* and its target genes *Gclm* and *Hmox1* peaked in the early morning (ZT0-3), whilst *Gsta3* peaked in the early evening (ZT15), suggesting that both transcriptional and post-transcriptional mechanisms may regulate tissue-specific antioxidant responses [130]). Further work in the mouse lung and fibroblasts demonstrated that *Nrf2* mRNA as well as its protein levels and nuclear activity are not only targets of the molecular clock under light/dark conditions, but are also regulated endogenously under constant darkness. Moreover, it was found that NRF2 itself rhythmically binds to promoters of antioxidant genes involved in glutathione synthesis (e.g., *Gclm/Gclc*) and utilisation (e.g., *Gsta3/4*), and that cyclical NRF2 levels were responsible for their rhythmicity [130]. This contrasts with an antioxidant gene *Prdx6*, which is cooperatively regulated by both BMAL1 and NRF2 in human lens epithelial cells [135].

Redox environment can also exert feedback on the molecular clock activity. The DNA binding activity of the CLOCK:BMAL1 heterodimer is dependent on the cellular redox status determined by the NADH/NAD^+ and NADPH/NADP ratios, whereby binding is enhanced under reducing conditions [136]. NRF2 can also modify circadian clock gene expression and rhythmicity in mouse fibroblasts, hepatocytes and liver, indicating that NRF2 is involved in fine-tuning the timekeeping mechanism through a circadian clock-NRF2 bi-directional loop [137], although the precise mechanisms of this regulation have not been studied and require further studies in other peripheral tissues such as skeletal muscle.

8. Circadian Regulation of Mitochondrial Structure, Function and Dynamics

Mitochondria are vital for the production of energy in the form ATP through a process of oxidative phosphorylation, which is required for the functioning of many biological processes, including contraction of skeletal muscle [138]. The first stage of oxidative phosphorylation involves the oxidation of Nicotinamide Adenine Dinucleotide (reduced) (NADH) and Flavin Adenine Dinucleotide (reduced) (FADH2) via a series of electron transport chain protein complexes embedded in the inner-mitochondrial membrane. This produces an electrochemical gradient, with a higher concentration of H+ protons outside the mitochondrial membrane. The second stage involves the movement of the protons through the protein ATP synthase, inducing the binding of ADP (Adenosine Diphosphate) and phosphate (Pi), and the formation of ATP [139].

A comprehensive circadian proteomics study of mitochondria isolated from mouse liver identified that ~38% of mitochondrial proteins are rhythmic and oscillate through the day, peaking in protein abundance from ZT0-4 [140]. Both post-transcriptional and post-translational circadian mechanisms have been implicated in mitochondrial function and dynamics [141]. Circadian regulation of mitochondrial oxidative phosphorylation is achieved through several pathways: deacetylation of mitochondrial proteins by SIRT3, a NAD-dependent mitochondrial deacetylase [142] and via AMPK, a redox-sensitive energy sensor [143]. A global analysis of the liver acetylome in mice revealed CLOCK-dependent acetylation of enzymes within the Krebs cycle, glutathione metabolism and mitochondrial proteins [144], which are altered in *Bmal1* KO mice [142,145,146].

Furthermore, mitochondrial morphology and distribution are in constant flux, undergoing cycles of fusion (joining of mitochondria) and fission (separation of mitochondria), with fused mitochondria being more metabolically active as they have higher capacity for respiration [147]. Fission of mitochondria is dependent upon dynamin-related protein-1 (DRP1), which is regulated through transcriptional and post-transcriptional mechanisms (phosphorylation) in a circadian manner as demonstrated in human skin fibroblasts as well as the mouse brain and liver, whilst mitochondrial fusion is regulated through circadian post-transcriptional regulation of fusion proteins MFN1/2 and OPA1 [148]; see Figure 2.

Mitochondrial membrane potential also follows a circadian pattern as demonstrated in murine macrophages, gradually increasing 4 h after serum shock and peaking at 12 h to 16 h before gradually depolarising for the rest of the circadian cycle [149]. The circadian clock also regulates the rhythmic synthesis of NAD+, thereby producing oscillations in the mitochondrial capacity for energy production [150].

The balance between mitochondrial biogenesis and mitophagy determines the mitochondrial content of the cell. PGC-1α/β are key regulators of mitochondrial biogenesis whilst PTEN-induced kinase 1 (PINK1) and Parkin RBR E3 ubiquitin ligase (PARKIN) are essential in mitophagy [151]. PGC-1α/β are rhythmically expressed in the liver and skeletal muscle in mice, with functioning PGC-1α being required for cell-autonomous clock function [152]. *Pink1* is a direct target of BMAL1, and in human skeletal muscle, PINK1 protein levels peak in the middle of the active phase [149]; see Figure 2. In synchronised human fibroblasts, transcription of *Park2* (coding for PARKIN) and *Park6* (coding for PINK1) displayed oscillations in expression, with ~24 h periods [153]. The oxygen consumption rate, a readout of mitochondrial respiration, exhibits circadian oscillation, which is lost in the liver mitochondria of *Per1/2* KO and *Bmal1* KO mice [140,142,148]. Additionally, PER1 and PER2 regulate the key rate-limiting mitochondrial enzymes, such as carnitine palmitoyl-transferase 1 (CPT1) in mouse liver, which are required for the processing of nutrients in a time-of-day manner [140]. Circadian clock components also affect lifespan determination in *Drosophila* by altering cellular respiration via mitochondrial uncoupling [154], suggesting that the mitochondrial metabolism and circadian rhythms are coupled at a systemic level [155].

Bmal1-/- and *Clock*$^{\Delta 19}$ mice show a ~40% reduction in mitochondrial volume in skeletal muscles, and of the mitochondria present, most were abnormal in shape, impairing their ability to undertake respiration and produce ATP. These mice also show altered expression levels of *Pgc1α/β* [66]. Exercise and melatonin were able to restore myogenesis and altered mitochondrial dynamics associated with sarcopenia in *Bmal1* KO mice [156]. *Nr1d1-/-* mice show a similar reduction in mitochondrial volume in skeletal muscles [93]. Deficiency of *Rev-erbα* in mice induced increased mitophagy through a lack of inhibition of *Park2* and Unc-51 like autophagy activating kinase 1 (*Ulk1*), shifting the mitochondrial dynamics, reducing energy capacity and resulting in a decline of total mitochondrial content [93]. Mice with a double knock-out of *Cry1/2* are capable of running further and faster than WT littermates, enhancing exercise capacity due to increased peroxisome proliferator-activated receptor δ (PPARδ) activity and fatty acid oxidation [157]. Furthermore, muscle-specific *Per2* deletion in mice led to a lower glucose tolerance during the day compared with WT littermates but no difference was seen during night-time. Night-time activity levels were also significantly reduced in *Per2* KO mice but this did not affect exercise capacity [158]. Circadian clock dysregulation is observed in primary myocytes and skeletal muscle biopsies in individuals with type 2 diabetes and is linked to altered rhythmic mitochondrial respiration and rhythmic transcripts of the mitochondrial inner membrane [159]. However, the circadian mechanisms of mitochondrial structure, function and dynamics in skeletal muscle have not been studied in detail, and whilst some of the circadian insights from other peripheral tissues are likely to operate in skeletal muscle, further studies are warranted, especially when it comes to responses to exercise and ageing.

9. Extracellular Matrix Homeostasis Regulation by the Molecular Clock

The ECM plays a major role in determining the stiffness of the skeletal muscle microenvironment during rest and in mechanical responses to exercise. Excessive accumulation of ECM in the skeletal muscle with ageing leads to increased stiffness, which inhibits the myogenic differentiation capacity of muscle-resident stem cells. The ECM components required for maintaining the SC niche are altered in the skeletal muscles of aged mice. Old muscles show increased collagen content by ~35–40% of the extracellular space with a higher proportion of type-I collagen to type-III collagen as well as increased proportions of collagens type IV and VI alongside increased levels of the non-enzymatic crosslinking of collagen fibres such as glycosylation. This leads to changes in the elastic modulus of the ECM from ~12 to ~ 418 kPa in older skeletal muscles, demonstrating much stiffer microenvironment, that affects both stem cell activity and generation of skeletal muscle force [10].

Skeletal muscle stem cells, known as satellite cells, reside in the niche within the ECM, and they are vital for the regeneration of skeletal muscle, maintaining strong oscillations of the circadian clock machinery [160]. Circadian time-series proteomics revealed that many ECM proteins show daily oscillations in their abundance including collagens and various components of the matrisome in the mouse femoral head cartilage [161]. Research by Chang and colleagues identified that key regulators of collagen type I secretion, including transport and Golgi organisation protein 1 (TANG01) and SEC61, are rhythmic in their expression, resulting in nocturnal procollagen-1 synthesis with daytime collagen type I fibril assembly, thus affecting oscillations of a pool of newly synthesised collagen type I. Arrhythmic *Clock*$^{\Delta 19}$ and *Bmal1* KO mice both accumulate *Col1α1* mRNA and have a disorganised and structurally and mechanically abnormal collagen matrix [162]. Matrix metalloproteinase-14 (*Mmp14*), a key protease involved in the remodelling of ECM through the degradation of macromolecules such as type I collagen, is also rhythmically expressed at both mRNA and protein levels. A loss of *Mmp14* leads to arrhythmic tendon fibroblasts and severe fibrosis of the ECM [163]. Dysregulation of various ECM proteins and tissue fibrosis have been seen in a wide range of peripheral tissues in both *Bmal1* KO and *Clock*$^{\Delta 19}$ mice including: lungs [130,164–167], heart [168–172], skeletal muscle [75], kidney [173–175], liver [176–180], pancreas [181] and adipose tissue [182,183]; see Figure 3. Inducible skeletal muscle-specific *Bmal1* deletion also resulted in increased muscle fibrosis, together with increased bone calcification and alterations in joint knee collagens IV and V [75]. Moreover, ageing of several peripheral tissues leads to diminished circadian regulation, with changes in the ECM content and associated fibrosis [184–188]. Therefore, the circadian clock regulates the function of connective tissue components within the microenvironment such as residing tissue-specific stem cells as well as the abundance and activity of collagens and ECM proteins, although the circadian regulation of ECM in skeletal muscle, and how it changes with ageing specifically, has not been studied and warrants further research.

10. Chronotyping: Personalisation of Skeletal Muscle Performance

A 'chronotype' refers to the phase of an individual's inner biological clock. It differs between individuals such that the behavioural preference for one individual might be to be most active in the morning, whilst for another it may be in the evening [189]. A chronotype is influenced by physiological, genetic and behavioural factors and can be categorised into three groups, with morning types or 'early birds' on the one end, evening types or 'night owls' on the other, and an intermediate type in between. Morning types are typically phase advanced, being most active in the morning and early hours of the day, whilst the evening types are most active and alert in the later hours of the day. The intermediate types have the most flexibility, having a greater capacity to adapt to a required delay or advance depending on lifestyle requirements. A study by Facer-Childs and Brandstaetter (2015) involved early-, intermediate- and late- types undertaking a progressive aerobic endurance test at 7 a.m., 10 a.m., 1 p.m. and 10 p.m. The early types performed best at ~12.10 p.m., the intermediates at ~15.50 p.m. and the late types at ~19.40 p.m., and there was performance variation between the groups over the day; ~7–10% for the early and intermediates, but up to 26% for the late types [190].

A systematic review by van der Merwe et al. (2022) investigated how different chronotypes affect body composition and biomarker outcomes. They found that the evening chronotypes had increased body mass index and increased weight gain over time when compared to early chronotypes [191]. Moreover, individuals with late chronotype may have better cognitive and motor performances in the evening compared to earlier times of day [192]. Evening type swimmers, for example, showed a reduced performance in the morning compared to the evening, with both morning and evening types displaying higher stress levels when training at times opposite to their chronotypes [193]. Similar observations were seen in morning chronotype cyclists performing in the evening who reported higher exertion rates [194].

Chronotypes can be entrained to a degree; morning exercise can help to advance a late chronotype's phase to improve their morning performance [195]. Indeed, Facer-Childs and colleagues investigated if a combination of timed light exposure, sleep, meals, caffeine and exercise could shift the phase of 'night owls' in a real-world setting. Using a carefully formulated regimen, the researchers were able to achieve a phase advance of ~2 h in the participants, who also reported reduced morning sleepiness and decreased rates of depression and stress alongside improved cognitive (reaction times) and physical (grip strength) performance [196]. Interestingly, whilst late chronotypes may benefit from phase advances induced by exercise in the morning or evening, evening exercise may exacerbate circadian misalignment in early chronotypes [197]. Additionally, for athletes participating in team sports, there is evidence that chronotype distribution within teams can influence team performance [190]. Thus, whilst chronotypes can influence many physiological processes, new studies are required to understand how best to optimise this knowledge, especially if chronotypes are to be incorporated into personalised exercise regimens.

11. Time-Scheduled Exercise as a Lifestyle Intervention for Skeletal Muscle Loss with Ageing?

Currently, exercise is one of the key promising treatments for skeletal muscle loss with ageing as it can target both the metabolic and functional deficits of skeletal muscle [198].

Acute and chronic exercise increase mitochondrial function in skeletal muscle of rats [199] and humans [200]. Whilst contracting skeletal muscles during exercise produce more RO/NS, it has been determined that RO/NS itself are required to trigger and mediate the adaptive responses to exercise [201]. However, a single acute bout of exercise can generate excessive amounts of RO/NS which can lead to oxidative damage within the skeletal muscle [201]. Conversely, regular exercise can induce adaptations to resist oxidative damage through the activity of antioxidant pathways [201].

Aerobic exercise training in young adults and older individuals was capable of increasing *Nrf2* signalling in both groups as well as of lowering the basal level of *Nrf2* in the older group [202]. Moreover, in humans, baseline NRF2 expression in skeletal muscle correlates with maximal oxygen uptake and high-intensity exercise performance [203]. A systematic review investigating exercise-inducing adaptations identified that exercise training also promoted significant increase in mRNA and protein expression of various collagens, including types I, III, IV and VI collagens, in young adults and older participants [204]. Older participants experienced a higher fold change in ECM content compared to young adults during high intensity interval training (HIIT), including changes in collagens IV and XIV, elastin and periostin [205].

Optimal redox responses to exercise require functional NRF2 signalling. Mice with a *Nrf2* loss in muscles perform significantly worse in treadmill exercises and show impaired oxidative/glycolytic muscle contractility, whilst loss of *Keap1*, a repressor of *Nrf2*, enhanced exercise capacity and muscle contractility [206]. In line with this, treatment of rats with *Nrf2* activator, sulforaphane, enhanced running capacity and diminished muscle fatigue by upregulating *Nrf2* signalling and increasing downstream antioxidant gene expression [207]. Conversely, in a model of chronic *Nrf2* activation through systemic loss of *Nrf2* regulator *Keap1*, exercise endurance capacity in mice was increased [208].

Exercise is an external cue capable of synchronising clock gene expression in skeletal muscle. Recent work has demonstrated that daily variations in the expression of human core clock genes can predict fluctuations in exercise performance across the day [209]. Moreover, the timing of aerobic exercise impacts the levels of BMAL1 and NRF2 protein expression. Mice that exercised at Zeitgeber time 12 (ZT12, ZT0 = lights on, ZT12 = lights off) exhibited higher total antioxidant capacity and lower RO/NS in skeletal muscles than mice that exercised at ZT24. Antioxidant proteins NQO1 and HMOX-1 were both significantly upregulated at ZT12 compared to ZT24, suggesting a more adaptable skeletal muscle clock and timed antioxidant response to exercise [210].

Daily voluntary running also dramatically remodels the mouse skeletal muscle circadian transcriptome, with skeletal muscle and liver glycogen exhibiting robust daily rhythms [111]. These authors determined that daily voluntary wheel running of male mice resulted in an increased maximum speed at ZT21, at the end of the active phase, but not at ZT13, at the start of the active phase. The active mice also gained less weight than their age- and sex-matched sedentary counterparts. Interestingly, this time-of-day difference was seen in male mice but was not observed in age-matched females, though females were overall more active and had greater maximum sprint speed than males [111]. When investigating metabolite levels at ZT13 and ZT21, there were stable levels of glucose and β-hydroxybutyrate (major energy sources particularly when the body is low on carbohydrates) at ZT21 along with increased blood lactate, but a significant reduction in glucose at ZT13. The authors suggested that maximum speed at ZT21 is associated with increased selective use of carbohydrates for energy production [111]. This is consistent with a human study by Iwayama et al. (2021) where participants showed increased respiratory exchange ratio (measure of utilisation of carbohydrates and fats during energy production) when undertaking a single bout of 60 minute cardiovascular exercise in the afternoon (4 p.m.) when compared to the morning (7 a.m.) [211]. Interestingly, afternoon HIIT was more efficacious than morning HIIT at improving blood glucose in men with type 2 diabetes [212], and it also led to increased skeletal muscle lipids and mitochondrial content, the clinical implications of which warrant further research [213]. In contrast, exercise responses during the early active phase (ZT12) are characterised by HIF1α-dependent activation of glycolytic pathways in the mouse skeletal muscle, resulting in carbohydrate exhaustion, usage of alternative energy sources and adaptation of systemic energy expenditure [214].

Daily scheduled voluntary exercise can also affect SCN's molecular and neuronal activities, thus promoting cellular clock synchrony and robust ~24 h rhythms in behaviour [215].

In mice lacking neuropeptide vasoactive intestinal polypeptide (VIP) in the SCN, which leads to blunted and temporally disorganised molecular clocks, daily voluntary exercise promoted SCN clock cell synchrony and robustness of ~24 h rhythms in behaviour [215].

Furthermore, specific clock proteins can modify exercise capacity in a time-dependent manner, likely through differences in food consumption and liver glycogen stores, with BMAL1 impairing and PERIODs enhancing exercise capacity in a time-dependent manner [216]. Moreover, muscle-specific *Bmal1* knock-out mice present with diminished transcriptional responses to acute exercise as demonstrated through limited upregulation of exercise-responsive transcription factors Neuron-derived orphan receptor 1 (*Nr4 a3*) and *Pgc1α* [108]. Research has demonstrated that contraction of skeletal muscle through exercise leads to increases in cAMP levels, which activates cAMP response element binding protein (CREB) that, in turn, activates circadian PERIOD1/2 proteins, whilst the generated RO/NS activate a NAD-dependent protein deacetylase Sirtuin-1 (SIRT1) and PGC-1α to bind to circadian transcription factor RORα/γ, both pathways improving exercise performance [67]. Moreover, synchronisation of aerobic exercise with the circadian rhythm of NAD^+-SIRT1 boosts MFN2-mediated mitochondrial fusion by activating the BMAL1-PER2-SIRT1-PPARα axis in the skeletal muscle of diabetic mice, leading to more effective glycaemic control and improved insulin resistance [217].

It is important to note that different forms of exercise rely on enhancing distinct physiological processes and functions of skeletal muscle. Resistance exercise (e.g., weightlifting) promotes anabolic responses and muscle hypertrophy by increasing muscle mass through the activation of mTOR pathway which regulates the balance between protein synthesis and degradation [218]. Endurance exercise (e.g., marathon running) improves skeletal muscle function without substantial hypertrophy by improving maximum volume of oxygen (VO_2 max) and stimulating mitochondrial biogenesis to optimise aerobic metabolism efficiency [219]. Therefore, different exercise regimens may be more optimal to undertake at specific times-of-day. For example, human skeletal muscle torque, strength and power is increased during the late afternoon (between 16 and 18 p.m.) compared to the morning [190]. In a study of Olympic swimmers, athletes produced more competitive results when competing in the afternoon races, particularly ~17 p.m. [220]. Semi-professional female volleyball players exhibited increased standing spike velocity, flexibility, dynamic balance and agility at 19 p.m. in comparison to 9 a.m. [221]. Male professional tennis players jumped higher and were faster during the afternoon [222], and male Olympic weight-lifters [223] and male professional rugby players [224] both demonstrated a better performance and a higher output during the afternoon. Conversely, professional marathon runners produced improved personal record times at 6.30 a.m. in the morning [225]. For a comprehensive review on the influence of circadian rhythms on different sports performances, please refer to [226].

Many studies have demonstrated regular exercise as a strategy to improve sarcopenia and physical functions in the elderly. Landi et al. (2014) provided a comprehensive review of different exercise regimens used as treatments for sarcopenia and concluded that physical exercise increases aerobic capacity, skeletal muscle strength and endurance in older individuals [227]. Indeed, a search for 'sarcopenia' and 'exercise' on the Clinical Trials website (www.clinicaltrials.gov) produced 424 trials. A total of 409 studies were targeting older adults (65+ years) and 392 classed 'exercise' as an intervention. There was a variety of exercise interventions being investigated, including high/low loads, pilates, resistance, high intensity and home-based physical activities. Exercise is therefore a very promising intervention for skeletal muscle loss with ageing as it can improve functional and metabolic aspects of skeletal muscle and ameliorate the deterioration of energy metabolism with age. Resistance training is potentially of most benefit to the elderly as it increases skeletal muscle mass and strength to improve quality of life; however, the elderly typically have earlier chronotypes. Therefore, it can be postulated that the afternoon resistance exercise may be better adapted to earlier times in the day for the elderly populations. However, how time-of-day exercise responses change with ageing and age-related chronotype shifts is currently unknown and further research is warranted to maximise the therapeutic potential of chrono-based exercise interventions.

12. Sex-Tailored Exercise Interventions for Sarcopenia and the Role of Circadian Hormone Melatonin

Whilst the above studies demonstrated that circadian rhythms are paramount in optimising athletic performance, it is worth noting that most of these studies were undertaken by male participants. Inclusion of women in research studies was not promoted by the National Institutes of Health (NIH) in the USA until 1986, therefore prior to this date only male organisms and isolated cells were used in research. Pre-1993, the US Food and Drug Administration banned women of childbearing age to be included in clinical trials; thus, men were the main subjects of research. Indeed, historical bias, costs, concerns for decreasing fertility and arguments of hormonal fluctuations impacting findings are some of the reasons behind scarcity of studies on the female population [228]. This sex bias was also present during the animal testing stages of research. Between 1990 and 2009, 80% of all rodent studies used exclusively males. Some arguments against the use of female animal models were similar to human, mainly related to challenges in controlling hormonal fluctuations. However, there were additional reasons including female animals being required as breeders, making males surplus and more readily available for experiments [229]. In 2014, the NIH issued a mandate that all studies which were awarded grants were to include sex as a biological variable in cell lines, tissues and organisms, unless biological sex was not relevant, though this needed to be proved through strict logic and analysis [229].

Systematic reviews by Ose et al. and Paul et al. highlighted this sex imbalance in human studies [230, 231]. Paul and colleagues reviewed 669 studies from 2017 to 2021 across six top sports medicine journals and found that 70.7% of the studies focused on male-only athletes, whilst 8.8% were female-only, with males dominating in studies on sports such as football, rugby and basketball, and women more commonly studied in softball and volleyball studies [230]. Ose and colleagues investigated 1441 original and epidemiological research studies published in three major sports and exercise medicine journals between 2021 and 2023, totalling ~40.1 million participants. The mean proportion of female participants was 40.22%. However, only 103 (7.15%) of studies were female-only, and of them, a total of 66 (5.6%) included menstrual-status considerations in their study design [231]. Whilst women are 2:1 to 9:1 more likely than men to experience anterior cruciate ligament (ACL) injuries during sport activities, male athletes still predominate research population in this research field [232].

Thus, an important consideration when investigating the effects of ageing on skeletal muscle is the sex of the participants as there are many sex-specific differences in skeletal muscle maintenance, many of which stem from hormonal regulation at the onset of puberty. For example, males typically have more muscle mass than females, regardless of activity levels, and females have a greater proportion of type I muscle fibres whereases males have a larger proportion of type II fibres [233], possibly due to greater hypertrophy of type II fibres induced by testosterone [234]. Some metabolic properties of skeletal muscle also differ between males and females. Women use more lipids during exercise than men do, with higher levels of intramuscular triacylglycerol and fatty acid translocase proteins present in women than men [234]. Sex also influences skeletal muscle calcium ($Ca^{2+)}$ regulation; Ca^{2+}-

ATPase activity was reduced with exercise in male participants but increased in women. This may be due to differences in proportions of fibre types, with type I fibres having lower maximum sarcoplasmic reticulum Ca^{2+}-ATPase activity, suggesting that female skeletal muscle is more likely to resynthesise ATP from oxidative phosphorylation during exercise [235]. Moreover, myofibre size is also influenced by sex. Across all myofibre types, cross-sectional area is greater in males than females, and this difference appears during the transition from childhood to adulthood [236]. Recent research has shown that type II myofibres in the *Vastus lateralis* muscle of untrained adult males contained more satellite cells than those of their female counterparts [237]. This was supported by murine studies where the number of satellite cells in the *Extensor digitorum longus* muscle of male mice, predominantly composed on type II fibres, was greater than in age-matched females [238].

A key difference between male and female skeletal muscle structural/functional properties lies with the sex-specific hormones present in the body, and/or circulating hormones levels at very different concentrations. Both sex hormones regulate key physiological processes including postnatal muscle growth, promoting contractility and governing the hypertrophic response to resistance exercise [233]. Testosterone binds to nuclear androgen receptors, promoting development of type II over type I fibres, upregulating glucose clearance and glycogen synthesis and enhancing muscle mass and strength. The cellular mechanism for the activity of testosterone involves the upregulation of follistatin, a protein which inhibits the activity of myostatin, restricting skeletal muscle growth [233]. Testosterone also impacts skeletal muscle contractile function, with supplementation of the hormone increasing isometric knee extension peak torque in older male humans [239].

When oestrogen binds to the oestrogen-receptor complex, it promotes preferential development of type I over type II myofibres, mitochondrial dynamics and bioenergetics. Additionally, it maintains the pool of satellite cells, skeletal muscle contractility and protects against muscle damage and inflammation [233]. Whilst the action of testosterone in skeletal muscle is well described in the literature, the actions of oestrogen in the same context remain to be elucidated. Oestrogen has a positive effect on skeletal muscle regeneration after injury whilst decreasing exercise-mediated skeletal muscle injury through inhibition of inflammatory responses and increased growth of satellite cells [240]. Oestrogen is important in the regulation of mitochondrial function in skeletal muscle. Female mice with a muscle-specific knock-out of oestrogen receptor-α have reduced basal and stimulated oxygen consumption rates, excessive RO/NS production and mitochondrial morphological abnormalities including enlargement, elongation and hyper-fusion [241].

Melatonin is a highly conserved circadian 'dark' hormone synthesised in the pineal gland and most tissues and organs of the body, including skeletal muscle. Melatonin produced by the pineal gland oscillates in synchrony with the light/dark cycle and is controlled by the master clock in the SCN, though extrapineal melatonin does not follow a circadian rhythm [242]. Melatonin is a very versatile molecule, with pineal melatonin serving chronobiological functions and extrapineal melatonin acting as an antioxidant and anti-inflammatory molecule [242]. Both pineal and extrapineal melatonin decrease with age, the former leading to a decline in circadian control of peripheral tissues and the latter linked to mitochondrial dysfunction [242], suggesting that melatonin administration could be used as a potential therapy against sarcopenia.

Young male footballers treated with melatonin prior to intense exercise showed reduced levels of lipid peroxidation when compared to controls and also showed increased total antioxidant activity for up to 60 minutes post-exercise [243]. In rats treated with melatonin and subjected to exercise, melatonin led to enhanced exercise tolerance, increased expression of PGC-1α, decreased skeletal muscle triglyceride content and increased muscle glycogen levels 3 hours post-exercise [244]. Treatment of aged mice with melatonin improved function and structure of the skeletal muscles whilst reducing damage to mitochondria [245]. In muscle-specific *Bmal1* knock-out mice, which express hallmark features of sarcopenia (circadian and mitochondrial disruption and impaired skeletal muscle function), combined exercise and melatonin treatment reversed these effects independently of *Bmal1*. In addition, mitochondrial dynamics was improved and myogenesis was restored [156]. In male rugby players, core clock genes *Bmal1*, *Rorα*, *Cry1*, *Per1*, *Per2* and *Nr1d1* were upregulated compared with sedentary controls, and melatonin levels oscillated over the day in athletes, but remained constant in sedentary males [246]. Thus, a potential dual therapy of exercise and melatonin supplementation could be a therapeutic strategy to reduce the effects of sarcopenia in the elderly. Currently, there are 969 clinical trials being undertaken with melatonin as the intervention, many of which are for sleeping problems; of these, 15 are investigating melatonin's anti-ageing effects, though most are either for neurodegenerative disorders or osteoporosis. Only one study investigated melatonin in relation to sarcopenia (NCT03784495), and no studies investigated sarcopenia, exercise and melatonin. Thus, more preclinical and clinical research is warranted before dual melatonin/exercise therapy could be tested as a therapeutic treatment for sarcopenia.

A recent UK biobank study revealed that whilst there was a significant age-related decline in skeletal muscle strength in both males and females, loss of muscle mass was more pronounced in males, whilst females displayed substantially greater age-related decline in quality of skeletal muscle [247]. The cross-sectional area of the *Quadriceps* has been shown to decrease in both sexes, but more significantly in women [248]. Older adults have decreased intermyofibrillar mitochondrial size, however, the age effect was mostly seen in women. Additionally, the slowing of myosin cross-bridge kinetics is particularly observed in elderly women [249]. The review by Della Peruta et al. (2023) details the sex differences in skeletal muscle in adulthood and in relation to ageing [250].

Furthermore, elderly men and women both benefit from prolonged (six months, three times per week) resistance training, which leads to increases in leg lean mass and *Quadriceps* cross-sectional area [251]. There are currently 244 clinical trials registered on Clinical Trials website (clinicaltrials.gov) when searching the parameters such as 'ageing', 'exercise' and 'skeletal muscle'. However, when 'circadian' is added to the search parameters, this number drops to three studies only. Of the three studies, only two, NCT06136013 and NCT07075133, specify the use of exercise in the morning or afternoon as the intervention. This demonstrates that whilst exercise is considered an important therapeutic intervention for the treatment of sarcopenia, there is a great need for further research into time-of-day exercise interventions in the elderly. Therefore, ageing studies using exercise as an intervention warrant inclusion of both circadian timing and sex of participants as important variables to be studied as circadian timing of exercise may affect the responses of different types of exercise whilst sex of participants may influence benefits from interventions targeting fibre size and quality, rather than quantity alone.

13. Summary

Both the circadian molecular clock and *Nrf2* redox signalling regulate skeletal muscle function, metabolism and mitochondrial activity, with growing evidence demonstrating the importance of these systems in age-related skeletal muscle decline and systemic loss of tissue resilience. Many tissues, including skeletal muscle, display both circadian dysfunction and redox imbalance during ageing [252–255], therefore future exercise regimens may maximise skeletal muscle functional and systemic health benefits by incorporating chronotherapy-based strategies to boost both the molecular clock robustness (i.e., circadian alignment) and *Nrf2*–mediated antioxidant system through timed exercise, diet or pharmacological interventions. Figure 4 depicts a proposed model of how boosting circadian alignment, through time-scheduled lifestyle approaches such as exercise, may delay the onset of and/or ameliorate skeletal muscle ageing and frailty when coupled with other lifestyle improvements such as diet and pharmacological supplementation (e.g., melatonin).

Recently, Zhang et al. identified that Achilles tendinopathy (AT) disrupts the *Achilles* tissue clock, reducing both *Bmal1* and *Nrf2*, which further exacerbated tissue inflammation. Enhancing *Bmal1* expression mitigated tendinopathy and increased AT relief [256]. Moreover, the timing of aerobic exercise modulates the levels of BMAL1 and NRF2 protein expression alongside the total antioxidant capacity and RO/NS levels. Furthermore, research has highlighted that AMPK signalling in adipocytes impacts exercise-induced beneficial effects on skeletal muscle function in a time-of-day manner by affecting skeletal muscle metabolites, thus implicating an important role for circadian inter-tissue cross-talk in exercise responses [257]. Despite this promising research, Palmese and colleagues (2025) report in a systematic review a relative lack of research undertaking circadian timing approaches in combating sarcopenia. They concluded that whilst available studies supported the role of the circadian timing system in sarcopenia, the research was limited [258]. Their suggestion is that researchers take advantage of published studies to optimise and refine experimental and clinical protocols, using clinical and preclinical models to address the wide knowledge gaps in the field.

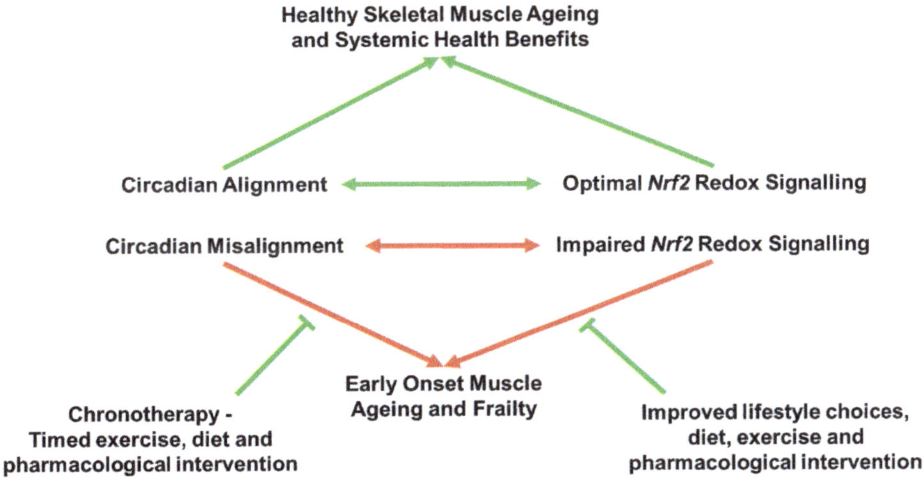

Figure 4. Schematic showing proposed bi-directional interactions between the circadian clock and *Nrf2*-based redox signalling to regulate the rate of skeletal muscle ageing and systemic health. Under appropriate circadian alignment, the *Nrf2* redox signalling can function optimally, which supports long-term maintenance of systemic health and healthier ageing of skeletal muscle. Circadian misalignment or impaired *Nrf2* redox signalling leads to pathophysiological consequences in the skeletal muscle, such as the accelerated development of sarcopenia and systemic loss of tissue resilience to stressors leading to frailty. Using chronotherapy-based approaches, including timed exercise, diet or pharmacological interventions, along with improved lifestyle choices, circadian alignment can be achieved alongside optimal redox homeostasis to promote healthy ageing of the skeletal muscle and systemic health benefits.

Exercise remains the most promising intervention for sarcopenia, yet its effectiveness may vary depending on many external factors, including feeding/fasting cycles and the time of day exercise is performed. A total of ~20% of people do not improve their respiratory capacity when they exercise regularly; they are referred to as "non-respondents". Additionally, ~7 to 15% of people experience adverse reactions when exercising on a regular basis [259], including increased systolic blood pressure, increased blood levels of triglycerides and insulin, and lower HDL cholesterol [260]. Despite this variability in exercise responses, few experimental studies and clinical trials account for circadian timing and personal chronotypes as important variables. Yet, appropriately timed exercise may improve health outcomes and reduce the frequency and/or duration of exercise needed, potentially lowering heterogeneity in exercise responses, especially given that circadian regulation is often phase-shifted in older adults.

Whilst outside the scope of this review, diet is an important factor to consider in future studies of skeletal muscle exercise, ageing and circadian biology. Lipid, glucose and amino acid metabolism are all regulated by the circadian clock and change with age; see reviews [261,262]. The timing and nutritional composition of food intake are important Zeitgebers for the circadian clocks in humans [262]. Meal frequency and distribution within a day are highly related to metabolic functions, and optimal time-restricted feeding has the potential to prevent several metabolic dysfunctions. Thus, eating at an inappropriate time (e.g., during the night) may have a desynchronizing effect on the circadian clocks and, in the long term, may result in adverse health outcomes (e.g., weight gain, obesity and poor metabolic function). The digestive, absorptive and metabolic capacities of nutrients also show the day–night variations in several peripheral tissues such as small intestine and liver. For example, the postprandial response of blood triacylglycerol to a specific diet and glucose tolerance exhibit clear time-of-day effects. Ingestion of carbohydrates increases secretion of insulin which, together with insulin-like growth factor 1 (IGF-1), can entrain cellular clocks by increasing the synthesis of PERIOD proteins alongside mechanistic target of rapamycin (mTOR) activation [263].

Diet and exercise are intrinsically linked. The nutritional composition of food intake influences the exercise responses. High protein diets enhance muscular performance and skeletal muscle mass in resistance-trained males, irrespective of protein intake immediately prior to exercise or 3 h post exercise [264]. Dietary protein is also required to repair damaged cells and tissues, synthesise hormones and support metabolic activities. Protein ingestion pre- or post-exercise has been shown to enhance recovery, immune function as well as growth and maintenance of lean body mass [265]. Beneficial synergistic effects of exercise and time-restricted feeding have been reported for mice fed a high-fat diet [266].

Older adults are at increased risk of malnutrition due to a ~25% decline in food intake with age, leading to energy supplies lower than the recommended values [267]. Insufficient protein intake is a common nutritional deficit seen in the elderly leading to lower skeletal muscle protein synthesis. It is currently suggested that the elderly ingest 1–1.3 g/kg/day of protein to sustain skeletal muscle mass and function [268]. For building muscle mass, overall protein intake should be 1.4–2 g/kg/day, and if this is not possible through food consumption, supplementation is recommended [269]. Thus, optimising nutritional intake could be a vital piece of puzzle in designing improved exercise regimens, especially for the elderly.

To maximise the benefits of exercise-based lifestyle interventions for musculoskeletal health, future studies should consider circadian timing of specific exercise regimens and diet intake/composition alongside sex, age and chronotype, as these factors interact in complex ways to influence health outcomes and are prone to misalignments between the circadian system and environmental cues (e.g., reduced physical activity, consuming energy-dense nutriments, irregular eating, disruptions in sleep patterns, shift work, light pollution). For example, resistance training appears optimal in the afternoon for skeletal muscle growth but may be better tailored to earlier times of day for individuals with an early chronotype. Recognising the complex interplay between physiological, chronobiological and nutritional factors will be essential for developing more effective personalised lifestyle and therapeutic strategies to prevent and/or treat age-related skeletal muscle function decline and its associated systemic health effects.

Funding: We thank MRC UK (MR/P003311/1); MRC-VA UK as part of CIMA (MR/R502182/1, MR/P020941/1); The Wellcome Trust Institutional Strategic Support Fund (097826/Z/11/A); The Royal Society (RGS\R1\211432); MD UK (21GRO-PG12-0532); BBSRC UK (BB/W010801/1, BB/W018314/1, BB/S020772/1); Rosetrees Trust (M709, CF-2021-2/133); and Science Fund of the Republic of Serbia (451-03-47/2023-01/200007).

Conflicts of Interest: The authors declare that they have no known competing financial interests or personal relationships that could have appeared to influence the work reported in this paper.

References

1. Frontera, W.R.; Ochala, J. Skeletal Muscle: A Brief Review of Structure and Function. *Calcif. Tissue Int.* **2015**, *96*, 183–195. [CrossRef]
2. Blaauw, B.; Schiaffino, S.; Reggiani, C. Mechanisms modulating skeletal muscle phenotype. *Compr. Physiol.* **2013**, *3*, 1645–1687. [CrossRef]
3. Mukund, K.; Subramaniam, S. Skeletal muscle: A review of molecular structure and function, in health and disease. *WIREs Syst. Biol. Med.* **2020**, *12*, e1462. [CrossRef]
4. Fernández Costa, J.M.; Fernández-Garibay, X.; Velasco, F.; Ramón-Azcón, J. Bioengineered in vitro skeletal muscles as new tools for muscular dystrophies preclinical studies. *J. Tissue Eng.* **2021**, *12*, 2041731420981339. [CrossRef] [PubMed]
5. Birch, H.L. Extracellular matrix and ageing. *Subcell. Biochem.* **2018**, *90*, 169–190. [CrossRef]
6. Csapo, R.; Gumpenberger, M.; Wessner, B. Skeletal Muscle Extracellular Matrix—What Do We Know About Its Composition, Regulation, and Physiological Roles? A Narrative Review. *Front. Physiol.* **2020**, *11*, 253. [CrossRef] [PubMed]
7. Dave, H.D.; Shook, M.; Varacallo, M. Anatomy, skeletal muscle. In *StatPearls [Internet]*; StatPearls Publishing: Treasure Island, FL, USA, 2021.
8. Kurtz, A.; Oh, S.-J. Age related changes of the extracellular matrix and stem cell maintenance. *Prev. Med.* **2012**, *54*, S50–S56. [CrossRef] [PubMed]
9. Gillies, A.R.; Lieber, R.L. Structure and function of the skeletal muscle extracellular matrix. *Muscle Nerve* **2011**, *44*, 318–331. [CrossRef]
10. Cai, L.; Shi, L.; Peng, Z.; Sun, Y.; Chen, J. Ageing of skeletal muscle extracellular matrix and mitochondria: Finding a potential link. *Ann. Med.* **2023**, *55*, 2240707. [CrossRef]
11. Gutierrez-Monreal, M.A.; Harmsen, J.; Schrauwen, P.; Esser, K.A. Ticking for Metabolic Health: The Skeletal-Muscle Clocks. *Obesity* **2020**, *28*, S46–S54. [CrossRef]
12. Wolfe, R.R. The underappreciated role of muscle in health and disease. *Am. J. Clin. Nutr.* **2006**, *84*, 475–482. [CrossRef]
13. Goldstein, M.S. Humoral nature of the hypoglycemic factor of muscular work. *Diabetes* **1961**, *10*, 232–234. [CrossRef]
14. Pedersen, B.K.; Steensberg, A.; Fischer, C.; Keller, C.; Ostrowski, K.; Schjerling, P. Exercise and cytokines with particular focus on muscle derived IL-6. *Exerc. Immunol. Rev.* **2001**, *7*, 18–31.
15. Pedersen, B.K.; Steensberg, A.; Keller, P.; Keller, C.; Fischer, C.; Hiscock, N.; Van Hall, G.; Plomgaard, P.; Febbraio, M.A. Muscle-derived interleukin-6: Lipolytic, anti-inflammatory and immune regulatory effects. *Pflügers Arch.* **2003**, *446*, 9–16. [CrossRef] [PubMed]
16. Hojman, P.; Dethlefsen, C.; Brandt, C.; Hansen, J.; Pedersen, L.; Pedersen, B.K. Exercise-induced muscle-derived cytokines inhibit mammary cancer cell growth. *Am. J. Physiol. Metab.* **2011**, *301*, E504–E510. [CrossRef]
17. Shero, J.A.; Lindholm, M.E.; Sandri, M.; Stanford, K.I. Skeletal muscle as a mediator of interorgan crosstalk during exercise: Implications for aging and obesity. *Circ. Res.* **2025**, *136*, 1407–1432. [CrossRef] [PubMed]
18. Hartwig, S.; Raschke, S.; Knebel, B.; Scheler, M.; Irmler, M.; Passlack, W.; Muller, S.; Hanisch, F.-G.; Franz, T.; Li, X. Secretome profiling of primary human skeletal muscle cells. *Biochim. Biophys. Acta (BBA)-Proteins Proteom.* **2014**, *1844*, 1011–1017. [CrossRef]
19. Pedersen, B.K.; Febbraio, M.A. Muscle as an endocrine organ: Focus on muscle-derived interleukin-6. *Physiol. Rev.* **2008**, *88*, 1379–1406. [CrossRef] [PubMed]

20. Schnyder, S.; Handschin, C. Skeletal muscle as an endocrine organ: PGC-1α, myokines and exercise. *Bone* **2015**, *80*, 115–125. [CrossRef]
21. Leal, L.G.; Lopes, M.A.; Batista, M.L., Jr. Physical exercise-induced myokines and muscle-adipose tissue crosstalk: A review of current knowledge and the implications for health and metabolic diseases. *Front. Physiol.* **2018**, *9*, 1307. [CrossRef]
22. Vechetti, I.J., Jr.; Valentino, T.; Mobley, C.B.; McCarthy, J.J. The role of extracellular vesicles in skeletal muscle and systematic adaptation to exercise. *J. Physiol.* **2021**, *599*, 845–861. [CrossRef]
23. Guescini, M.; Guidolin, D.; Vallorani, L.; Casadei, L.; Gioacchini, A.M.; Tibollo, P.; Battistelli, M.; Falcieri, E.; Battistin, L.; Agnati, L.F. C2C12 myoblasts release micro-vesicles containing mtDNA and proteins involved in signal transduction. *Exp. Cell Res.* **2010**, *316*, 1977–1984. [CrossRef] [PubMed]
24. Aoi, W.; Tanimura, Y. Roles of skeletal muscle-derived exosomes in organ metabolic and immunological communication. *Front. Endocrinol.* **2021**, *12*, 697204. [CrossRef] [PubMed]
25. Guescini, M.; Canonico, B.; Lucertini, F.; Maggio, S.; Annibalini, G.; Barbieri, E.; Luchetti, F.; Papa, S.; Stocchi, V. Muscle releases alpha-sarcoglycan positive extracellular vesicles carrying miRNAs in the bloodstream. *PLoS ONE* **2015**, *10*, e0125094.
26. Kirby, T.J.; McCarthy, J.J. MicroRNAs in skeletal muscle biology and exercise adaptation. *Free Radic. Biol. Med.* **2013**, *64*, 95–105. [CrossRef]
27. Lananna, B.V.; Musiek, E.S. The wrinkling of time: Aging, inflammation, oxidative stress, and the circadian clock in neurodegeneration. *Neurobiol. Dis.* **2020**, *139*, 104832. [CrossRef]
28. Soendenbroe, C.; Heisterberg, M.F.; Schjerling, P.; Karlsen, A.; Kjaer, M.; Andersen, J.L.; Mackey, A.L. Molecular indicators of denervation in aging human skeletal muscle. *Muscle Nerve* **2019**, *60*, 453–463. [CrossRef]
29. Kirkeby, S.; Garbarsch, C. Aging affects different human muscles in various ways. An image analysis of the histomorphometric characteristics of fiber types in human masseter and vastus lateralis muscles from young adults and the very old. *Histol. Histopathol.* **2000**, *15*, 61–72.
30. Barnouin, Y.; McPhee, J.S.; Butler-Browne, G.; Bosutti, A.; De Vito, G.; Jones, D.A.; Narici, M.; Behin, A.; Hogrel, J.; Degens, H. Coupling between skeletal muscle fiber size and capillarization is maintained during healthy aging. *J. Cachexia. Sarcopenia Muscle* **2017**, *8*, 647–659. [CrossRef]
31. Power, G.A.; Dalton, B.H.; Rice, C.L. Human neuromuscular structure and function in old age: A brief review. *J. Sport Health Sci.* **2013**, *2*, 215–226. [CrossRef]
32. D'Antona, G.; Pellegrino, M.A.; Adami, R.; Rossi, R.; Carlizzi, C.N.; Canepari, M.; Saltin, B.; Bottinelli, R. The effect of ageing and immobilization on structure and function of human skeletal muscle fibres. *J. Physiol.* **2003**, *552*, 499–511. [CrossRef]
33. Wilkinson, D.J.; Piasecki, M.; Atherton, P. The age-related loss of skeletal muscle mass and function: Measurement and physiology of muscle fibre atrophy and muscle fibre loss in humans. *Ageing Res. Rev.* **2018**, *47*, 123–132. [CrossRef]
34. Ewald, C.Y. The matrisome during aging and longevity: A systems-level approach toward defining matreotypes promoting healthy aging. *Gerontology* **2020**, *66*, 266–274. [CrossRef] [PubMed]
35. Singh, V.P.; Bali, A.; Singh, N.; Jaggi, A.S. Advanced glycation end products and diabetic complications. *Korean J. Physiol. Pharmacol. Off. J. Korean Physiol. Soc. Korean Soc. Pharmacol.* **2014**, *18*, 1. [CrossRef]
36. Ren, L.; Fu, Y.; Deng, Y.; Qi, L.; Jin, L. Advanced glycation end products inhibit the expression of collagens type I and III by human gingival fibroblasts. *J. Periodontol.* **2009**, *80*, 1166–1173. [CrossRef]
37. Stearns-Reider, K.M.; D'Amore, A.; Beezhold, K.; Rothrauff, B.; Cavalli, L.; Wagner, W.R.; Vorp, D.A.; Tsamis, A.; Shinde, S.; Zhang, C.; et al. Aging of the skeletal muscle extracellular matrix drives a stem cell fibrogenic conversion. *Aging Cell* **2017**, *16*, 518–528. [CrossRef] [PubMed]
38. Marzetti, E.; Calvani, R.; Cesari, M.; Buford, T.W.; Lorenzi, M.; Behnke, B.J.; Leeuwenburgh, C. Mitochondrial dysfunction and sarcopenia of aging: From signaling pathways to clinical trials. *Int. J. Biochem. Cell Biol.* **2013**, *45*, 2288–2301. [CrossRef]
39. Chistiakov, D.A.; Sobenin, I.A.; Revin, V.V.; Orekhov, A.N.; Bobryshev, Y.V. Mitochondrial aging and age-related dysfunction of mitochondria. *Biomed Res. Int.* **2014**, *2014*, 238463. [CrossRef] [PubMed]
40. Calvani, R.; Joseph, A.-M.; Adhihetty, P.J.; Miccheli, A.; Bossola, M.; Leeuwenburgh, C.; Bernabei, R.; Marzetti, E. Mitochondrial pathways in sarcopenia of aging and disuse muscle atrophy. *Biol. Chem.* **2013**, *394*, 393–414. [CrossRef]
41. Ma, Q. Role of Nrf2 in Oxidative Stress and Toxicity. *Annu. Rev. Pharmacol. Toxicol.* **2013**, *53*, 401–426. [CrossRef]
42. Uruno, A.; Yagishita, Y.; Katsuoka, F.; Kitajima, Y.; Nunomiya, A.; Nagatomi, R.; Pi, J.; Biswal, S.S.; Yamamoto, M.; Uruno, C.A. Nrf2-Mediated Regulation of Skeletal Muscle Glycogen Metabolism Downloaded from. *Mol. Cell. Biol.* **2016**, *36*, 1655–1672. [CrossRef] [PubMed]
43. Cuadrado, A.; Rojo, A.I.; Wells, G.; Hayes, J.D.; Cousin, S.P.; Rumsey, W.L.; Attucks, O.C.; Franklin, S.; Levonen, A.L.; Kensler, T.W.; et al. Therapeutic targeting of the NRF2 and KEAP1 partnership in chronic diseases. *Nat. Rev. Drug Discov.* **2019**, *18*, 295–317. [CrossRef] [PubMed]

44. Noh, Y.H.; Kim, K.Y.; Shim, M.S.; Choi, S.H.; Choi, S.; Ellisman, M.H.; Weinreb, R.N.; Perkins, G.A.; Ju, W. Inhibition of oxidative stress by coenzyme Q10 increases mitochondrial mass and improves bioenergetic function in optic nerve head astrocytes. *Cell Death Dis.* **2013**, *4*, e820. [CrossRef]
45. Chow, C.K. Vitamin E regulation of mitochondrial superoxide generation. *Neurosignals* **2001**, *10*, 112–124. [CrossRef]
46. Vasilaki, A.; Jackson, M.J. Role of reactive oxygen species in the defective regeneration seen in aging muscle. *Free Radic. Biol. Med.* **2013**, *65*, 317–323. [CrossRef]
47. Bellezza, I.; Giambanco, I.; Minelli, A.; Donato, R. Nrf2-Keap1 signaling in oxidative and reductive stress. *Biochim. Biophys. Acta (BBA)-Mol. Cell Res.* **2018**, *1865*, 721–733. [CrossRef]
48. Kitaoka, Y.; Tamura, Y.; Takahashi, K.; Takeda, K.; Takemasa, T.; Hatta, H. Effects of Nrf2 deficiency on mitochondrial oxidative stress in aged skeletal muscle. *Physiol. Rep.* **2019**, *7*, e13998. [CrossRef]
49. Huang, D.-D.; Fan, S.-D.; Chen, X.-Y.; Yan, X.-L.; Zhang, X.-Z.; Ma, B.-W.; Yu, D.-Y.; Xiao, W.-Y.; Zhuang, C.-L.; Yu, Z. Nrf2 deficiency exacerbates frailty and sarcopenia by impairing skeletal muscle mitochondrial biogenesis and dynamics in an age-dependent manner. *Exp. Gerontol.* **2019**, *119*, 61–73. [CrossRef]
50. Miller, C.J.; Gounder, S.S.; Kannan, S.; Goutam, K.; Muthusamy, V.R.; Firpo, M.A.; Symons, J.D.; Paine, R.; Hoidal, J.R.; Rajasekaran, N.S. Disruption of Nrf2/ARE signaling impairs antioxidant mechanisms and promotes cell degradation pathways in aged skeletal muscle. *Biochim. Biophys. Acta—Mol. Basis Dis.* **2012**, *1822*, 1038–1050. [CrossRef] [PubMed]
51. Narasimhan, M.; Hong, J.; Atieno, N.; Muthusamy, V.R.; Davidson, C.J.; Abu-Rmaileh, N.; Richardson, R.S.; Gomes, A.V.; Hoidal, J.R.; Rajasekaran, N.S. Nrf2 deficiency promotes apoptosis and impairs PAX7/MyoD expression in aging skeletal muscle cells. *Free Radic. Biol. Med.* **2014**, *71*, 402–414. [CrossRef]
52. Ahn, B.; Pharaoh, G.; Premkumar, P.; Huseman, K.; Ranjit, R.; Kinter, M.; Szweda, L.; Kiss, T.; Fulop, G.; Tarantini, S.; et al. Nrf2 deficiency exacerbates age-related contractile dysfunction and loss of skeletal muscle mass. *Redox Biol.* **2018**, *17*, 47–58. [CrossRef]
53. Safdar, A.; deBeer, J.; Tarnopolsky, M.A. Dysfunctional Nrf2-Keap1 redox signaling in skeletal muscle of the sedentary old. *Free Radic. Biol. Med.* **2010**, *49*, 1487–1493. [CrossRef]
54. Schroder, E.A.; Esser, K.A. Circadian rhythms, skeletal muscle molecular clocks, and exercise. *Exerc. Sport Sci. Rev.* **2013**, *41*, 224–229. [CrossRef]
55. Serin, Y.; Tek, N.A. Effect of circadian rhythm on metabolic processes and the regulation of energy balance. *Ann. Nutr. Metab.* **2019**, *74*, 322–330. [CrossRef]
56. Aoyama, S.; Shibata, S. The role of circadian rhythms in muscular and osseous physiology and their regulation by nutrition and exercise. *Front. Neurosci.* **2017**, *11*, 63. [CrossRef]
57. Dibner, C.; Schibler, U.; Albrecht, U. The mammalian circadian timing system: Organization and coordination of central and peripheral clocks. *Annu. Rev. Physiol.* **2010**, *72*, 517–549. [CrossRef]
58. Wolff, C.A.; Esser, K.A. Exercise timing and circadian rhythms. *Curr. Opin. Physiol.* **2019**, *10*, 64–69. [CrossRef] [PubMed]
59. Mistlberger, R.E. Food as circadian time cue for appetitive behavior. *F1000Research* **2020**, *9*, F1000-Faculty. [CrossRef] [PubMed]
60. Rey, G.; Reddy, A.B. Interplay between cellular redox oscillations and circadian clocks. *Diabetes Obes. Metab.* **2015**, *17*, 55–64. [CrossRef] [PubMed]
61. Murphy, B.A.; Wagner, A.L.; McGlynn, O.F.; Kharazyan, F.; Browne, J.A.; Elliott, J.A. Exercise influences circadian gene expression in equine skeletal muscle. *Vet. J.* **2014**, *201*, 39–45. [CrossRef]
62. Hannibal, J. Hirota, T., Hatori, M., Panda, S., Eds.; *Neuroanatomy of the Suprachiasmatic Nucleus (SCN): Multicolor Immunohistochemistry and 3D Reconstruction BT—Circadian Clocks*; Springer: New York, NY, USA, 2022; pp. 85–98. ISBN 978-1-0716-2577-4.
63. Ma, M.; Morrison, E. Neuroanatomy, Nucleus Suprachiasmatic. In *Neuroanatomy, Nucleus Suprachiasmatic*; StatPearls Publishing LLC: Treasure Island Florida USA, 2025. [PubMed]
64. Hastings, M.H.; Maywood, E.S.; Brancaccio, M. Generation of circadian rhythms in the suprachiasmatic nucleus. *Nat. Rev. Neurosci.* **2018**, *19*, 453–469. [CrossRef]
65. Jewett, B.; Sharma, S. Physiology, GABA. In *Physiology, GABA*; StatPearls Publishing Ltd: Treasure Island Florida USA, 2023. [PubMed]
66. Andrews, J.L.; Zhang, X.; Mccarthy, J.J.; Mcdearmon, E.L.; Hornberger, T.A.; Russell, B.; Campbell, K.S.; Arbogast, S.; Reid, M.B.; Walker, J.R.; et al. CLOCK and BMAL1 regulate MyoD and are necessary for maintenance of skeletal muscle phenotype and function. *Proc. Natl. Acad. Sci. USA* **2010**, *107*, 19090–19095. [CrossRef] [PubMed]
67. Mansingh, S.; Handschin, C. Time to Train: The Involvement of the Molecular Clock in Exercise Adaptation of Skeletal Muscle. *Front. Physiol.* **2022**, *13*, 902031. [CrossRef]
68. Nakao, R.; Nikawa, T.; Oishi, K. The skeletal muscle circadian clock: Current insights. *ChronoPhysiology Ther.* **2017**, *7*, 47–57. [CrossRef]

69. Narasimamurthy, R.; Virshup, D.M. The phosphorylation switch that regulates ticking of the circadian clock. *Mol. Cell* **2021**, *81*, 1133–1146. [CrossRef]
70. Zhang, R.; Lahens, N.F.; Ballance, H.I.; Hughes, M.E.; Hogenesch, J.B. A circadian gene expression atlas in mammals: Implications for biology and medicine. *Proc. Natl. Acad. Sci. USA* **2014**, *111*, 16219–16224. [CrossRef]
71. Panda, S.; Sato, T.K.; Castrucci, A.M.; Rollag, M.D.; DeGrip, W.J.; Hogenesch, J.B.; Provencio, I.; Kay, S.A. Melanopsin (Opn4) requirement for normal light-induced circadian phase shifting. *Science* **2002**, *298*, 2213–2216. [CrossRef]
72. Riley, L.A.; Zhang, X.; Douglas, C.M.; Mijares, J.M.; Hammers, D.W.; Wolff, C.A.; Wood, N.B.; Olafson, H.R.; Du, P.; Labeit, S. The skeletal muscle circadian clock regulates titin splicing through RBM20. *Elife* **2022**, *11*, e76478. [CrossRef]
73. Kelu, J.J.; Hughes, S.M. Muscle peripheral circadian clock drives nocturnal protein degradation via raised Ror/Rev-erb balance and prevents premature sarcopenia. *Proc. Natl. Acad. Sci. USA* **2025**, *122*, e2422446122. [CrossRef]
74. Kahn, R.E.; Dayanidhi, S.; Lacham-Kaplan, O.; Hawley, J.A. Molecular clocks, satellite cells, and skeletal muscle regeneration. *Am. J. Physiol. Physiol.* **2023**, *324*, C1332–C1340. [CrossRef] [PubMed]
75. Schroder, E.A.; Harfmann, B.D.; Zhang, X.; Srikuea, R.; England, J.H.; Hodge, B.A.; Wen, Y.; Riley, L.A.; Yu, Q.; Christie, A.; et al. Intrinsic muscle clock is necessary for musculoskeletal health. *J. Physiol.* **2015**, *593*, 5387–5404. [CrossRef] [PubMed]
76. Hodge, B.A.; Wen, Y.; Riley, L.A.; Zhang, X.; England, J.H.; Harfmann, B.D.; Schroder, E.A.; Esser, K.A. The endogenous molecular clock orchestrates the temporal separation of substrate metabolism in skeletal muscle. *Skelet. Muscle* **2015**, *5*, 17. [CrossRef] [PubMed]
77. Ehlen, J.C.; Brager, A.J.; Baggs, J.; Pinckney, L.; Gray, C.L.; DeBruyne, J.P.; Esser, K.A.; Takahashi, J.S.; Paul, K.N. Bmal1 function in skeletal muscle regulates sleep. *Elife* **2017**, *6*, e26557. [CrossRef]
78. Zylka, M.J.; Shearman, L.P.; Weaver, D.R.; Reppert, S.M. Three period homologs in mammals: Differential light responses in the suprachiasmatic circadian clock and oscillating transcripts outside of brain. *Neuron* **1998**, *20*, 1103–1110. [CrossRef]
79. Miller, B.H.; McDearmon, E.L.; Panda, S.; Hayes, K.R.; Zhang, J.; Andrews, J.L.; Antoch, M.P.; Walker, J.R.; Esser, K.A.; Hogenesch, J.B. Circadian and CLOCK-controlled regulation of the mouse transcriptome and cell proliferation. *Proc. Natl. Acad. Sci. USA* **2007**, *104*, 3342–3347. [CrossRef]
80. Pizarro, A.; Hayer, K.; Lahens, N.F.; Hogenesch, J.B. CircaDB: A database of mammalian circadian gene expression profiles. *Nucleic Acids Res.* **2013**, *41*, D1009-13. [CrossRef] [PubMed]
81. McCarthy, J.J.; Andrews, J.L.; McDearmon, E.L.; Campbell, K.S.; Barber, B.K.; Miller, B.H.; Walker, J.R.; Hogenesch, J.B.; Takahashi, J.S.; Esser, K.A. Identification of the circadian transcriptome in adult mouse skeletal muscle. *Physiol. Genom.* **2007**, *31*, 86–95. [CrossRef]
82. Harfmann, B.D.; Schroder, E.A.; Esser, K.A. Circadian rhythms, the molecular clock, and skeletal muscle. *J. Biol. Rhythms* **2015**, *30*, 84–94. [CrossRef]
83. Dyar, K.A.; Ciciliot, S.; Tagliazucchi, G.M.; Pallafacchina, G.; Tothova, J.; Argentini, C.; Agatea, L.; Abraham, R.; Ahdesmäki, M.; Forcato, M. The calcineurin-NFAT pathway controls activity-dependent circadian gene expression in slow skeletal muscle. *Mol. Metab.* **2015**, *4*, 823–833. [CrossRef]
84. Perrin, L.; Loizides-Mangold, U.; Chanon, S.; Gobet, C.; Hulo, N.; Isenegger, L.; Weger, B.D.; Migliavacca, E.; Charpagne, A.; Betts, J.A. Transcriptomic analyses reveal rhythmic and CLOCK-driven pathways in human skeletal muscle. *Elife* **2018**, *7*, e34114. [CrossRef]
85. Potter, G.D.M.; Skene, D.J.; Arendt, J.; Cade, J.E.; Grant, P.J.; Hardie, L.J. Circadian rhythm and sleep disruption: Causes, metabolic consequences, and countermeasures. *Endocr. Rev.* **2016**, *37*, 584–608. [CrossRef]
86. Choi, Y.I.; Park, D.K.; Chung, J.-W.; Kim, K.O.; Kwon, K.A.; Kim, Y.J. Circadian rhythm disruption is associated with an increased risk of sarcopenia: A nationwide population-based study in Korea. *Sci. Rep.* **2019**, *9*, 12015. [CrossRef]
87. Matsumoto, T.; Tanizawa, K.; Tachikawa, R.; Murase, K.; Minami, T.; Inouchi, M.; Handa, T.; Oga, T.; Hirai, T.; Chin, K. Associations of obstructive sleep apnea with truncal skeletal muscle mass and density. *Sci. Rep.* **2018**, *8*, 6550. [CrossRef]
88. Nedeltcheva, A.V.; Kilkus, J.M.; Imperial, J.; Schoeller, D.A.; Penev, P.D. Insufficient sleep undermines dietary efforts to reduce adiposity. *Ann. Intern. Med.* **2010**, *153*, 435–441. [CrossRef]
89. Morrison, M.; Halson, S.L.; Weakley, J.; Hawley, J.A. Sleep, circadian biology and skeletal muscle interactions: Implications for metabolic health. *Sleep Med. Rev.* **2022**, *66*, 101700. [CrossRef]
90. van Moorsel, D.; Hansen, J.; Havekes, B.; Scheer, F.A.J.L.; Jörgensen, J.A.; Hoeks, J.; Schrauwen-Hinderling, V.B.; Duez, H.; Lefebvre, P.; Schaper, N.C. Demonstration of a day-night rhythm in human skeletal muscle oxidative capacity. *Mol. Metab.* **2016**, *5*, 635–645. [CrossRef] [PubMed]
91. Wefers, J.; Connell, N.J.; Fealy, C.E.; Andriessen, C.; de Wit, V.; van Moorsel, D.; Moonen-Kornips, E.; Jörgensen, J.A.; Hesselink, M.K.C.; Havekes, B. Day-night rhythm of skeletal muscle metabolism is disturbed in older, metabolically compromised individuals. *Mol. Metab.* **2020**, *41*, 101050. [CrossRef] [PubMed]

92. Kondratov, R.V.; Kondratova, A.A.; Gorbacheva, V.Y.; Vykhovanets, O.V.; Antoch, M.P. Early aging and age-related pathologies in mice deficient in BMAL1, the core component of the circadian clock. *Genes Dev.* **2006**, *20*, 1868–1873. [CrossRef] [PubMed]
93. Woldt, E.; Sebti, Y.; Solt, L.A.; Duhem, C.; Lancel, S.; Eeckhoute, J.; Hesselink, M.K.C.; Paquet, C.; Delhaye, S.; Shin, Y. Rev-erb-α modulates skeletal muscle oxidative capacity by regulating mitochondrial biogenesis and autophagy. *Nat. Med.* **2013**, *19*, 1039–1046. [CrossRef]
94. Gao, H.; Xiong, X.; Lin, Y.; Chatterjee, S.; Ma, K. The clock regulator Bmal1 protects against muscular dystrophy. *Exp. Cell Res.* **2020**, *397*, 112348. [CrossRef]
95. Chatterjee, S.; Yin, H.; Li, W.; Lee, J.; Yechoor, V.K.; Ma, K. The nuclear receptor and clock repressor Rev-erbα suppresses myogenesis. *Sci. Rep.* **2019**, *9*, 4585. [CrossRef]
96. Welch, R.D.; Billon, C.; Valfort, A.-C.; Burris, T.P.; Flaveny, C.A. Pharmacological inhibition of REV-ERB stimulates differentiation, inhibits turnover and reduces fibrosis in dystrophic muscle. *Sci. Rep.* **2017**, *7*, 17142. [CrossRef] [PubMed]
97. Xiong, X.; Gao, H.; Lin, Y.; Yechoor, V.; Ma, K. Inhibition of Rev-erbα ameliorates muscular dystrophy. *Exp. Cell Res.* **2021**, *406*, 112766. [CrossRef] [PubMed]
98. Hao, Y.; Xue, T.; Liu, S.; Geng, S.; Shi, X.; Qian, P.; He, W.; Zheng, J.; Li, Y.; Lou, J. Loss of CRY2 promotes regenerative myogenesis by enhancing PAX7 expression and satellite cell proliferation. *MedComm* **2023**, *4*, e202. [CrossRef]
99. Rossi, R.; Falzarano, M.S.; Osman, H.; Armaroli, A.; Scotton, C.; Mantuano, P.; Boccanegra, B.; Cappellari, O.; Schwartz, E.; Yuryev, A. Circadian genes as exploratory biomarkers in DMD: Results from both the Mdx mouse model and patients. *Front. Physiol.* **2021**, *12*, 678974. [CrossRef]
100. Betts, C.A.; Jagannath, A.; van Westering, T.L.E.; Bowerman, M.; Banerjee, S.; Meng, J.; Falzarano, M.S.; Cravo, L.; McClorey, G.; Meijboom, K.E. Dystrophin involvement in peripheral circadian SRF signalling. *Life Sci. Alliance* **2021**, *4*, e202101014. [CrossRef]
101. Scotton, C.; Bovolenta, M.; Schwartz, E.; Falzarano, M.S.; Martoni, E.; Passarelli, C.; Armaroli, A.; Osman, H.; Rodolico, C.; Messina, S. Deep RNA profiling identified CLOCK and molecular clock genes as pathophysiological signatures in collagen VI myopathy. *J. Cell Sci.* **2016**, *129*, 1671–1684. [CrossRef]
102. Kiperman, T.; Ma, K. Circadian clock in muscle disease etiology and therapeutic potential for duchenne muscular dystrophy. *Int. J. Mol. Sci.* **2024**, *25*, 4767. [CrossRef] [PubMed]
103. Savarese, M.; Sarparanta, J.; Vihola, A.; Udd, B.; Hackman, P. Increasing role of titin mutations in neuromuscular disorders. *J. Neuromuscul. Dis.* **2016**, *3*, 293–308. [CrossRef]
104. Dyar, K.A.; Ciciliot, S.; Wright, L.E.; Biensø, R.S.; Tagliazucchi, G.M.; Patel, V.R.; Forcato, M.; Paz, M.I.P.; Gudiksen, A.; Solagna, F. Muscle insulin sensitivity and glucose metabolism are controlled by the intrinsic muscle clock. *Mol. Metab.* **2014**, *3*, 29–41. [CrossRef]
105. Perrin, L.; Loizides-Mangold, U.; Skarupelova, S.; Pulimeno, P.; Chanon, S.; Robert, M.; Bouzakri, K.; Modoux, C.; Roux-Lombard, P.; Vidal, H.; et al. Human skeletal myotubes display a cell-autonomous circadian clock implicated in basal myokine secretion. *Mol. Metab.* **2015**, *4*, 834–845. [CrossRef] [PubMed]
106. Zhao, X.; Karpac, J. Muscle directs diurnal energy homeostasis through a myokine-dependent hormone module in Drosophila. *Curr. Biol.* **2017**, *27*, 1941–1955. [CrossRef]
107. de Souza Teixeira, A.A.; Biondo, L.; Silveira, L.S.; Lima, E.A.; Diniz, T.A.; Lira, F.S.; Seelaender, M.; Rosa Neto, J.C. Exercise training induces alteration of clock genes and myokines expression in tumor-bearing mice. *Cell Biochem. Funct.* **2023**, *41*, 1383–1394. [CrossRef]
108. Viggars, M.R.; Berko, H.E.; Hesketh, S.J.; Wolff, C.A.; Gutierrez-Monreal, M.A.; Martin, R.A.; Jennings, I.G.; Huo, Z.; Esser, K.A. Skeletal muscle BMAL1 is necessary for transcriptional adaptation of local and peripheral tissues in response to endurance exercise training. *Mol. Metab.* **2024**, *86*, 101980. [CrossRef]
109. Gutierrez-Monreal, M.A.; Wolff, C.A.; Rijos, E.E.; Viggars, M.R.; Douglas, C.M.; Pagala, V.; Peng, J.; Hunt, L.C.; Ding, H.; Huo, Z. Targeted Bmal1 restoration in muscle prolongs lifespan with systemic health effects in aging model. *JCI Insight* **2024**, *9*, e174007. [CrossRef] [PubMed]
110. Sato, S.; Dyar, K.A.; Treebak, J.T.; Jepsen, S.L.; Ehrlich, A.M.; Ashcroft, S.P.; Trost, K.; Kunzke, T.; Prade, V.M.; Small, L. Atlas of exercise metabolism reveals time-dependent signatures of metabolic homeostasis. *Cell Metab.* **2022**, *34*, 329–345. [CrossRef] [PubMed]
111. Casanova-Vallve, N.; Duglan, D.; Vaughan, M.E.; Pariollaud, M.; Handzlik, M.K.; Fan, W.; Yu, R.T.; Liddle, C.; Downes, M.; Delezie, J. Daily running enhances molecular and physiological circadian rhythms in skeletal muscle. *Mol. Metab.* **2022**, *61*, 101504. [CrossRef]
112. Wolff, C.A.; Gutierrez-Monreal, M.A.; Meng, L.; Zhang, X.; Douma, L.G.; Costello, H.M.; Douglas, C.M.; Ebrahimi, E.; Pham, A.; Oliveira, A.C. Defining the age-dependent and tissue-specific circadian transcriptome in male mice. *Cell Rep.* **2023**, *42*, 111982. [CrossRef]

113. Sies, H.; Belousov, V.V.; Chandel, N.S.; Davies, M.J.; Jones, D.P.; Mann, G.E.; Murphy, M.P.; Yamamoto, M.; Winterbourn, C. Defining roles of specific reactive oxygen species (ROS) in cell biology and physiology. *Nat. Rev. Mol. Cell Biol.* **2022**, *23*, 499–515. [CrossRef]
114. Wood, Z.A.; Poole, L.B.; Karplus, P.A. Peroxiredoxin evolution and the regulation of hydrogen peroxide signaling. *Science* **2003**, *300*, 650–653. [CrossRef]
115. Jackson, M.J.; Pollock, N.; Staunton, C.A.; Stretton, C.; Vasilaki, A.; McArdle, A. Oxidative stress in skeletal muscle: Unraveling the potential beneficial and deleterious roles of reactive oxygen species. *Oxidative Stress* **2020**, *2020*, 713–733.
116. Pizzino, G.; Irrera, N.; Cucinotta, M.; Pallio, G.; Mannino, F.; Arcoraci, V.; Squadrito, F.; Altavilla, D.; Bitto, A. Oxidative stress: Harms and benefits for human health. *Oxid. Med. Cell. Longev.* **2017**, *2017*, 8416763. [CrossRef]
117. Milev, N.B.; Reddy, A.B. Circadian redox oscillations and metabolism. *Trends Endocrinol. Metab.* **2015**, *26*, 430–437. [CrossRef]
118. O'Neill, J.S.; Van Ooijen, G.; Dixon, L.E.; Troein, C.; Corellou, F.; Bouget, F.Y.; Reddy, A.B.; Millar, A.J. Circadian rhythms persist without transcription in a eukaryote. *Nature* **2011**, *469*, 554–558. [CrossRef]
119. Edgar, R.S.; Green, E.W.; Zhao, Y.; Van Ooijen, G.; Olmedo, M.; Qin, X.; Xu, Y.; Pan, M.; Valekunja, U.K.; Feeney, K.A.; et al. Peroxiredoxins are conserved markers of circadian rhythms. *Nature* **2012**, *485*, 459–464. [CrossRef]
120. Wang, T.A.; Yu, Y.V.; Govindaiah, G.; Ye, X.; Artinian, L.; Coleman, T.P.; Sweedler, J.V.; Cox, C.L.; Gillette, M.U. Circadian rhythm of redox state regulates excitability in suprachiasmatic nucleus neurons. *Science* **2012**, *337*, 839–842. [CrossRef] [PubMed]
121. Geyfman, M.; Kumar, V.; Liu, Q.; Ruiz, R.; Gordon, W.; Espitia, F.; Cam, E.; Millar, S.E.; Smyth, P.; Ihler, A.; et al. Brain and muscle Arnt-like protein-1 (BMAL1) controls circadian cell proliferation and susceptibility to UVB-induced DNA damage in the epidermis. *Proc. Natl. Acad. Sci. USA* **2012**, *109*, 11758–11763. [CrossRef]
122. Lai, A.G.; Doherty, C.J.; Mueller-Roeber, B.; Kay, S.A.; Schippers, J.H.M.; Dijkwel, P.P. Circadian Clock-Associated 1 regulates ROS homeostasis and oxidative stress responses. *Proc. Natl. Acad. Sci. USA* **2012**, *109*, 17129–17134. [CrossRef]
123. Krishnan, N.; Davis, A.J.; Giebultowicz, J.M. Circadian regulation of response to oxidative stress in Drosophila melanogaster. *Biochem. Biophys. Res. Commun.* **2008**, *374*, 299–303. [CrossRef] [PubMed]
124. O'Neill, J.S.; Reddy, A.B. Circadian clocks in human red blood cells. *Nature* **2011**, *469*, 498–503. [CrossRef] [PubMed]
125. Itoh, K.; Wakabayashi, N.; Katoh, Y.; Ishii, T.; Igarashi, K.; Engel, J.D.; Yamamoto, M. Keap1 represses nuclear activation of antioxidant responsive elements by Nrf2 through binding to the amino-terminal Neh2 domain. *Genes Dev.* **1999**, *13*, 76–86. [CrossRef]
126. McMahon, M.; Itoh, K.; Yamamoto, M.; Hayes, J.D. Keap1-dependent proteasomal degradation of transcription factor Nrf2 contributes to the negative regulation of antioxidant response element-driven gene expression. *J. Biol. Chem.* **2003**, *278*, 21592–21600. [CrossRef]
127. Chen, W.; Sun, Z.; Wang, X.-J.; Jiang, T.; Huang, Z.; Fang, D.; Zhang, D.D. Direct interaction between Nrf2 and p21Cip1/WAF1 upregulates the Nrf2-mediated antioxidant response. *Mol. Cell* **2009**, *34*, 663–673. [CrossRef]
128. Komatsu, M.; Kurokawa, H.; Waguri, S.; Taguchi, K.; Kobayashi, A.; Ichimura, Y.; Sou, Y.-S.; Ueno, I.; Sakamoto, A.; Tong, K.I. The selective autophagy substrate p62 activates the stress responsive transcription factor Nrf2 through inactivation of Keap1. *Nat. Cell Biol.* **2010**, *12*, 213–223. [CrossRef] [PubMed]
129. Chowdhry, S.; Zhang, Y.; McMahon, M.; Sutherland, C.; Cuadrado, A.; Hayes, J.D. Nrf2 is controlled by two distinct β-TrCP recognition motifs in its Neh6 domain, one of which can be modulated by GSK-3 activity. *Oncogene* **2013**, *32*, 3765–3781. [CrossRef] [PubMed]
130. Pekovic-Vaughan, V.; Gibbs, J.; Yoshitane, H.; Yang, N.; Pathiranage, D.; Guo, B.; Sagami, A.; Taguchi, K.; Bechtold, D.; Loudon, A.; et al. The circadian clock regulates rhythmic activation of the NRF2/glutathionemediated antioxidant defense pathway to modulate pulmonary fibrosis. *Genes Dev.* **2014**, *28*, 548–560. [CrossRef]
131. Early, J.O.; Menon, D.; Wyse, C.A.; Cervantes-Silva, M.P.; Zaslona, Z.; Carroll, R.G.; Palsson-McDermott, E.M.; Angiari, S.; Ryan, D.G.; Corcoran, S.E.; et al. Circadian clock protein BMAL1 regulates IL-1β in macrophages via NRF2. *Proc. Natl. Acad. Sci. USA* **2018**, *115*, E8460–E8468. [CrossRef]
132. Xu, Y.Q.; Zhang, D.; Jin, T.; Cai, D.J.; Wu, Q.; Lu, Y.; Liu, J.; Klaassen, C.D. Diurnal Variation of Hepatic Antioxidant Gene Expression in Mice. *PLoS ONE* **2012**, *7*, e44237. [CrossRef]
133. Musiek, E.S.; Lim, M.M.; Yang, G.; Bauer, A.Q.; Qi, L.; Lee, Y.; Roh, J.H.; Ortiz-Gonzalez, X.; Dearborn, J.T.; Culver, J.P.; et al. Circadian clock proteins regulate neuronal redox homeostasis and neurodegeneration. *J. Clin. Investig.* **2013**, *123*, 5389–5400. [CrossRef] [PubMed]
134. Lee, S.; Hur, E.; Ryoo, I.; Jung, K.-A.; Kwak, J.; Kwak, M.-K. Involvement of the Nrf2-proteasome pathway in the endoplasmic reticulum stress response in pancreatic β-cells. *Toxicol. Appl. Pharmacol.* **2012**, *264*, 431–438. [CrossRef]
135. Chhunchha, B.; Kubo, E.; Singh, D.P. Clock Protein Bmal1 and Nrf2 Cooperatively Control Aging or Oxidative Response and Redox Homeostasis by Regulating Rhythmic Expression of Prdx6. *Cells* **2020**, *9*, 1861. [CrossRef]

136. Rutter, J.; Reick, M.; Wu, L.C.; McKnight, S.L. Regulation of clock and NPAS2 DNA binding by the redox state of NAD cofactors. *Science* **2001**, *293*, 510–514. [CrossRef]
137. Wible, R.S.; Ramanathan, C.; Sutter, C.H.; Olesen, K.M.; Kensler, T.W.; Liu, A.C.; Sutter, T.R. NRF2 regulates core and stabilizing circadian clock loops, coupling redox and timekeeping in mus musculus. *Elife* **2018**, *7*, e31656. [CrossRef]
138. Dirksen, R.T. Sarcoplasmic reticulum–mitochondrial through-space coupling in skeletal muscle. *Appl. Physiol. Nutr. Metab.* **2009**, *34*, 389–395. [CrossRef]
139. Cooper, G.M. The Mechanism of Oxidative Phosphorylation. In *The Cell: A Molecular Approach*, 2nd ed.; Sinauer Associates: Sunderland Massachusetts USA, 2000; pp. 396–402.
140. Neufeld-Cohen, A.; Robles, M.S.; Aviram, R.; Manella, G.; Adamovich, Y.; Ladeuix, B.; Nir, D.; Rousso-Noori, L.; Kuperman, Y.; Golik, M. Circadian control of oscillations in mitochondrial rate-limiting enzymes and nutrient utilization by PERIOD proteins. *Proc. Natl. Acad. Sci. USA* **2016**, *113*, E1673–E1682. [CrossRef]
141. Mezhnina, V.; Ebeigbe, O.P.; Poe, A.; Kondratov, R.V. Circadian control of mitochondria in reactive oxygen species homeostasis. *Antioxid. Redox Signal.* **2022**, *37*, 647–663. [CrossRef] [PubMed]
142. Peek, C.B.; Affinati, A.H.; Ramsey, K.M.; Kuo, H.-Y.; Yu, W.; Sena, L.A.; Ilkayeva, O.; Marcheva, B.; Kobayashi, Y.; Omura, C. Circadian clock NAD+ cycle drives mitochondrial oxidative metabolism in mice. *Science* **2013**, *342*, 1243417. [CrossRef]
143. Hardie, D.G.; Ross, F.A.; Hawley, S.A. AMPK: A nutrient and energy sensor that maintains energy homeostasis. *Nat. Rev. Mol. Cell Biol.* **2012**, *13*, 251–262. [CrossRef] [PubMed]
144. Masri, S.; Patel, V.R.; Eckel-Mahan, K.L.; Peleg, S.; Forne, I.; Ladurner, A.G.; Baldi, P.; Imhof, A.; Sassone-Corsi, P. Circadian acetylome reveals regulation of mitochondrial metabolic pathways. *Proc. Natl. Acad. Sci. USA* **2013**, *110*, 3339–3344. [CrossRef]
145. Cela, O.; Scrima, R.; Pazienza, V.; Merla, G.; Benegiamo, G.; Augello, B.; Fugetto, S.; Menga, M.; Rubino, R.; Fuhr, L. Clock genes-dependent acetylation of complex I sets rhythmic activity of mitochondrial OxPhos. *Biochim. Biophys. Acta (BBA)-Mol. Cell Res.* **2016**, *1863*, 596–606. [CrossRef]
146. Mauvoisin, D.; Atger, F.; Dayon, L.; Galindo, A.N.; Wang, J.; Martin, E.; Da Silva, L.; Montoliu, I.; Collino, S.; Martin, F.-P. Circadian and feeding rhythms orchestrate the diurnal liver acetylome. *Cell Rep.* **2017**, *20*, 1729–1743. [CrossRef] [PubMed]
147. Mishra, P.; Chan, D.C. Metabolic regulation of mitochondrial dynamics. *J. Cell Biol.* **2016**, *212*, 379–387. [CrossRef] [PubMed]
148. Jacobi, D.; Liu, S.; Burkewitz, K.; Kory, N.; Knudsen, N.H.; Alexander, R.K.; Unluturk, U.; Li, X.; Kong, X.; Hyde, A.L. Hepatic Bmal1 regulates rhythmic mitochondrial dynamics and promotes metabolic fitness. *Cell Metab.* **2015**, *22*, 709–720. [CrossRef]
149. Oliva-Ramírez, J.; Moreno-Altamirano, M.M.B.; Pineda-Olvera, B.; Cauich-Sánchez, P.; Sánchez-García, F.J. Crosstalk between circadian rhythmicity, mitochondrial dynamics and macrophage bactericidal activity. *Immunology* **2014**, *143*, 490–497. [CrossRef]
150. Aguilar-López, B.A.; Moreno-Altamirano, M.M.B.; Dockrell, H.M.; Duchen, M.R.; Sánchez-García, F.J. Mitochondria: An integrative hub coordinating circadian rhythms, metabolism, the microbiome, and immunity. *Front. Cell Dev. Biol.* **2020**, *8*, 51. [CrossRef] [PubMed]
151. De Goede, P.; Wefers, J.; Brombacher, E.C.; Schrauwen, P.; Kalsbeek, A. Circadian rhythms in mitochondrial respiration. *J. Mol. Endocrinol.* **2018**, *60*, R115–R130. [CrossRef]
152. Liu, C.; Li, S.; Liu, T.; Borjigin, J.; Lin, J.D. Transcriptional coactivator PGC-1α integrates the mammalian clock and energy metabolism. *Nature* **2007**, *447*, 477–481. [CrossRef]
153. Pacelli, C.; Rotundo, G.; Lecce, L.; Menga, M.; Bidollari, E.; Scrima, R.; Cela, O.; Piccoli, C.; Cocco, T.; Vescovi, A.L. Parkin mutation affects clock gene-dependent energy metabolism. *Int. J. Mol. Sci.* **2019**, *20*, 2772. [CrossRef]
154. Ulgherait, M.; Chen, A.; McAllister, S.F.; Kim, H.X.; Delventhal, R.; Wayne, C.R.; Garcia, C.J.; Recinos, Y.; Oliva, M.; Canman, J.C. Circadian regulation of mitochondrial uncoupling and lifespan. *Nat. Commun.* **2020**, *11*, 1927. [CrossRef]
155. Kim, J.; Sun, W. Circadian coordination: Understanding interplay between circadian clock and mitochondria. *Anim. Cells Syst.* **2024**, *28*, 228–236. [CrossRef] [PubMed]
156. Ramírez-Casas, Y.; Fernández-Martínez, J.; Martín-Estebané, M.; Aranda-Martínez, P.; López-Rodríguez, A.; Esquivel-Ruiz, S.; Yang, Y.; Escames, G.; Acuña-Castroviejo, D. Melatonin and Exercise Restore Myogenesis and Mitochondrial Dynamics Deficits Associated With Sarcopenia in iMS-Bmal1$^{-/-}$ Mice. *J. Pineal Res.* **2025**, *77*, e70049. [CrossRef] [PubMed]
157. Jordan, S.D.; Kriebs, A.; Vaughan, M.; Duglan, D.; Fan, W.; Henriksson, E.; Huber, A.-L.; Papp, S.J.; Nguyen, M.; Afetian, M. CRY1/2 selectively repress PPARδ and limit exercise capacity. *Cell Metab.* **2017**, *26*, 243–255. [CrossRef]
158. Mansingh, S.; Maier, G.; Delezie, J.; Westermark, P.O.; Ritz, D.; Duchemin, W.; Santos, G.; Karrer-Cardel, B.; Steurer, S.A.; Albrecht, U. More than the clock: Distinct regulation of muscle function and metabolism by PER2 and RORα. *J. Physiol.* **2024**, *602*, 6373–6402. [CrossRef]
159. Gabriel, B.M.; Altıntaş, A.; Smith, J.A.B.; Sardon-Puig, L.; Zhang, X.; Basse, A.L.; Laker, R.C.; Gao, H.; Liu, Z.; Dollet, L. Disrupted circadian oscillations in type 2 diabetes are linked to altered rhythmic mitochondrial metabolism in skeletal muscle. *Sci. Adv.* **2021**, *7*, eabi9654. [CrossRef]

160. Solanas, G.; Peixoto, F.O.; Perdiguero, E.; Jardí, M.; Ruiz-Bonilla, V.; Datta, D.; Symeonidi, A.; Castellanos, A.; Welz, P.-S.; Caballero, J.M. Aged stem cells reprogram their daily rhythmic functions to adapt to stress. *Cell* **2017**, *170*, 678–692. [CrossRef]
161. Dudek, M.; Angelucci, C.; Pathiranage, D.; Wang, P.; Mallikarjun, V.; Lawless, C.; Swift, J.; Kadler, K.E.; Boot-Handford, R.P.; Hoyland, J.A. Circadian time series proteomics reveals daily dynamics in cartilage physiology. *Osteoarthr. Cartil.* **2021**, *29*, 739–749. [CrossRef]
162. Chang, J.; Garva, R.; Pickard, A.; Yeung, C.Y.C.; Mallikarjun, V.; Swift, J.; Holmes, D.F.; Calverley, B.; Lu, Y.; Adamson, A.; et al. Circadian control of the secretory pathway maintains collagen homeostasis. *Nat. Cell Biol.* **2020**, *22*, 74–86. [CrossRef]
163. Garva, R.; Yeung, C.-Y.C.; Pickard, A.; Lu, Y.; Mallikarjun, V.; Swift, J.; Taylor, S.H.; Rai, J.; Eyre, D.R.; Chaturvedi, M. Mmp14-controlled extracellular matrix homeostasis is required for circadian rhythm. *bioRxiv* **2022**. [CrossRef]
164. Cunningham, P.S.; Meijer, P.; Nazgiewicz, A.; Anderson, S.G.; Borthwick, L.A.; Bagnall, J.; Kitchen, G.B.; Lodyga, M.; Begley, N.; Venkateswaran, R.V.; et al. The circadian clock protein REVERBα inhibits pulmonary fibrosis development. *Proc. Natl. Acad. Sci. USA* **2020**, *117*, 1139–1147. [CrossRef] [PubMed]
165. Prasad, C.; Hahn, K.; Duraisamy, S.K.; Salathe, M.A.; Huang, S.K.; Burris, T.P.; Sundar, I.K. Rev-erbα agonists suppresses TGFβ1-induced fibroblast-to-myofibroblast transition and pro-fibrotic phenotype in human lung fibroblasts. *Biochem. Biophys. Res. Commun.* **2023**, *669*, 120–127. [CrossRef] [PubMed]
166. Dong, C.; Gongora, R.; Sosulski, M.L.; Luo, F.; Sanchez, C.G. Regulation of transforming growth factor-beta1 (TGF-β1)-induced pro-fibrotic activities by circadian clock gene BMAL1. *Respir. Res.* **2016**, *17*, 4. [PubMed]
167. Wang, Q.; Sundar, I.K.; Lucas, J.H.; Park, J.-G.; Nogales, A.; Martinez-Sobrido, L.; Rahman, I. Circadian clock molecule REV-ERBα regulates lung fibrotic progression through collagen stabilization. *Nat. Commun.* **2023**, *14*, 1295. [CrossRef]
168. Ingle, K.A.; Kain, V.; Goel, M.; Prabhu, S.D.; Young, M.E.; Halade, G.V. Cardiomyocyte-specific Bmal1 deletion in mice triggers diastolic dysfunction, extracellular matrix response, and impaired resolution of inflammation. *Am. J. Physiol. Circ. Physiol.* **2015**, *309*, H1827–H1836. [CrossRef]
169. Mia, S.; Kane, M.S.; Latimer, M.N.; Reitz, C.J.; Sonkar, R.; Benavides, G.A.; Smith, S.R.; Frank, S.J.; Martino, T.A.; Zhang, J. Differential effects of REV-ERBα/β agonism on cardiac gene expression, metabolism, and contractile function in a mouse model of circadian disruption. *Am. J. Physiol. Circ. Physiol.* **2020**, *318*, H1487–H1508. [CrossRef]
170. Liang, Q.; Xu, H.; Liu, M.; Qian, L.; Yan, J.; Yang, G.; Chen, L. Postnatal deletion of Bmal1 in cardiomyocyte promotes pressure overload induced cardiac remodeling in mice. *J. Am. Heart Assoc.* **2022**, *11*, e025021. [CrossRef]
171. Durgan, D.J.; Pulinilkunnil, T.; Villegas-Montoya, C.; Garvey, M.E.; Frangogiannis, N.G.; Michael, L.H.; Chow, C.-W.; Dyck, J.R.B.; Young, M.E. Ischemia/reperfusion tolerance is time-of-day–dependent: Mediation by the cardiomyocyte circadian clock. *Circ. Res.* **2010**, *106*, 546–550. [CrossRef]
172. Yoshida, Y.; Matsunaga, N.; Nakao, T.; Hamamura, K.; Kondo, H.; Ide, T.; Tsutsui, H.; Tsuruta, A.; Kurogi, M.; Nakaya, M. Alteration of circadian machinery in monocytes underlies chronic kidney disease-associated cardiac inflammation and fibrosis. *Nat. Commun.* **2021**, *12*, 2783. [CrossRef] [PubMed]
173. Chen, W.; Zhao, S.; Xing, J.; Yu, W.; Rao, T.; Zhou, X.; Ruan, Y.; Li, S.; Xia, Y.; Song, T. BMAL1 inhibits renal fibrosis and renal interstitial inflammation by targeting the ERK1/2/ELK-1/Egr-1 axis. *Int. Immunopharmacol.* **2023**, *125*, 111140. [CrossRef]
174. Chen, W.-D.; Yeh, J.-K.; Peng, M.-T.; Shie, S.-S.; Lin, S.-L.; Yang, C.-H.; Chen, T.-H.; Hung, K.-C.; Wang, C.-C.; Hsieh, I.-C. Circadian CLOCK mediates activation of transforming growth factor-β signaling and renal fibrosis through cyclooxygenase 2. *Am. J. Pathol.* **2015**, *185*, 3152–3163. [CrossRef]
175. Zhang, J.; Liu, C.; Liang, Q.; Zheng, F.; Guan, Y.; Yang, G.; Chen, L. Postnatal deletion of Bmal1 in mice protects against obstructive renal fibrosis via suppressing Gli2 transcription. *FASEB J.* **2021**, *35*, e21530. [CrossRef]
176. Crouchet, E.; Dachraoui, M.; Jühling, F.; Roehlen, N.; Oudot, M.A.; Durand, S.C.; Ponsolles, C.; Gadenne, C.; Meiss-Heydmann, L.; Moehlin, J. Targeting the liver clock improves fibrosis by restoring TGF-β signaling. *J. Hepatol.* **2025**, *82*, 120–133. [CrossRef]
177. Jokl, E.; Llewellyn, J.; Simpson, K.; Adegboye, O.; Pritchett, J.; Zeef, L.; Donaldson, I.; Athwal, V.S.; Purssell, H.; Street, O. Circadian disruption primes myofibroblasts for accelerated activation as a mechanism underpinning fibrotic progression in non-alcoholic fatty liver disease. *Cells* **2023**, *12*, 1582. [CrossRef] [PubMed]
178. Chen, P.; Han, Z.; Yang, P.; Zhu, L.; Hua, Z.; Zhang, J. Loss of clock gene mPer2 promotes liver fibrosis induced by carbon tetrachloride. *Hepatol. Res.* **2010**, *40*, 1117–1127. [CrossRef] [PubMed]
179. Wang, J.; Wang, Y.; Lin, L.; Pei, W.; Li, Y. Rev-erbα: The circadian guardian against NLRP3-driven liver fibrosis. *Mol. Med. Rep.* **2025**, *32*, 270.
180. Chen, P.; Kakan, X.; Wang, S.; Dong, W.; Jia, A.; Cai, C.; Zhang, J. Deletion of clock gene Per2 exacerbates cholestatic liver injury and fibrosis in mice. *Exp. Toxicol. Pathol.* **2013**, *65*, 427–432. [CrossRef] [PubMed]
181. Jiang, W.; Jin, L.; Ju, D.; Lu, Z.; Wang, C.; Guo, X.; Zhao, H.; Shen, S.; Cheng, Z.; Shen, J. The pancreatic clock is a key determinant of pancreatic fibrosis progression and exocrine dysfunction. *Sci. Transl. Med.* **2022**, *14*, eabn3586. [CrossRef]

182. Hunter, A.L.; Pelekanou, C.E.; Barron, N.J.; Northeast, R.C.; Grudzien, M.; Adamson, A.D.; Downton, P.; Cornfield, T.; Cunningham, P.S.; Billaud, J.-N. Adipocyte NR1D1 dictates adipose tissue expansion during obesity. *Elife* **2021**, *10*, e63324. [CrossRef]
183. Xiong, X.; Lin, Y.; Lee, J.; Paul, A.; Yechoor, V.; Figueiro, M.; Ma, K. Chronic circadian shift leads to adipose tissue inflammation and fibrosis. *Mol. Cell. Endocrinol.* **2021**, *521*, 111110. [CrossRef]
184. Gossan, N.; Zeef, L.; Hensman, J.; Hughes, A.; Bateman, J.F.; Rowley, L.; Little, C.B.; Piggins, H.D.; Rattray, M.; Boot-Handford, R.P. The circadian clock in murine chondrocytes regulates genes controlling key aspects of cartilage homeostasis. *Arthritis Rheum.* **2013**, *65*, 2334–2345. [CrossRef]
185. Yang, N.; Williams, J.; Pekovic-Vaughan, V.; Wang, P.; Olabi, S.; McConnell, J.; Gossan, N.; Hughes, A.; Cheung, J.; Streuli, C.H.; et al. Cellular mechano-environment regulates the mammary circadian clock. *Nat. Commun.* **2017**, *8*, 14287. [CrossRef]
186. Ding, S.-L.; Zhang, T.-W.; Zhang, Q.-C.; Ding, W.; Li, Z.-F.; Han, G.-J.; Bai, J.-S.; Li, X.-L.; Dong, J.; Wang, H.-R. Excessive mechanical strain accelerates intervertebral disc degeneration by disrupting intrinsic circadian rhythm. *Exp. Mol. Med.* **2021**, *53*, 1911–1923. [CrossRef]
187. Dudek, M.; Gossan, N.; Yang, N.; Im, H.-J.; Ruckshanthi, J.P.D.; Yoshitane, H.; Li, X.; Jin, D.; Wang, P.; Boudiffa, M. The chondrocyte clock gene Bmal1 controls cartilage homeostasis and integrity. *J. Clin. Investig.* **2016**, *126*, 365–376. [CrossRef]
188. Dudek, M.; Morris, H.; Rogers, N.; Pathiranage, D.R.J.; Raj, S.S.; Chan, D.; Kadler, K.E.; Hoyland, J.; Meng, Q.-J. The clock transcription factor BMAL1 is a key regulator of extracellular matrix homeostasis and cell fate in the intervertebral disc. *Matrix Biol.* **2023**, *122*, 1–9. [CrossRef]
189. Lastella, M.; Miller, D.J.; Quilelli, M.; Roberts, S.; Aisbett, B.; Condo, D. The impact of chronotype on the sleep and training responses of elite female Australian footballers. *Clocks & Sleep* **2021**, *3*, 528–535. [CrossRef]
190. Facer-Childs, E.; Brandstaetter, R. The impact of circadian phenotype and time since awakening on diurnal performance in athletes. *Curr. Biol.* **2015**, *25*, 518–522. [CrossRef] [PubMed]
191. van der Merwe, C.; Münch, M.; Kruger, R. Chronotype differences in body composition, dietary intake and eating behavior outcomes: A scoping systematic review. *Adv. Nutr.* **2022**, *13*, 2357–2405. [CrossRef] [PubMed]
192. Facer-Childs, E.R.; Boiling, S.; Balanos, G.M. The effects of time of day and chronotype on cognitive and physical performance in healthy volunteers. *Sport. Med.* **2018**, *4*, 47. [CrossRef] [PubMed]
193. Anderson, A.; Murray, G.; Herlihy, M.; Weiss, C.; King, J.; Hutchinson, E.; Albert, N.; Ingram, K.K. Circadian effects on performance and effort in collegiate swimmers. *J. Circadian Rhythms* **2018**, *16*, 8. [CrossRef]
194. Kunorozva, L.; Roden, L.C.; Rae, D.E. Perception of effort in morning-type cyclists is lower when exercising in the morning. *J. Sports Sci.* **2014**, *32*, 917–925. [CrossRef]
195. Lang, C.; Richardson, C.; Short, M.A.; Gradisar, M. Low-intensity scheduled morning exercise for adolescents with a late chronotype: A novel treatment to advance circadian phase? *Sleep Adv.* **2022**, *3*, zpac021. [CrossRef]
196. Facer-Childs, E.R.; Middleton, B.; Skene, D.J.; Bagshaw, A.P. Resetting the late timing of 'night owls' has a positive impact on mental health and performance. *Sleep Med.* **2019**, *60*, 236–247. [CrossRef] [PubMed]
197. Thomas, J.M.; Kern, P.A.; Bush, H.M.; McQuerry, K.J.; Black, W.S.; Clasey, J.L.; Pendergast, J.S. Circadian rhythm phase shifts caused by timed exercise vary with chronotype. *JCI Insight* **2020**, *5*, e134270. [CrossRef]
198. Distefano, G.; Goodpaster, B.H. Effects of exercise and aging on skeletal muscle. *Cold Spring Harb. Perspect. Med.* **2018**, *8*, a029785. [CrossRef]
199. Yoo, S.-Z.; No, M.-H.; Heo, J.-W.; Park, D.-H.; Kang, J.-H.; Kim, J.-H.; Seo, D.-Y.; Han, J.; Jung, S.-J.; Kwak, H.-B. Effects of acute exercise on mitochondrial function, dynamics, and mitophagy in rat cardiac and skeletal muscles. *Int. Neurourol. J.* **2019**, *23*, S22–S31. [CrossRef] [PubMed]
200. Lundby, C.; Jacobs, R.A. Adaptations of skeletal muscle mitochondria to exercise training. *Exp. Physiol.* **2016**, *101*, 17–22. [CrossRef]
201. Yavari, A.; Javadi, M.; Mirmiran, P.; Bahadoran, Z. Exercise-induced oxidative stress and dietary antioxidants. *Asian, J. Sports Med.* **2015**, *6*, e24898. [CrossRef]
202. Ostrom, E.L.; Traustadóttir, T. Aerobic exercise training partially reverses the impairment of Nrf2 activation in older humans. *Free Radic. Biol. Med.* **2020**, *160*, 418–432. [CrossRef]
203. Martinez-Canton, M.; Galvan-Alvarez, V.; Martin-Rincon, M.; Calbet, J.A.L.; Gallego-Selles, A. Unlocking peak performance: The role of Nrf2 in enhancing exercise outcomes and training adaptation in humans. *Free Radic. Biol. Med.* **2024**, *224*, 168–181. [CrossRef] [PubMed]
204. Kritikaki, E.; Asterling, R.; Ward, L.; Padget, K.; Barreiro, E.; Simoes, D.C.M. Exercise training-induced extracellular matrix protein adaptation in locomotor muscles: A systematic review. *Cells* **2021**, *10*, 1022. [CrossRef]

205. Robinson, M.M.; Dasari, S.; Konopka, A.R.; Johnson, M.L.; Manjunatha, S.; Esponda, R.R.; Carter, R.E.; Lanza, I.R.; Nair, K.S. Enhanced protein translation underlies improved metabolic and physical adaptations to different exercise training modes in young and old humans. *Cell Metab.* **2017**, *25*, 581–592. [CrossRef]
206. Gao, L.; Kumar, V.; Vellichirammal, N.N.; Park, S.-Y.; Rudebush, T.L.; Yu, L.; Son, W.-M.; Pekas, E.J.; Wafi, A.M.; Hong, J.; et al. Functional, proteomic and bioinformatic analyses of Nrf2- and Keap1- null skeletal muscle. *J. Physiol.* **2020**, *598*, 5427–5451. [CrossRef]
207. Oh, S.; Komine, S.; Warabi, E.; Akiyama, K.; Ishii, A.; Ishige, K.; Mizokami, Y.; Kuga, K.; Horie, M.; Miwa, Y. Nuclear factor (erythroid derived 2)-like 2 activation increases exercise endurance capacity via redox modulation in skeletal muscles. *Sci. Rep.* **2017**, *7*, 12902. [CrossRef]
208. Onoki, T.; Izumi, Y.; Takahashi, M.; Murakami, S.; Matsumaru, D.; Ohta, N.; Wati, S.M.; Hatanaka, N.; Katsuoka, F.; Okutsu, M. Skeletal muscle-specific Keap1 disruption modulates fatty acid utilization and enhances exercise capacity in female mice. *Redox Biol.* **2021**, *43*, 101966. [CrossRef]
209. Basti, A.; Yalçin, M.; Herms, D.; Hesse, J.; Aboumanify, O.; Li, Y.; Aretz, Z.; Garmshausen, J.; El-Athman, R.; Hastermann, M. Diurnal variations in the expression of core-clock genes correlate with resting muscle properties and predict fluctuations in exercise performance across the day. *BMJ Open Sport Exerc. Med.* **2021**, *7*, e000876. [CrossRef] [PubMed]
210. Xu, L.; Jia, J.; Yu, J.; Miao, S.; Zhang, Y. The impact of aerobic exercise timing on BMAL1 protein expression and antioxidant responses in skeletal muscle of mice. *Free Radic. Res.* **2024**, *58*, 311–322. [CrossRef] [PubMed]
211. Iwayama, K.; Tanabe, Y.; Tanji, F.; Ohnishi, T.; Takahashi, H. Diurnal variations in muscle and liver glycogen differ depending on the timing of exercise. *J. Physiol. Sci.* **2021**, *71*, 35. [CrossRef] [PubMed]
212. Savikj, M.; Gabriel, B.M.; Alm, P.S.; Smith, J.; Caidahl, K.; Björnholm, M.; Fritz, T.; Krook, A.; Zierath, J.R.; Wallberg-Henriksson, H. Afternoon exercise is more efficacious than morning exercise at improving blood glucose levels in individuals with type 2 diabetes: A randomised crossover trial. *Diabetologia* **2019**, *62*, 233–237. [CrossRef]
213. Savikj, M.; Stocks, B.; Sato, S.; Caidahl, K.; Krook, A.; Deshmukh, A.S.; Zierath, J.R.; Wallberg-Henriksson, H. Exercise timing influences multi-tissue metabolome and skeletal muscle proteome profiles in type 2 diabetic patients–A randomized crossover trial. *Metabolism* **2022**, *135*, 155268. [CrossRef]
214. Sato, S.; Basse, A.L.; Schönke, M.; Chen, S.; Samad, M.; Altıntaş, A.; Laker, R.C.; Dalbram, E.; Barrès, R.; Baldi, P. Time of exercise specifies the impact on muscle metabolic pathways and systemic energy homeostasis. *Cell Metab.* **2019**, *30*, 92–110. [CrossRef]
215. Hughes, A.T.L.; Samuels, R.E.; Baño-Otálora, B.; Belle, M.D.C.; Wegner, S.; Guilding, C.; Northeast, R.C.; Loudon, A.S.I.; Gigg, J.; Piggins, H.D. Timed daily exercise remodels circadian rhythms in mice. *Commun. Biol.* **2021**, *4*, 761. [CrossRef] [PubMed]
216. Adamovich, Y.; Dandavate, V.; Ezagouri, S.; Manella, G.; Zwighaft, Z.; Sobel, J.; Kuperman, Y.; Golik, M.; Auerbach, A.; Itkin, M. Clock proteins and training modify exercise capacity in a daytime-dependent manner. *Proc. Natl. Acad. Sci. USA* **2021**, *118*, e2101115118. [CrossRef]
217. Pourabdi, R.; Shahidi, F.; Tabandeh, M.R.; Salehpour, M. Aerobic exercise timing affects mitochondrial dynamics and insulin resistance by regulating the circadian clock protein expression and NAD+-SIRT1-PPARα-MFN2 pathway in the skeletal muscle of high-fat-diet-induced diabetes mice. *J. Physiol. Biochem.* **2025**, *81*, 199–214. [CrossRef]
218. Ferraro, E.; Giammarioli, A.M.; Chiandotto, S.; Spoletini, I.; Rosano, G. Exercise-induced skeletal muscle remodeling and metabolic adaptation: Redox signaling and role of autophagy. *Antioxid. Redox Signal.* **2014**, *21*, 154–176. [CrossRef]
219. Milanović, Z.; Sporiš, G.; Weston, M. Effectiveness of high-intensity interval training (HIT) and continuous endurance training for VO2max improvements: A systematic review and meta-analysis of controlled trials. *Sport. Med.* **2015**, *45*, 1469–1481. [CrossRef]
220. Lok, R.; Zerbini, G.; Gordijn, M.C.M.; Beersma, D.G.M.; Hut, R.A. Gold, silver or bronze: Circadian variation strongly affects performance in Olympic athletes. *Sci. Rep.* **2020**, *10*, 16088. [CrossRef] [PubMed]
221. Martín-López, J.; Sedliak, M.; Valadés, D.; Muñoz, A.; Buffet-García, J.; García-Oviedo, R.; Rodríguez-Aragón, M.; Pérez-López, A.; López-Samanes, Á. Impact of time-of-day and chronotype on neuromuscular performance in semi-professional female volleyball players. *Chronobiol. Int.* **2022**, *39*, 1006–1014. [CrossRef]
222. López-Samanes, Á.; Moreno-Pérez, D.; Maté-Muñoz, J.L.; Domínguez, R.; Pallarés, J.G.; Mora-Rodriguez, R.; Ortega, J.F. Circadian rhythm effect on physical tennis performance in trained male players. *J. Sports Sci.* **2017**, *35*, 2121–2128. [CrossRef]
223. Ammar, A.; Chtourou, H.; Trabelsi, K.; Padulo, J.; Turki, M.; El Abed, K.; Hoekelmann, A.; Hakim, A. Temporal specificity of training: Intra-day effects on biochemical responses and Olympic-Weightlifting performances. *J. Sports Sci.* **2015**, *33*, 358–368. [CrossRef] [PubMed]
224. West, D.J.; Cook, C.J.; Beaven, M.C.; Kilduff, L.P. The influence of the time of day on core temperature and lower body power output in elite rugby union sevens players. *J. Strength Cond. Res.* **2014**, *28*, 1524–1528. [CrossRef]
225. Henst, R.H.P.; Jaspers, R.T.; Roden, L.C.; Rae, D.E. A chronotype comparison of South African and Dutch marathon runners: The role of scheduled race start times and effects on performance. *Chronobiol. Int.* **2015**, *32*, 858–868. [CrossRef]

226. Ayala, V.; Martínez-Bebia, M.; Latorre, J.A.; Gimenez-Blasi, N.; Jimenez-Casquet, M.J.; Conde-Pipo, J.; Bach-Faig, A.; Mariscal-Arcas, M. Influence of circadian rhythms on sports performance. *Chronobiol. Int.* **2021**, *38*, 1522–1536. [PubMed]
227. Landi, F.; Marzetti, E.; Martone, A.M.; Bernabei, R.; Onder, G. Exercise as a remedy for sarcopenia. *Curr. Opin. Clin. Nutr. Metab. Care* **2014**, *17*, 25–31. [CrossRef]
228. Meignié, A.; Toussaint, J.-F.; Antero, J. Dealing with menstrual cycle in sport: Stop finding excuses to exclude women from research. *Eur. J. Appl. Physiol.* **2022**, *122*, 2489–2490. [CrossRef]
229. Justice, M.J. Sex matters in preclinical research. *Dis. Model. Mech.* **2024**, *17*, dmm050759. [CrossRef]
230. Paul, R.W.; Sonnier, J.H.; Johnson, E.E.; Hall, A.T.; Osman, A.; Connors, G.M.; Freedman, K.B.; Bishop, M.E. Inequalities in the evaluation of male versus female athletes in sports medicine research: A systematic review. *Am. J. Sports Med.* **2023**, *51*, 3335–3342. [PubMed]
231. Ose, B.M.; Eisenhauer, J.; Roepe, I.; Herda, A.A.; Vopat, B.G.; Vopat, L.M. Where are all the female participants in sports and exercise medicine research? A decade later. *Am. J. Sports Med.* **2025**, *53*, 2022–2028. [CrossRef]
232. Ellison, T.M.; Flagstaff, I.; Johnson, A.E. Sexual dimorphisms in anterior cruciate ligament injury: A current concepts review. *Orthop. J. Sport. Med.* **2021**, *9*, 23259671211025304.
233. Emmert, M.E.; Emmert, A.S.; Goh, Q.; Cornwall, R. Sexual dimorphisms in skeletal muscle: Current concepts and research horizons. *J. Appl. Physiol.* **2024**, *137*, 274–299. [CrossRef]
234. Schiaffino, S.; Reggiani, C. Fiber types in mammalian skeletal muscles. *Physiol. Rev.* **2011**, *91*, 1447–1531. [CrossRef]
235. Harmer, A.R.; Ruell, P.A.; Hunter, S.K.; McKenna, M.J.; Thom, J.M.; Chisholm, D.J.; Flack, J.R. Effects of type 1 diabetes, sprint training and sex on skeletal muscle sarcoplasmic reticulum Ca2+ uptake and Ca2+-ATPase activity. *J. Physiol.* **2014**, *592*, 523–535. [CrossRef] [PubMed]
236. Esbjörnsson, M.E.; Dahlström, M.S.; Gierup, J.W.; Jansson, E.C. Muscle fiber size in healthy children and adults in relation to sex and fiber types. *Muscle Nerve* **2021**, *63*, 586–592. [CrossRef] [PubMed]
237. Moesgaard, L.; Jessen, S.; Mackey, A.L.; Hostrup, M. Myonuclear addition is associated with sex-specific fiber hypertrophy and occurs in relation to fiber perimeter not cross-sectional area. *J. Appl. Physiol.* **2022**, *133*, 732–741. [CrossRef] [PubMed]
238. Neal, A.; Boldrin, L.; Morgan, J.E. The satellite cell in male and female, developing and adult mouse muscle: Distinct stem cells for growth and regeneration. *PLoS ONE* **2012**, *7*, e37950. [CrossRef] [PubMed]
239. Srinivas-Shankar, U.; Roberts, S.A.; Connolly, M.J.; O'Connell, M.D.L.; Adams, J.E.; Oldham, J.A.; Wu, F.C.W. Effects of testosterone on muscle strength, physical function, body composition, and quality of life in intermediate-frail and frail elderly men: A randomized, double-blind, placebo-controlled study. *J. Clin. Endocrinol. Metab.* **2010**, *95*, 639–650. [CrossRef]
240. Ikeda, K.; Horie-Inoue, K.; Inoue, S. Functions of estrogen and estrogen receptor signaling on skeletal muscle. *J. Steroid Biochem. Mol. Biol.* **2019**, *191*, 105375. [CrossRef]
241. Ribas, V.; Drew, B.G.; Zhou, Z.; Phun, J.; Kalajian, N.Y.; Soleymani, T.; Daraei, P.; Widjaja, K.; Wanagat, J.; de Aguiar Vallim, T.Q. Skeletal muscle action of estrogen receptor α is critical for the maintenance of mitochondrial function and metabolic homeostasis in females. *Sci. Transl. Med.* **2016**, *8*, 334ra54. [CrossRef]
242. Fernández-Martínez, J.; Ramírez-Casas, Y.; Yang, Y.; Aranda-Martínez, P.; Martínez-Ruiz, L.; Escames, G.; Acuña-Castroviejo, D. From Chronodisruption to Sarcopenia: The Therapeutic Potential of Melatonin. *Biomolecules* **2023**, *13*, 1779. [CrossRef]
243. Maldonado, M.D.; Manfredi, M.; Ribas-Serna, J.; Garcia-Moreno, H.; Calvo, J.R. Melatonin administrated immediately before an intense exercise reverses oxidative stress, improves immunological defenses and lipid metabolism in football players. *Physiol. Behav.* **2012**, *105*, 1099–1103. [CrossRef]
244. Faria, V.S.; Manchado-Gobatto, F.B.; Scariot, P.P.M.; Zagatto, A.M.; Beck, W.R. Melatonin Potentiates Exercise-Induced Increases in Skeletal Muscle PGC-1 α and Optimizes Glycogen Replenishment. *Front. Physiol.* **2022**, *13*, 803126. [CrossRef] [PubMed]
245. Sayed, R.K.A.; Fernández-Ortiz, M.; Diaz-Casado, M.E.; Rusanova, I.; Rahim, I.; Escames, G.; López, L.C.; Mokhtar, D.M.; Acuña-Castroviejo, D. The protective effect of melatonin against age-associated, sarcopenia-dependent tubular aggregate formation, lactate depletion, and mitochondrial changes. *J. Gerontol. Ser. A* **2018**, *73*, 1330–1338.
246. Song, Y.; Choi, G.; Jang, L.; Kim, S.-W.; Jung, K.-H.; Park, H. Circadian rhythm gene expression and daily melatonin levels vary in athletes and sedentary males. *Biol. Rhythm Res.* **2018**, *49*, 237–245. [CrossRef]
247. Fieldsend, T.W.; O'Neill, C.R.; Shrivastava, A.; Ogden, H.E.; Dand, N.; Hughes, S.M. Sexual dimorphism in human muscle ageing. *MedRxiv* **2025**, 2001–2025. [CrossRef]
248. Kasai, T.; Ishiguro, N.; Matsui, Y.; Harada, A.; Takemura, M.; Yuki, A.; Kato, Y.; Otsuka, R.; Ando, F.; Shimokata, H. Sex-and age-related differences in mid-thigh composition and muscle quality determined by computed tomography in middle-aged and elderly J apanese. *Geriatr. Gerontol. Int.* **2015**, *15*, 700–706. [CrossRef]

249. Callahan, D.M.; Bedrin, N.G.; Subramanian, M.; Berking, J.; Ades, P.A.; Toth, M.J.; Miller, M.S. Age-related structural alterations in human skeletal muscle fibers and mitochondria are sex specific: Relationship to single-fiber function. *J. Appl. Physiol.* **2014**, *116*, 1582–1592. [CrossRef] [PubMed]
250. Della Peruta, C.; Lozanoska-Ochser, B.; Renzini, A.; Moresi, V.; Sanchez Riera, C.; Bouché, M.; Coletti, D. Sex differences in inflammation and muscle wasting in aging and disease. *Int. J. Mol. Sci.* **2023**, *24*, 4651. [CrossRef]
251. Leenders, M.; Verdijk, L.B.; van der Hoeven, L.; Van Kranenburg, J.; Nilwik, R.; van Loon, L.J.C. Elderly men and women benefit equally from prolonged resistance-type exercise training. *J. Gerontol. Ser. A Biomed. Sci. Med. Sci.* **2013**, *68*, 769–779. [CrossRef]
252. Desvergne, A.; Ugarte, N.; Petropoulos, I.; Friguet, B. Circadian modulation of proteasome activities and removal of carbonylated proteins. *Free Radic. Biol. Med.* **2014**, *75*, S18.
253. Desvergne, A.; Ugarte, N.; Radjei, S.; Gareil, M.; Petropoulos, I.; Friguet, B. Circadian modulation of proteasome activity and accumulation of oxidized protein in human embryonic kidney HEK 293 cells and primary dermal fibroblasts. *Free Radic. Biol. Med.* **2016**, *94*, 195–207. [CrossRef]
254. Klichko, V.I.; Chow, E.S.; Kotwica-Rolinska, J.; Orr, W.C.; Giebultowicz, J.M.; Radyuk, S.N. Aging alters circadian regulation of redox in Drosophila. *Front. Genet.* **2015**, *6*, 83. [CrossRef]
255. Zhang, H.; Davies, K.J.A.; Forman, H.J. Oxidative stress response and Nrf2 signaling in aging. *Free Radic. Biol. Med.* **2015**, *88*, 314–336. [CrossRef]
256. Zhang, Y.; Wu, Y.; Wang, Y.; Lu, J.; Lu, Y.; Wang, P.; Li, L.; Yan, W.; Cai, H.; Leigh, W.H.; et al. Achilles tendinopathy treatment via circadian rhythm regulation. *J. Adv. Res.* **2025**, *75*, 505–520. [CrossRef] [PubMed]
257. Chen, J.; Xiang, J.; Zhou, M.; Huang, R.; Zhang, J.; Cui, Y.; Jiang, X.; Li, Y.; Zhou, R.; Xin, H. Dietary timing enhances exercise by modulating fat-muscle crosstalk via adipocyte AMPKα2 signaling. *Cell Metab.* **2025**, *37*, 1364–1380. [CrossRef] [PubMed]
258. Palmese, F.; Druda, Y.; Del Toro, R.; Bedogni, G.; Domenicali, M.; Silvani, A. The role of the circadian timing system in sarcopenia in old age: A scoping review. *Eur. Geriatr. Med.* **2025**, *16*, 447–460. [CrossRef]
259. Bouchard, C.; Blair, S.N.; Church, T.S.; Earnest, C.P.; Hagberg, J.M.; Häkkinen, K.; Jenkins, N.T.; Karavirta, L.; Kraus, W.E.; Leon, A.S. Adverse metabolic response to regular exercise: Is it a rare or common occurrence? *PLoS ONE* **2012**, *7*, e37887. [CrossRef]
260. Bherer, L.; Juneau, M. Exerkines: Mediators of the Health Benefits of Exercise. 2023. Available online: https://observatoireprevention.org/en/2023/10/30/exerkines-mediators-of-the-health-benefits-of-exercise/ (accessed on 31 July 2025).
261. Aoyama, S.; Shibata, S. Time-of-day-dependent physiological responses to meal and exercise. *Front. Nutr.* **2020**, *7*, 18. [CrossRef]
262. Tahara, Y.; Shibata, S. Entrainment of the mouse circadian clock: Effects of stress, exercise, and nutrition. *Free Radic. Biol. Med.* **2018**, *119*, 129–138. [CrossRef]
263. Crosby, P.; Hamnett, R.; Putker, M.; Hoyle, N.P.; Reed, M.; Karam, C.J.; Maywood, E.S.; Stangherlin, A.; Chesham, J.E.; Hayter, E.A. Insulin/IGF-1 drives PERIOD synthesis to entrain circadian rhythms with feeding time. *Cell* **2019**, *177*, 896–909. [CrossRef]
264. Lak, M.; Bagheri, R.; Ghobadi, H.; Campbell, B.; Wong, A.; Shahrbaf, A.; Shariatzadeh, M.; Dutheil, F. Timing matters? The effects of two different timing of high protein diets on body composition, muscular performance, and biochemical markers in resistance-trained males. *Front. Nutr.* **2024**, *11*, 1397090. [CrossRef]
265. Kreider, R.B.; Campbell, B. Protein for exercise and recovery. *Phys. Sportsmed.* **2009**, *37*, 13–21. [CrossRef] [PubMed]
266. Raza, G.S.; Kaya, Y.; Stenbäck, V.; Sharma, R.; Sodum, N.; Mutt, S.J.; Gagnon, D.D.; Tulppo, M.; Järvelin, M.; Herzig, K. Effect of Aerobic Exercise and Time-Restricted Feeding on Metabolic Markers and Circadian Rhythm in Mice Fed with the High-Fat Diet. *Mol. Nutr. Food Res.* **2024**, *68*, 2300465. [CrossRef] [PubMed]
267. Calvani, R.; Picca, A.; Coelho-Júnior, H.J.; Tosato, M.; Marzetti, E.; Landi, F. Diet for the prevention and management of sarcopenia. *Metabolism* **2023**, *146*, 155637. [CrossRef] [PubMed]
268. Rogeri, P.S.; Zanella, R., Jr.; Martins, G.L.; Garcia, M.D.A.; Leite, G.; Lugaresi, R.; Gasparini, S.O.; Sperandio, G.A.; Ferreira, L.H.B.; Souza-Junior, T.P. Strategies to prevent sarcopenia in the aging process: Role of protein intake and exercise. *Nutrients* **2021**, *14*, 52. [CrossRef]
269. Jäger, R.; Kerksick, C.M.; Campbell, B.I.; Cribb, P.J.; Wells, S.D.; Skwiat, T.M.; Purpura, M.; Ziegenfuss, T.N.; Ferrando, A.A.; Arent, S.M. International society of sports nutrition position stand: Protein and exercise. *J. Int. Soc. Sports Nutr.* **2017**, *14*, 20. [CrossRef] [PubMed]

antioxidants MDPI

Article

Nitrosative Stress in Astronaut Skeletal Muscle in Spaceflight

Dieter Blottner [1,2,*,†], Manuela Moriggi [3,†], Gabor Trautmann [1], Sandra Furlan [4], Katharina Block [1], Martina Gutsmann [1], Enrica Torretta [5], Pietro Barbacini [3], Daniele Capitanio [3], Joern Rittweger [6], Ulrich Limper [6,7], Pompeo Volpe [8], Cecilia Gelfi [3,5,‡] and Michele Salanova [1,2,‡]

1 Institute of Integrative Neuroanatomy, Charité—Universitätsmedizin Berlin, Corporate Member of Freie Universität Berlin, Humboldt-Universität zu Berlin, and Berlin Institute of Health, 10115 Berlin, Germany; gabor.trautmann@charite.de (G.T.); katharina.block@charite.de (K.B.); martina.gutsmann@charite.de (M.G.); michele.salanova@charite.de (M.S.)
2 NeuroMuscular System and Signaling Group, Center of Space Medicine and Extreme Environments, 10115 Berlin, Germany
3 Department of Biomedical Sciences for Health, University of Milan, 20133 Milan, Italy; manuela.moriggi@unimi.it (M.M.); pietro.barbacini@unimi.it (P.B.); daniele.capitanio@unimi.it (D.C.); cecilia.gelfi@unimi.it (C.G.)
4 C.N.R. Neuroscience Institute, I-35121 Padova, Italy; sfurlan@mail.bio.unipd.it
5 Laboratory of Proteomics and Lipidomics, IRCCS Orthopedic Institute Galeazzi, Via R. Galeazzi 4, 20161 Milan, Italy; enrica.torretta@grupposandonato.it
6 Institute of Aerospace Medicine, German Aerospace Center (DLR), 51147 Cologne, Germany; joern.rittweger@dlr.de (J.R.); ulrich.limper@dlr.de (U.L.)
7 Anesthesiology and Intensive Care Medicine, Merheim Medical Center, Witten/Herdecke University, 51109 Cologne, Germany
8 Department of Biomedical Sciences, Università di Padova, I-35121 Padova, Italy; pompeo.volpe@unipd.it
* Correspondence: dieter.blottner@charite.de
† These authors (first authors) contributed equally to this work.
‡ These authors (last authors) contributed equally to this work.

Abstract: Long-duration mission (LDM) astronauts from the International Space Station (ISS) (>180 ISS days) revealed a close-to-normal sarcolemmal nitric oxide synthase type-1 (NOS1) immunoexpression in myofibers together with biochemical and quantitative qPCR changes in deep calf soleus muscle. Nitro-DIGE analyses identified functional proteins (structural, metabolic, mitochondrial) that were over-nitrosylated post- vs. preflight. In a short-duration mission (SDM) astronaut (9 ISS days), s-nitrosylation of a nodal protein of the glycolytic flux, specific proteins in tricarboxylic acid (TCA) cycle, respiratory chain, and over-nitrosylation of creatine kinase M-types as signs of impaired ATP production and muscle contraction proteins were seen. S-nitrosylation of serotransferrin (TF) or carbonic anhydrase 3 (CA3b and 3c) represented signs of acute response microgravity muscle maladaptation. LDM nitrosoprofiles reflected recovery of mitochondrial activity, contraction proteins, and iron transporter TF as signs of muscle adaptation to microgravity. Nitrosated antioxidant proteins, alcohol dehydrogenase 5/S-nitrosoglutathione reductase (ADH5/GSNOR), and selenoprotein thioredoxin reductase 1 (TXNRD1) levels indicated signs of altered redox homeostasis and reduced protection from nitrosative stress in spaceflight. This work presents a novel spaceflight-generated dataset on s-nitrosylated muscle protein signatures from astronauts that helps both to better understand the structural and molecular networks associated to muscular nitrosative stress and to design countermeasures to dysfunction and impaired performance control in human spaceflight missions.

Keywords: spaceflight; nitric oxide synthase; microgravity; RONS; nitrosated proteins

1. Introduction

In human space life sciences, the typical (and still not fully understood) loss of muscle mass and function in astronauts was recently highlighted by new research priorities. These priorities focus on fundamental molecular mechanisms of space-related musculoskeletal disuse and adaptation [1], muscle and bone loss [2], nutrition and metabolism [3], and the need for optimization of exercise as countermeasures (CMs), as exemplified by the targeting of certain muscle groups of the human body as prime "executers" of motion in Earth gravity (1G) or variable gravity (ΔG) vectors (e.g., Moon, Mars) in future deep space explorations [4–6].

Changes in redox genes and signal-related pathways are signatures of long-term exposure of astronauts to extreme space conditions [7] with as-yet unknown crew health risks. Attempts to trigger exercise-induced antioxidant responses in normal muscle [8] or by therapeutical denitrosylation in various human diseases [9] are promising, as are new strategies to combat antioxidant capacity, such as by exercise-driven interventions in normal conditions [10], chronic disuse in bed rest (a ground-based analogue to spaceflight [11]), and dietary nutritional intervention [12]. Taken together, these approaches inspired the hope for a better management of oxidative/nitrosative stress in normal and clinical populations but also for astronaut musculoskeletal health, in particular during long-duration mission (LDM) spaceflights (>6 months or more).

Reactive oxygen (ROS) and nitrogen species (RNS) induced by free radicals (e.g., ONOO peroxides), collectively termed 'reactive oxidative/nitrosative species' (RONS) [13,14], are known common stressors in disuse atrophy [15] and inter-organ communication [16], as, for example, seen in muscle wasting diseases and aging [17], that result from imbalanced production or expression in cells and tissues under various conditions. For example, nitrosative stress in dystrophic muscle disease [18] or in chronic disuse conditions with otherwise healthy volunteers in bed rest is characterized by an imbalance of nitric oxide synthase (NOS) expression and regulation, promoting mutable free radical NO signals associated to s-nitrosylation (SNO) of functional muscle proteins [19,20]. Moreover, three major sources of the biologic gaseous NO signals were found in brain neuronal cells (via neuronal NOS, nNOS/NOS1), immune system (macrophages, inducible NOS2), and vascular system (endothelial cells, endothelial NOS3), but also in skeletal muscle (NOS1, μNOS) [21].

Skeletal muscle (taken together, forming the largest body organ) is a major source of NO in the human body [22]. NO production is maintained via both the NOS-dependent pathway (e.g., L-arginine + O_2 -> NO) and the NOS-independent pathways (e.g., NO_3 -> NO_2 -> NO, -> mostly via leafy vegetable diet [23]). In muscle physiology, NO serves as a critical molecular signal during development, growth [24–27], neuromuscular formation [28], muscle contraction [29], microcirculation [30], and exercise tolerance [27]. In exercise physiology, dietary nitrate supplementation (e.g., sodium nitrate [$NaNO_3$], NO_3-rich beetroot juice) has been highlighted as potential nutritional "ergogenic aids" to promote 'physical performance for health' benefits [23]. In muscle pathophysiology, NO is involved in muscle repair [31], wound healing and inflammation [32], impaired muscle perfusion [33], and microcirculation [34] as well as in post-exercise muscle cramps [35] and in various muscle weakness diseases [36].

Robust inactivity-induced NOS changes in murine skeletal muscle have been demonstrated both in ground-based studies (tissue sharing) and in space-flown mice on the NASA/ASI Mouse Drawer System (MDS) mission (91 days on ISS, NASA STS-129), thus confirming microgravity-induced delocalization of membrane/cytosolic NOS1 immunosignals in postural murine hindlimb skeletal muscle [37]. A previous bed rest study has shown extensive NOS1 immunosignals at the skeletal muscle–myofiber outer membranes of the soleus muscle, suggesting robust NOS re-localization following resistive exercise as a countermeasure in otherwise healthy musculature [38]. In another bed rest study, we have demonstrated targeted SNO levels of functional proteins such as nitrosylated ryanodine receptor type 1 (SNO-RyR1) and several others involved in contraction coupling and related proteins that clearly showed aberrant SNO protein abundance as signatures of immobilization-induced muscle disuse attenuated by resistive vibration exercise [19]. We thus hypothesized that NOS/NO signalling and key nitrosative stress indices via RONS products are linked both to muscle activity/inactivity and to microgravity-unloaded skeletal muscle. If so, SNO signatures in skeletal muscle proteins could then serve as biomarkers and thus be reflecting efficacy of inflight exercise protocols as countermeasures (CMs) to offset or mitigate sarcolemmal NOS dislocation in myofibers and aberrant SNO patterns of key muscle proteins. Investigation of SNO post-translational protein modifications in long-term spaceflight helps to shed light on RONS-associated and space-related disuse atrophy based on disrupted redox homeostasis in skeletal muscle.

In the present study, we report novel results from the MUSCLE BIOPSY (ESA op.nom.) experiment onboard the International Space Station (ISS) using pre- and postflight biopsy tissue samples available from LDM astronauts' lower leg (calf) soleus muscle [39]. Our first study aim was to compare sarcolemmal NOS1 immuno-localization patterns directly in soleus muscle samples from a short-duration mission (SDM) astronaut (9 days on ISS, without inflight exercise) with those available from LDM astronauts (>6 month or more on ISS, with routine inflight exercise). In particular, pixel-intensity analyses of sarcolemmal NOS1 immuno-expression patterns were used as signs of a potential NO source at the muscle fiber sub-cellular level. Our second study aim was to investigate differential changes in SNO protein levels using small amounts of soleus muscle biopsy lysates from astronauts subjected to un-targeted Nitro-DIGE analysis [40]. In addition to global nitroso-profiles of canonical proteoforms involved in energy metabolism, transport, stress response, and muscle contraction, we were interested in more specific SNO signatures in disused skeletal muscle from two of the most prominent complex enzyme systems triggered by the Nuclear factor (erythroid-derived 2)-2 (Nrf2), such as the thyoredoxin (TRX) and S-Nitrosoglutatione reductase (GSNOR) systems, which are necessary for the maintenance of protein s-nitrosylation/denitrosylation homeostasis [41].

2. Materials and Methods

2.1. Study Participants

The present work (op. nom. MUSCLE BIOPSY; European Space Agency (ESA)'s Science & Exploration Research Plan, ILSRA-2004-155) includes data from six ($n = 6$ valid) male ISS crew members enrolled from the United States Orbital Segment (USOS) astronaut corps from the National Aeronautics and Space Agency (NASA), the Japanese Space Exploration Agency (JAXA), and the European Space Agency (ESA). One study participant, astronaut A, was assigned to a short-duration mission (SDM) (9 days on ISS), and five other astronauts, B to F ($n = 5$), were assigned to a long-duration mission (LDM, ~180 days/ 6 months on ISS). Anthropometric data (body mass indices), general study protocol, and total numbers of inflight physical exercise countermeasure (CM) days (walking/running on T2, cycling with CEVIS, weight lifting with aRED) for the participating astronauts are published elsewhere [39].

2.2. Muscle Biopsy

From the study participants of MUSCLE BIOPSY, we collected one pre-flight (L-90 ± 30 days prior to launch) and one post-flight biopsy (return day R+0/1, landing day) from the right soleus muscle, a major slow-type and postural deep calf skeletal muscle, by an established protocol procedure either at the European Astronaut Center (EAC)/Human Physiology Lab at the DLR Institute of Aerospace Medicine, German Aerospace Center, Deutsches Zentrum für Luft- und Raumfahrt (DLR:envihab, Cologne, Germany), or at NASA's Johnson Space Center (JSC, Houston, TX, USA), depending on the astronaut's return travel schedules, as previously reported [39]. Biopsy material was immediately frozen in liquid nitrogen on site and stored at minus 80 °C in a freezer until further use.

2.3. Immunohistochemistry

Serially cut cryosections (8–10 μm thickness, obtained by using a cryostat CM-1860, leica-microsystems.com, accessed on 21 March 2024) from frozen tissue blocks were mounted on glass slides (Superfrost, Thermo Fisher Scientific, Henningsdorf, Germany) and stored frozen (−80 °C) in sealed boxes until further use for analyses together with an established immunostaining protocol in parallel to avoid any bias in laboratory staining outcome across the study duration. Cryosections were fixed (4% buffered paraformaldehyde, 10 min, 4 °C), rinsed in physiological PBS buffer, and preincubated with mouse IgG blocking reagent (1:500 M.O.M., Vector Laboratories Inc., Newark, CA, USA) in buffer. We used monoclonal antibodies raised against nitric oxide synthase type-1 (anti-NOS, monoclonal, sc-5302, N-terminal amino acid 2-300, Santa Cruz Biotechnology, Dallas, TX, USA). Primary antibodies were visualized by using fluorochrome Alexa-488 and Alexa-555 conjugated secondary antibodies (affinity purified goat-anti-mouse IgG, Invitrogen, Waltham, MA, USA) using an established experimental protocol [38]. Confocal imaging was performed with a high-resolution three-channel laser confocal microscope (SP-8, leica-microsystems.com, accessed on 21 March 2024). For quantification of subcellular immunofluorescence signal intensities (pixel intensity per μm^3 and image volume given as arbitrary units, A.U.), each cryosection was subdivided into several non-overlapping regions of interest (ROIs), thus covering most of the cross-sectioned muscle fiber profiles identified with a sarcolemma biomarker (1:1000, monoclonal anti-dystrophin, NCL-DYS2 #107416, Novocastra Antibodies, Newcastle, UK) and with a myonuclear biomarker 4′,6-diamidino-2-phenylindole fluorescent DNA stain (1:1000, DAPI, Hoechst 33258 stain), and inspected with a 40× objective (SP8 laser scanning microscope, leica-microsystems.com, accessed on 21 March 2024). The number of fields (ROIs), however, varied between 4 and 16 fields among cryosections due to the size of the cross-sectional planes from each biopsy sample.

2.4. Proteomics with Nitro-DIGE

2.4.1. Protein Extraction

Muscle biopsies taken before and after the flight were suspended, sonicated, clarified, and quantified as described previously [39].

2.4.2. Identification of S-Nitrosated Proteins by 2-D CyDye-Maleimide DIGE (Nitro-DIGE)

A modified biotin switch method was used to identify SNO proteins, as previously described [42]. For each sample, two technical replicates were considered and labeled with Cy5, while a mixture containing an equal amount of all samples was labeled with Cy3 as the internal standard.

2.4.3. Image Acquisition

Images from CyDye-labeled gels were acquired by Typhoon 9200 Imager (GE HealthCare, Solingen, Germany), and image analysis was performed by DeCyder software (version 6.5, GE HealthCare). For each experimental group (post- and pre-SDM; post- and pre-LDM), proteins identified in at least 70% of samples were considered.

2.4.4. Protein Identification

Protein identification was carried out by matrix-assisted laser desorption/ionization time-of-flight (MALDI-ToF) mass spectrometry (MS), as previously described [42].

2.4.5. Immunoblotting

Protein extracts (50 µg) from one SDM (acute exposure) and two LDM astronauts (chronic exposure) were loaded in duplicate and resolved on 10–16% gradient polyacrylamide gels. Blots were incubated with primary antibodies as follows: mouse monoclonal anti-alcohol dehydrogenase 5/S-nitrosoglutathione reductase (ADH5/GSNOR, 40 kDa, Santa Cruz Biotechnology, Dallas, TX, USA, sc-293460, 1:500), mouse monoclonal anti-thioredoxin reductase 1 (TXNRD1, 55 kDa, Santa Cruz Biotechnology, Dallas, TX, USA, sc-28321, 1:500), rabbit polyclonal anti-nitric oxide synthase 1 (NOS1, 155KDa, Santa Cruz Biotechnology, Dallas, TX, USA, sc-8309, 1:500). The membranes were then incubated with anti-mouse secondary antibodies conjugated to horseradish peroxidase (KPL, Seracare,

Milford, MA, USA, 1:5000). The ECL Prime Detection Kit and the Image Quant LAS 4000 analysis system (GE Healthcare) were used to visualize the signals by chemiluminescence.

2.4.6. RNA Extraction and qPCR

Total RNA was extracted from small tissue pieces of soleus using a commercially available TRIzol Reagent (Thermo Fisher Scientific) following the manufacturer's instructions. The concentration and purity of the RNA were determined using NanoDrop Lite Spectrophotometer (Thermo Fisher Scientific) by measuring the absorbance at 260 (OD260) and 280 (OD280) nm. Following isolation, 1 μg of RNA was reverse transcribed into cDNA using a VILO cDNA synthesis kit (Thermo Fisher Scientific). Primers for reference genes B2M and PPIA were published elsewhere [42]. Specific primers for NOS1 (NOS1, *Homo sapiens* (human) gene ID:4842) were designed using Primer3 software (http://frodo.wi.mit.edu/, accessed on 1 February 2024, version 4.1.0, Whitehead Institute for Biomedical Research, Camebridge, MA, USA) and their thermodynamic specificity was determined using BLAST sequence alignment (NCBI) and vector NTI® software (Thermo Fisher Scientific). No Genomic DNA interference was allowed because cDNAs of interest were amplified using the pair of primers derived from different exons of the genes. Human NOS1 specific primers:

Fw: CAATGTGCCTGTCGTCCTCA; Rv: GTGCATCCCGTTTCCAATGT

Quantitative PCR was performed using SYBR green chemistry (Bio-Rad Laboratories GmbH, Neuried, Germany) in triplicate in a 96-well CFX Thermal Cycler (Bio-Rad). All samples were run in parallel with RNA- and RT-negative controls. Efficiency of each run was monitored by a standard curve. Normalization was performed by $2^{-\Delta\Delta Ct}$ algorithm to determine the relative fold change of gene expression of samples. B2M and PPIA were tested as putative reference genes, with B2M being the most stable gene to normalize cycle threshold (Ct) values.

2.5. Statistics

For NOS1 immunosignal analyses, we used the Pearson–D'Augustino normality test (pre vs. post) for Gaussian normal distribution, together with parametric unpaired *t*-test and Mann–Whitney test (non-Gaussian). Significance level was set at $p \leq 0.05$, or even higher ($p \leq 0.001$) as indicated. GraphPad Prism (v9.4.0) was used for graphical representation and statistical data analyses. Due to highly limited amount of tissue available, sarcolemmal NOS1 immunosignal intensities were pooled and determined in at least three different cryosections per each sample with at least 50 cross-sectioned myofibers (type 1 or 2) at two different time points (pre- vs. postflight). Image analysis of immunostained cryosections was performed at identical settings with Leica Application Suite (LAS) X co-localization/3D image analysis software (release #3.5.7.23225.3D, www.leica-microsystems.com, accessed on 21 March 2024).

For Nitro-DIGE data analysis, statistically significant differences were computed by paired Student's *t*-test with a *p*-value threshold of 0.05. False discovery rate was applied to correct for multiple tests to reduce the overall error.

For immunoblotting data analysis, band quantification was performed using the Image Quant TL v. 8.1(GE HealthCare) software followed by statistical analysis (Student's *t*-test, *p*-value ≤ 0.05). Band intensities were normalized against the total amount of proteins stained by Sypro ruby total protein stain.

3. Results

3.1. Quantitative PCR (qPCR) Analyses of NOS1 Transcripts in Astronaut Soleus Muscle Samples

We analyzed skeletal muscle biopsy samples obtained from short- (SDM) and long-duration mission (LDM) astronauts before and after spaceflight. Astronaut A belongs to a short-duration mission (SDM-A; 11-day spaceflight with 9 ISS days) and was not obliged to regular Medical Operational (MedOps) inflight exercise because of the short flight duration and stay onboard the ISS. In contrast, LDM astronauts (LDM-B, C, D, and E) performed regular inflight exercise as a countermeasure (1.5 to 2 h daily) during their entire mission onboard the ISS.

As shown in Figure 1, quantitative PCR analysis in soleus muscle samples pre- vs. post-flight showed that in two out of three LDM subjects (LDM-B and C), there was a 2- to 2.5-fold change in NOS1 transcripts that was not seen in SDM-A or LDM-E (trend only).

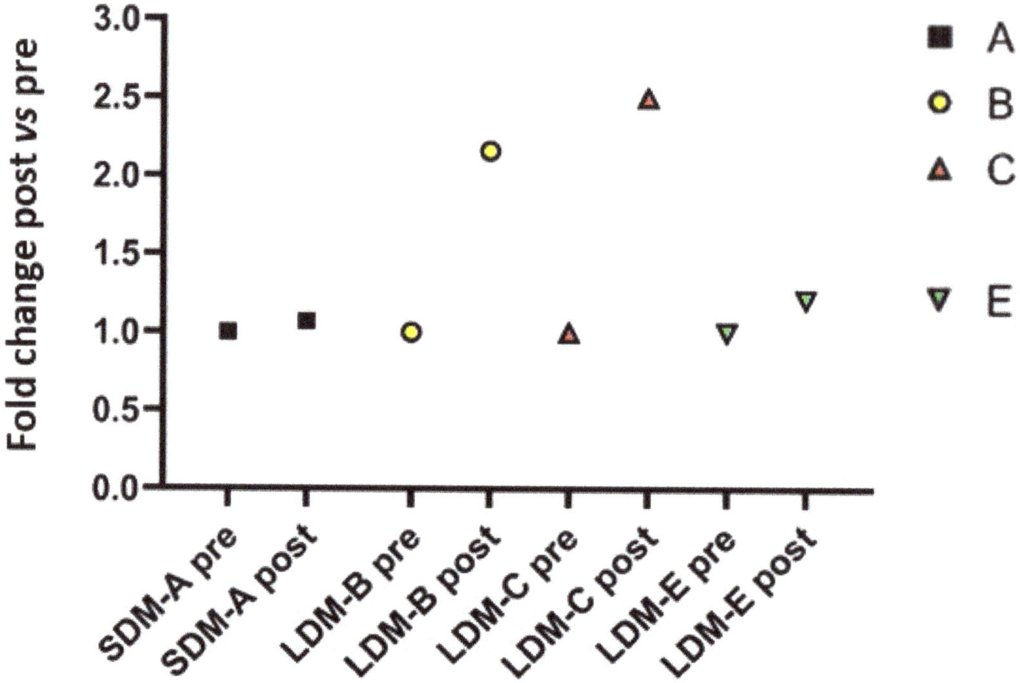

Figure 1. Scatter plot of NOS1 quantitative PCR (qPCR) data from soleus muscles from one ($n = 1$) SDM astronaut (SDM-A, 9 ISS days, black squares) and three ($n = 3$) LDM astronauts (LDM-B, yellow circle/C, red triangles/E, green triangles, 180 ISS days or more). NOS1 transcript levels were elevated in LDM-B and C (2–2.5-fold change) or in LDM-E (trend only) vs. SDM condition post- vs. preflight. Astronaut D values not available (tissue limitation).

3.2. NOS1 Immunolocalization in Astronaut Soleus Muscle Cryosections

Figures 2 and 3A show qualitative and quantitative analysis of subsarcolemma-associated NOS1 immunosignal intensity values in cryosections of the postural deep calf soleus muscle from SDM astronaut A (little or no exercise) before (pre-flight) and after spaceflight (post-flight). In the SDM astronaut, sarcolemma NOS1 immunosignal was reduced ($p = 0.001$) in muscle post-flight shortly (R0/+1) after landing vs. pre-flight (Figures 2 and 3A).

Quantitative analysis of NOS1 immunosignal pixel intensities confirmed that, in the soleus muscle of at least two LDM crew members (B, C; $p = 0.001$), subsarcolemmal NOS1 localization ($p = 0.001$) was increased, whereas no change was found in astronaut D and reduced NOS1 signal intensities were found in astronaut E post- vs. pre-flight (Figure 3A).

Eventually, a drop in sarcolemma NOS1 immunosignal found in the SDM astronaut was largely prevented in LDM astronauts (Figure 3A). With the exception of astronaut E, these findings support the notion that subsarcolemmal NOS1 expression likely reflects the positive muscle-type-specific effects induced by targeted inflight CM exercise protocols in three LDM crew members (astronauts B–D) in the deep calf soleus muscle.

Quantitative NOS1 biochemical analyses from soleus lysates of astronauts show that the amount of total NOS1 protein is changed in LDM B (trend) and LDM C (reduced) vs. SDM (no change) (see Figure 3B).

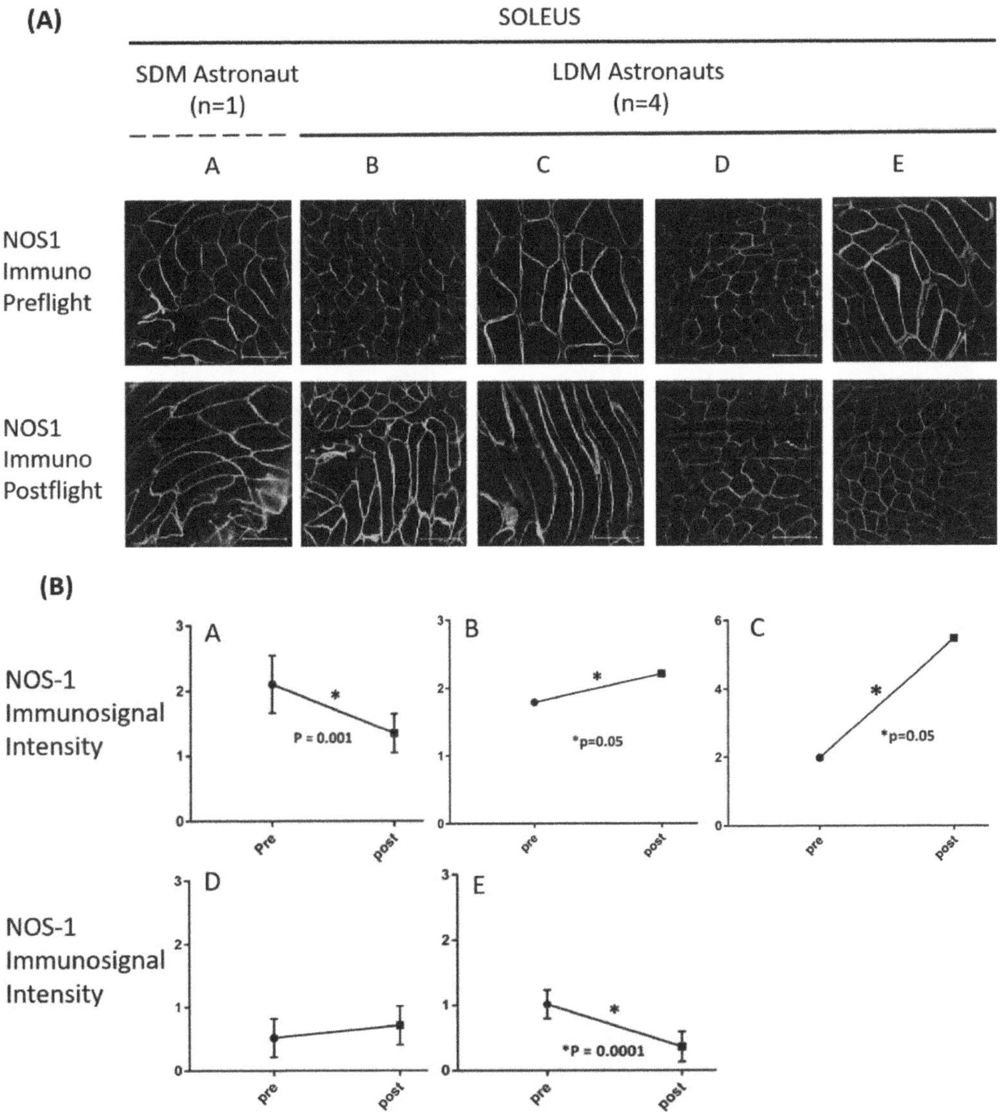

Figure 2. Nitric oxide synthase type-1 (NOS1) subsarcolemma immunoexpression (green fluorescence) in soleus muscles before (pre-flight) and after (post-flight) spaceflight. (**A**) SDM astronaut A is considered as a "non-exercise" control compared to the four LDM astronauts (B to E) with routine inflight exercise as a countermeasure. (**B**) NOS1 immunosignal intensity semi-quantification by three-dimensional (3D) confocal image analysis (mean pixel intensity per NOS1 positive volume/image size, in μm^3) in astronauts (y-axis: arbitrary units, a.u.) pre-flight (pre) vs. post-flight (post). Bar (images) = 150 µm. * *p*-values (graphs) ranged from significance level $p = 0.001$ to 0.0001.

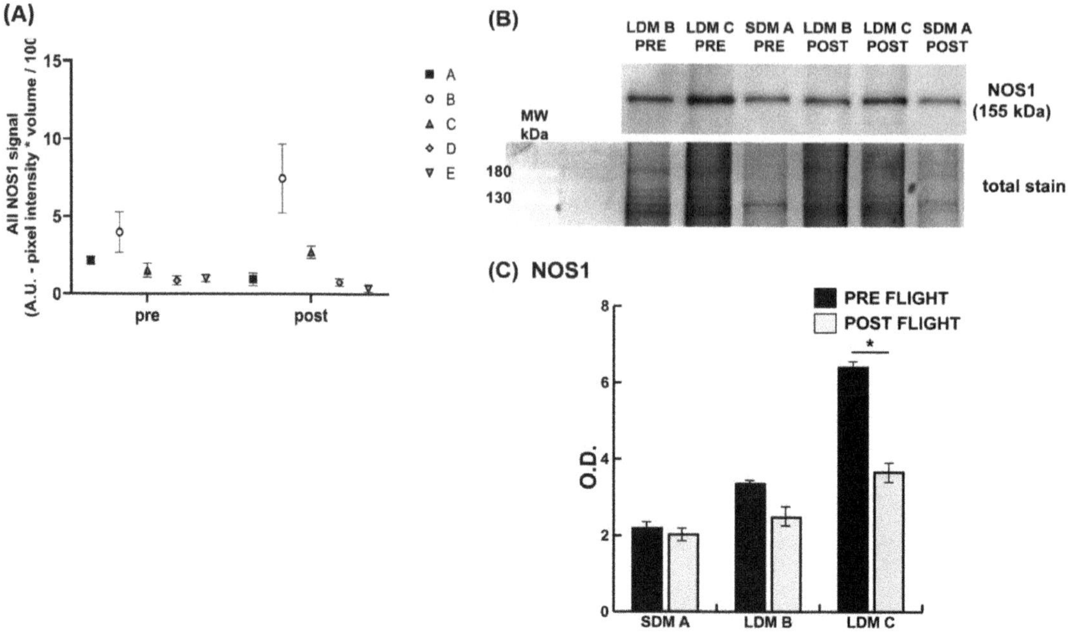

Figure 3. (**A**) Graph (scatter plot) showing sarcolemmal anti-NOS1 immunosignal pixel changes (pre- vs. postflight) in the soleus muscle (left panel) of LDM astronauts B, C, D, and E (n = 4) vs. one SDM astronaut (A, n = 1) as reference control. Color codes of astronaut values (as presented by square, circle, triangles, diamond symbols) highlight inter-subject variability (see astronaut B and E). (**B**) Representative NOS1 immunoblot images from SDM-A, LDM B, and LDM-C. (**C**) Representative bar graph (means ± SD) showing quantitative pre-flight (black bars) vs. post-flight (grey bars) NOS1 protein change in soleus lysates. Data were normalized against the total amount of loaded proteins stained with Sypro Ruby. O.D. = optical density; * significant difference; Student's t-test, $n = 2$, $p < 0.05$. Full-length images are available in the Supplementary Materials. Astronaut D and E immunoblot values not available (tissue limitation).

3.3. *Nitro-DIGE Analysis of Astronaut Soleus Biosamples*

We then assessed the abundance of proteoforms provided by the Nitro-DIGE technology, which generates typical "nitro-profiles" from SNO proteins based on a modified biotin switch method [40] in which the labile SNO group is selectively reduced with ascorbate and labelled with CyDyes. Analyses were performed on pre- and post-flight soleus biopsy samples from a single non-exercise SDM astronaut (n = 1) and four LDM astronauts (n = 4) performing daily routine countermeasure protocols during their stay aboard the ISS. Changes in nitrosylation levels were assessed for each protein by normalizing the Nitro-DIGE level over the label-free LC-MS/MS abundance dataset previously published based on the same subjects [39]. Proteins showing both decreased abundance and nitrosylation or vice versa, after microgravity exposure, were not considered, as we would like to identify proteins that are over- or under-nitrosylated compared to under- or over-expressed total protein. A total of 1822 proteins were identified from SDM and LDM astronaut soleus muscle in the label-free analysis, whereas 111 spots were found nitrosylated in Nitro-DIGE, among which 84 were present in the label-free dataset (Figure 4). Of the 84 nitrosylated proteins identified, in SDM exposure, eight showed statistically significant differential expression in both Nitro-DIGE and label-free data, two were significantly over-nitrosylated in Nitro-DIGE only, and 18 showed differential abundance in label-free data, although nitrosylation changes were not statistically significant. In LDM exposure, only one protein was differentially nitrosylated before and after exposure in Nitro-DIGE and label-free data, whereas 11 were not differentially nitrosylated but were differentially abundant (Figure 4). The list of proteins differentially expressed in SDM and LDM, together with statistical analyses, protein accession number, and gene name, is available in Supplementary Tables S1 (SDM) and S2 (LDM). A representative "nitroprofile" of soleus muscle protein extracts from SDM and LDM obtained by Nitro-DIGE is available in Supplementary Figure S1.

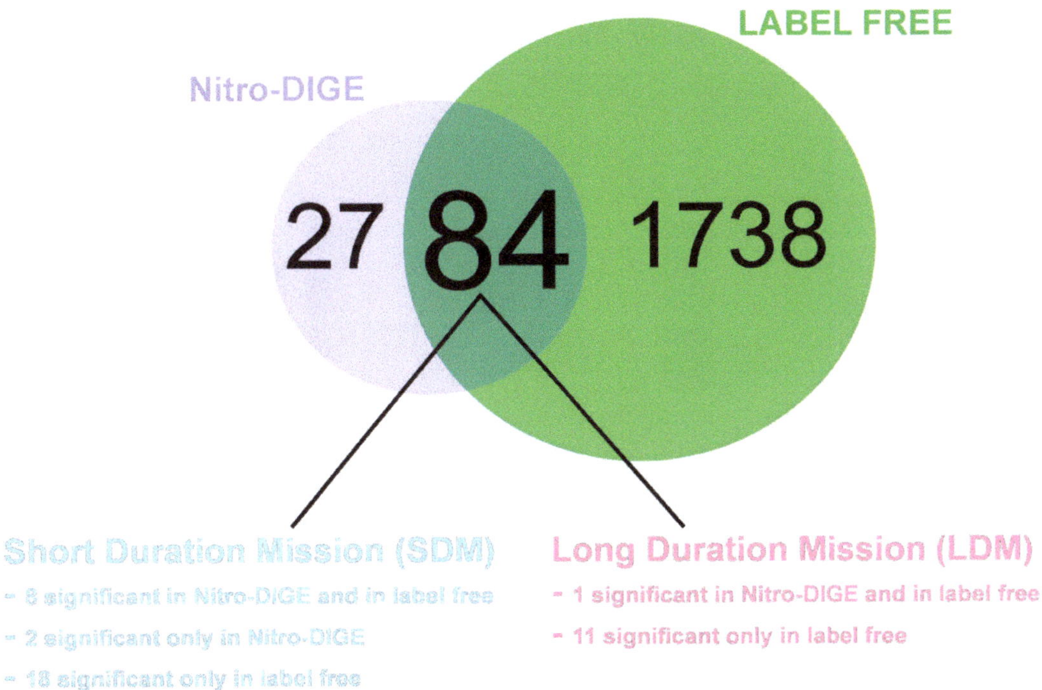

Figure 4. Venn diagram comparing the number of identified global proteins with posttranslational SNO modifications (total 111 proteins, Nitro-DIGE, violet circle) with total protein identified in label-free LC-MS/MS analysis (total 1822 proteins, LABEL FREE, green circle). A total of 84 were present both in Nitro-DIGE and label-free data (see overlapping area of circles), of which 28 or 12 were over-/under-nitrosylated in SDM or LDM. See also Supplementary Materials.

Figure 5 shows the correlation between level of nitrosylation and protein abundance in the SDM astronaut, and shows a 'statistically significant or only trend' increase in the level of nitrosylation despite a significant decrease in the total protein abundance of several proteoforms involved in energy metabolism, transport, stress response, and muscle contraction. Specifically, in the glycolytic pathway, fructose-bisphosphate aldolase A proteoforms (ALDOA a, b) were over-nitrosylated, while in the TCA cycle, OXPHOS, and malate/aspartate shuttle, increased nitrosylation was observed in aconitate hydratase proteoforms (ACO2 a, b, c), cytoplasmic aspartate aminotransferase proteoforms (GOT1 a, b), succinate dehydrogenase [ubiquinone] flavoprotein subunit proteoforms (SDHA a, b, c), 2-oxoglutarate dehydrogenase complex component E2 proteoforms (OGDH a, b), cytochrome b-c1 complex subunit 1 (UQCRC1), cytoplasmic malate dehydrogenase (MDH1), and creatine kinase M-type proteoforms (CKM a, b, c, d, e). Among the contractile proteins, myosin light chain 3 (MYL3) and slow skeletal muscle troponin T (TNNT1) were over-nitrosylated as serum transferrin proteoforms (TF a, b, c, d, e) and carbonic anhydrase 3 proteoforms (CA b, c).

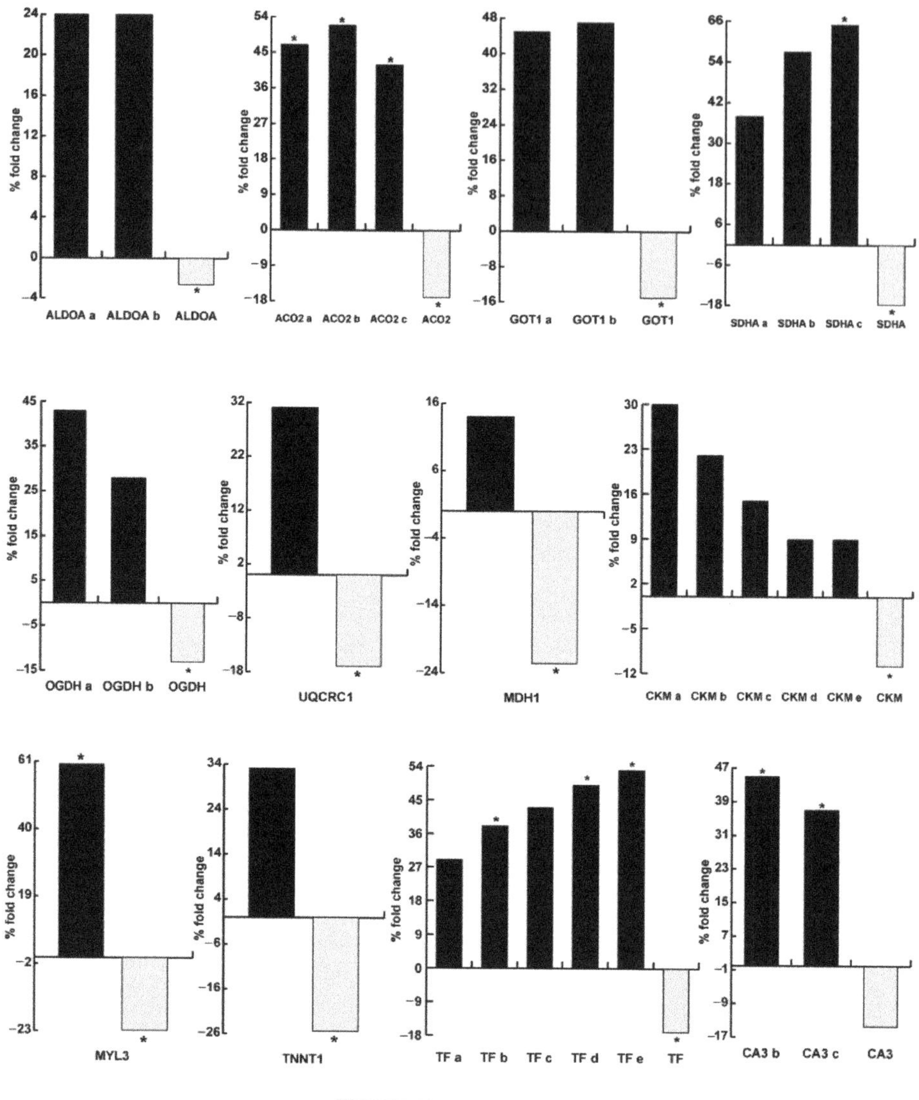

Figure 5. SDM Astronaut soleus nitrosoprofiles (s-nitrosylation) and abundance (total amount) of metabolic (glycolytic/TCA cycle), contractile muscle proteins (myosin light chain/troponins), and ion transporter serrotransferrin (TF). Total protein abundance (gray bars, label-free LC-MS/MS) and S-nitrosated (black bars, Nitro-DIGE) proteoforms (% fold change). Fructose-bisphosphate aldolase A (ALDOA), aconitate hydratase (ACO2), cytoplasmic aspartate aminotransferase (GOT1), succinate dehydrogenase [ubiquinone] flavoprotein subunit (SDHA), 2-oxoglutarate dehydrogenase complex component E2 (OGDH), cytochrome b-c1 complex subunit 1 (UQCRC1), cytoplasmic malate dehydrogenase (MDH1), creatine kinase M-type (CKM), myosin light chain 3 (MYL3), slow skeletal muscle troponin T (TNNT1), serotransferrin (TF), and carbonic anhydrase 3 (CA3) in SDM post vs. pre. (* = significant difference; POST vs. PRE SDM, Student's t-test and FDR, $n = 1$).

Figure 6 presents the relation between level of nitrosylation and protein abundance in LDM astronauts. A significant decrease in nitrosylation despite an increase in protein abundance was observed for UQCRC1. At variance, for isocitrate dehydrogenase [NADP] (IDH2), proteoforms of OGDH, mitochondrial aspartate aminotransferase proteoforms (GOT2 a, b), glutamine amidotransferase-like class 1 domain-containing protein 3 (GATD3), proteoforms of CA3, and MYL3, the total abundance was decreased, while levels of nitrosylation were increased, although not significantly changed.

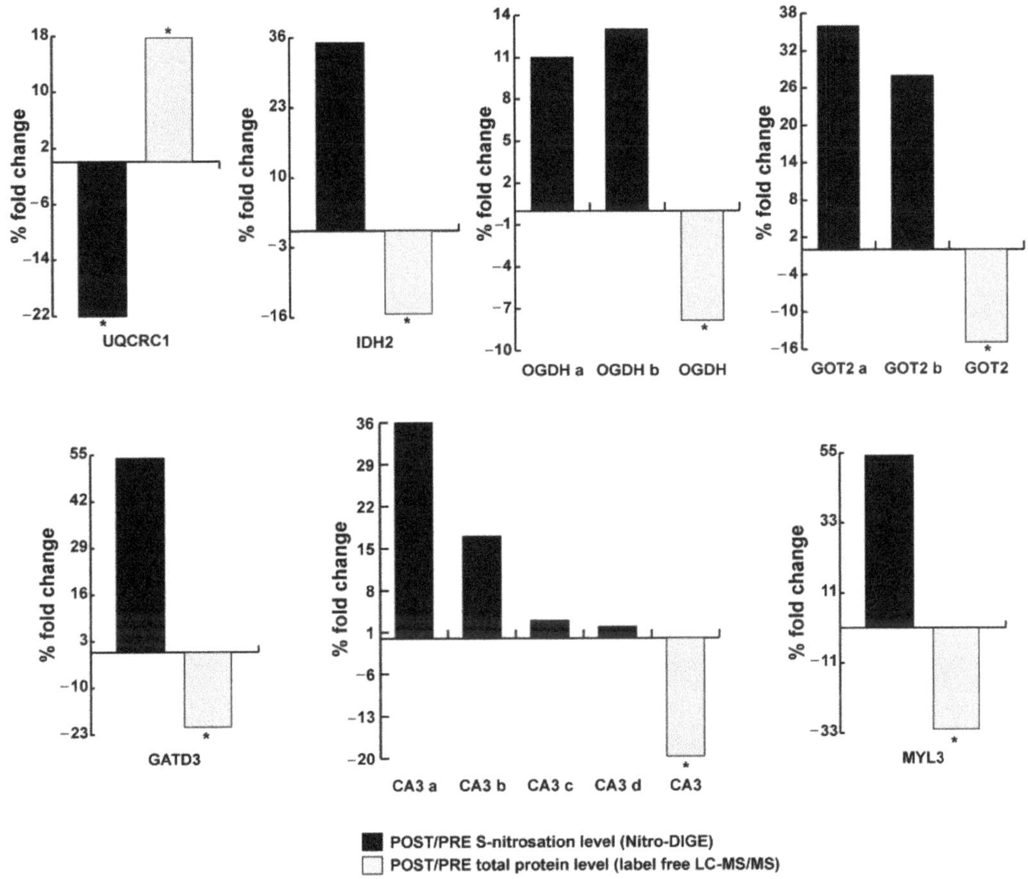

Figure 6. LDM astronaut soleus nitrosoprofiles (s-nitrosylation) and abundance (total amount) of mitochondria-related (respiratory chain) and structural (MYL3) proteins. Total protein abundance (gray bars, label-free LC-MS/MS) and S-nitrosated (black bars, Nitro-DIGE) level variations (% fold change) of cytochrome b-c1 complex subunit 1 (UQCRC1), isocitrate dehydrogenase [NADP] (IDH2), 2-oxoglutarate dehydrogenase complex component E2 (OGDH), mitochondrial aspartate aminotransferase (GOT2), glutamine amidotransferase-like class 1 domain-containing protein 3 (GATD3), carbonic anhydrase 3 (CA3), and myosin light chain 3 (MYL3) in LDM post vs. pre. (* = significant difference; POST vs. PRE LDM, paired Student's *t*-test and FDR, $n = 4$).

Figure 7 depicts the levels of molecules known to control nitrosative stress, such as alcohol dehydrogenase 5/S-nitrosoglutathione reductase (ADH5/GSNOR) and thioredoxin reductase 1 (TXNRD1), as assessed by immunoblotting in one SDM and two LDM astronauts. Results indicate that both ADH5 and TXNRD1 levels were decreased in all subjects, but only ADH5 in LDM (Figure 7B) was significant.

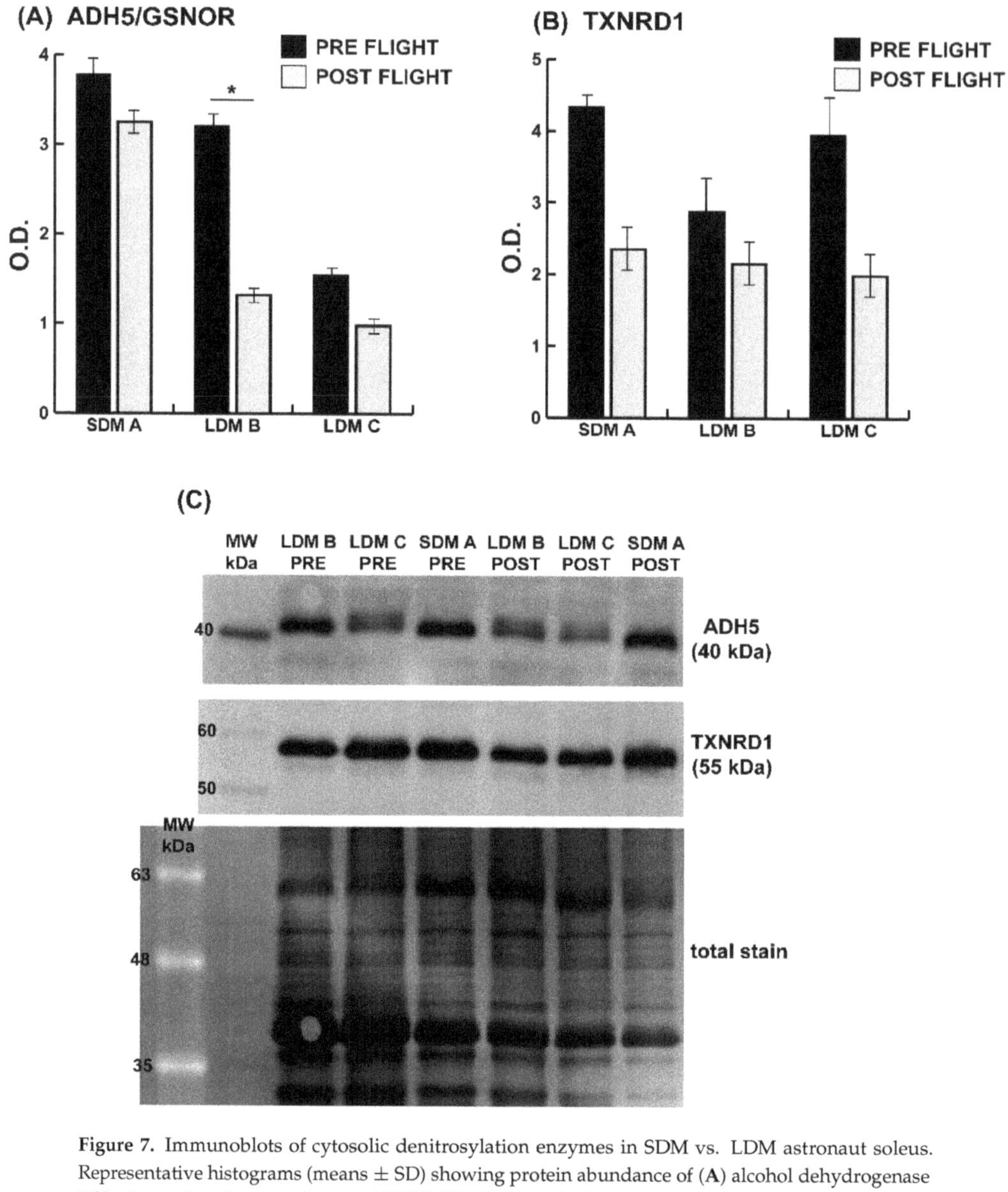

Figure 7. Immunoblots of cytosolic denitrosylation enzymes in SDM vs. LDM astronaut soleus. Representative histograms (means ± SD) showing protein abundance of (**A**) alcohol dehydrogenase 5/S-nitrosoglutathione reductase (ADH5/GSNOR) and (**B**) thioredoxin reductase 1 (TXNRD1) in preflight (black bars) vs. postflight (grey bars) in one (SDM A) astronaut, and two (LDM-B/-C) astronauts. (**C**) Representative immunoblot images from pre- vs. postflight SDM-A and LDM-B/-C. Data were normalized against the total amount of loaded proteins stained with Sypro Ruby. O.D. = optical density; * significant difference; Student's *t*-test, $n = 2$, $p < 0.05$. Full-length images are available in the Supplementary Materials.

4. Discussion

The present work reports that in contrast to the SDM condition in spaceflight (acute microgravity impact), LDM astronauts (chronic µG impact) showed that (i) sarcolemmal NOS-type 1 dislocation in soleus myofibers was not evident shortly after return from spaceflight, and (ii) a restricted set of functional muscle proteins (structural, metabolic, mitochondrial) were over-nitrosylated (s-nitrosylated vs. total protein abundance) after spaceflight compared to pre-flight controls. This work presents a novel dataset of s-nitrosylated skeletal muscle proteins of astronaut bio-samples (muscle biopsy). It aimed to gain insight into organ-based oxidative homeostasis in spaceflight [43] and to complement interdisciplinary space-related space-omics research databases on various space-flown biological organisms [44–46], development of countermeasures in personalized space medicine [47], and recent human space-omics reports [48,49]. Sarcolemmal NOS1 expression and protein s-nitrosylation in skeletal muscle are further keys to better understanding the molecular networks associated to the acute or chronic nitrosative stress management to defend skeletal muscle dysfunctions and impaired performance control in human spaceflight.

Altered sarcolemmal NOS1 expression is an oxidative stress-dependent early event and valuable cell marker linked not only to muscle activity/inactivity but also to various muscle pathologies [50]. Functionally distinct neuronal NOS (nNOS/NOS1) and its known splice variants (nNOSμ, nNOSß) are localized in muscle fibers to the sarcolemma (nNOS via PDZ domains of syntrophin to dystrophin glycoprotein complex), cis-Golgi complex (nNOSß), and cytoplasm (Balke et al., 2019), thus reflecting a complex interplay in various subsystems via NO/NOS signaling in normal muscle functions [51]. A fine regulation of myofiber function and homeostasis is apparently lost when nNOS (NOS1) dislocates from the sarcolemma to the cyto(sarco)solic compartment following disuse conditions [50]. Similar immuno-patterns of NOS1 were found in the apparently hypoactive or even passive soleus muscles of the SDM astronaut in the absence of a prescribed exercise training protocol because of the relatively short duration of space travel (1.5 weeks SDM vs. 24 weeks LDM) combined with a minimal risk for very low to moderate disuse atrophy with little impact on muscle health condition during short spaceflight and recovery thereafter. Increased s-nitrosylation of cytosolic, contractile proteins, and ion transporters was detectable in the SDM astronaut (Figure 5) despite similar NOS1 abundance in SDM pre- vs. post-flight (Figure 3B) measures, suggesting an interplay between alternative myofiber NO sources (NOS1, inducible NOS2) expressed in human skeletal muscle [38].

Reduced subsarcolemmal NOS1 is also an early event driving disuse-induced muscle atrophy in rats and short-term bed rest [50], as well as in long-term bed rest [38] and in space-flown mice [37], and thus is considered a "non-exercise" reference to the NOS1 immuno-expression levels found in our study of LDM astronauts (B to E) who performed up to 2.5 h of regular daily inflight exercise sessions during their 6-month missions onboard the ISS. All four LDM astronauts studied in this work showed preservation of sarcolemmal NOS1 immuno-reactivity around their soleus muscle myofibers post- vs. pre-flight, which is plausibly due to their routine inflight countermeasure prescriptions that included, in particular, daily inflight exercise with the advanced resistive exercise device (aRED) [52] as an integral part of the routine inflight countermeasure portfolio throughout their six-month missions onboard the ISS [53].

In the SDM astronaut, we found nitrosylation of (1) a nodal protein of the glycolytic flux, (2) specific proteins in TCA cycle and respiratory chain, (3) over-nitrosylation of proteins involved in muscle contraction, and (4) CK together with decrease in antioxidant protection as well as nitrosylation of TF that can promote ferroptosis (intracellular ion-dependent cell death [54]) as signs of microgravity muscle maladaptation.

In LDM astronauts, nitroso-profiles of some muscle proteins reflected a recovery of the glycolytic (ALDO) and energy flux (CKM), and of several enzymes of the TCA cycle (ACO2 and MDH1), while a recovery of mitochondrial activity was observed with no differential change in protein abundance and undetectable nitrosylation for SDHA. However, a significant increase in protein abundance was observed for UQCRC1 coupled with a decrease in nitrosylation. A decreased protein level with a tendency to retain an over-nitrosylated state was observed for mitochondrial GATD3, IDH2, GOT2, and OGDH. The same occurred with MYL3, whereas TNNT1 and TF were normalized as signs of muscle adaptation to microgravity. Notably, independent of the mission duration in space, levels of enzymes mediating denitrosylation of ADH5/GSNOR and TXNRD1 [41] were detectable as unmistakable signs of impaired redox homeostasis and reduced protection from nitrosative stress in muscle. Immunoblotting indicated decreased levels of NOS1 in muscle tissue that likely reflected a protective mechanism. Over-nitrosylation of enzymes in the TCA cycle as well as proteins in the electron transport chain can hinder the efficiency of oxidative phosphorylation and subsequently ATP synthesis [55]. We can hypothesize that over-nitrosylation of contractile proteins may reflect a protective mechanism to preserve muscle contraction homeostasis, reducing Ca^{2+} sensitivity and ATPase activity [56,57].

The use of Nitro-DIGE technology represents a reliable methodological tool that allowed us to identify variable expression patterns (loss vs. gain) detectable by individual nitrosoprofiles of functional proteins (structural/contractile, energy metabolism, transport, stress response, mitochondria) as SNO signatures from astronauts (short vs. long duration). Functional protein nitroso-profiles included members of primary contractile proteins, such as myosin light chains (MYL-3), Ca^{2+}-binding proteins, which modulate force transduction and cross-bridge kinetics, which were found over-nitrosylated in both SDM and LDM subjects (Supplementary Tables S1 and S2). Essential fast myosin light chains (MYL-3/MLC-3) belong to key contractile proteins like Myosin heavy chains (MyHCs), troponins (TnTs 1-3))/tropomyosins (TPM1-3), and α-actins (ACTA1) from skeletal muscle and are considered hallmarks of fiber type transition in the muscle aging process (sarcopenia) that leads to atrophy and frailty [58]. Earlier work on muscle contractile function with exercise also showed loss of peak forces in skinned myofibers from a slow soleus muscle after 17 d spaceflight [59] and increased MLC-3 subunit in soleus and extensor digitorum longus (EDL) muscles. More interestingly, protein content in MYL-3 was up-regulated in competing runners vs. recreational runners, suggesting MLC compositional changes in extreme loading conditions (endurance training) [59]. MYL-3 decreased in aged muscle, suggesting weakness in aging population [60]. In this respect, microgravity adaptation of inactive vs. active astronaut soleus muscle is linked to altered SNO patterns of contractile muscle proteins, and presumably also linked to myofiber type shifts (slow to fast).

We also found elevated SNO levels of carbonic anhydrases (CA), CA3b and CA3c, in both LDM and SDM astronauts. Cytosolic muscle proteins like CAs mediate reversible hydration of carbon dioxide. Among others, CAs are frequently found in the human proteome, including muscle [61]. Moreover, CA3 is enriched in slow fibers and an ideal marker for fiber shift in muscle adaptation [62], aging [63], and also in disused vastus lateralis muscles in bed rest, a human immobilization study as an analog to extended spaceflight [42]. These findings show that altered nitrosylation products are to be investigated in various cellular subsystems following skeletal muscle adaptation linked to various atrophy conditions including disuse, aging, and disease, but also spaceflight, with as-yet un-estimated consequences for dysfunctional human skeletal muscle on Earth and in space.

Aberrant s-nitrosylation is a hallmark of muscle wasting and neuromuscular dysfunction of human skeletal muscle [64]. Low levels of ADH/GSNOR are detected in muscular dystrophies, muscle atrophy, aging, amyotrophic lateral sclerosis (ALS), and neurodegenerative diseases, thus suggesting a key role in overall muscle homeostasis. Moreover, NOS1/GSNOR contributes to muscle differentiation and homeostasis [41]. We were able to identify decreased levels of the denitrosylation enzymes ADH5/GSNOR [41] and of the cytosolic selenoprotein TRXRD1, both linked to the transcription factor Nrf2, a potent regulator essential to the maintenance of redox homeostasis [65]. The level of TRXRD1 showed a trend in both SDM (non-exercise) and LDM (inflight exercise) conditions, and ADH5/GSNOR was significantly decreased in the LDM-B astronaut post- vs. pre-flight, suggesting a risk of insufficient mitigation effects of currently available inflight exercise training protocols, particularly regarding those affecting redox homeostasis and RONS-linked skeletal muscle properties in human muscle on Earth [9] and concerning microgravity-induced muscle hypoactivity/disuse in space.

NO is principally produced by NOS1 in contracting muscle fibers by the NADPH oxidase enzymes, NOX2 (plasma membrane) and NOX4 (intracellular membranes), but also by mitochondria [66,67]. Mitochondrial signaling also contributes to disuse muscle atrophy [68], and is common to heart and skeletal muscle pathologies, including sarcopenia [69]. We found robust recovery signs of mitochondrial activity in LDM astronaut bio-samples. Mitochondrial homeostasis and redox status appear to be highly challenged in spaceflight, as previously analyzed in hair-samples from LDM astronauts (collected pre-, in-, and post-flight) showing mitochondrial damage (reduced mtDNA/nDNA) together with other redox signal genes involved in oxidative stress observed post-flight [7].

In the skeletal muscle of astronauts, we found mitochondrial aconitate hydratase enzyme isoforms (i.e., ACO2a-c) as part of the TCA cycle highly nitrosylated in the SDM astronaut, while other mitochondrial enzymes, such as aspartate aminotransferases (GOT2), but also isocitrate dehydrogenase (NADP) (IDH2), were nitrosylated following LDM spaceflight, suggesting mitochondrial enzymes are likely targets for various mitochondrial functions related to spaceflight impact [70].

Previous other work with healthy subjects found elevated citrate as a potential stress response to preserve redox status of the cell during intense exercise [71] and aconitase post-translational modification as hallmarks in redox signaling and metabolism of ROS [72], further suggesting nitrosylation of mitochondrial proteins as another post-translational modification in skeletal muscle metabolism during spaceflight. Aconitase (ACOs) and oxoglutarate dehydrogenase (OGDH) inhibition via NO in rat soleus skeletal muscle strips in vitro suggests an impact of NO on catalytic enzyme functions in skeletal metabolism [73]. Mammalian GATD3 is a mitochondrial matrix enzyme with bacterial provenance that functions as a deglycase and interacts with the mitochondrial translation machinery to stabilize mitochondrial integrity involved in mitochondrial glycation status and thus dynamics [74]. It has been shown to be highly nitrosylated (54%-fold change) in LDM samples. Data from SDHA proteomic changes during bed rest [75,76], as well as conventional SDH histochemistry on muscle cross-sections after 11 days of spaceflight [77], support our results regarding the possible reduced metabolic support via mitochondria under the non-exercise SDM condition studied in this work.

Skeletal muscle function is linked to a well-balanced and coordinated redox production and detoxification system for physiological function [78]. Oxidative stresses are closely linked to mechanical loading stress for maximal force generation [27,79]. Hyper-nitrosylated RyR calcium release channels are leaky in dystrophic muscle [80]. Redox balance (cellular redox state) fine-tunes muscle hypertrophy and NO exerts a negative effect on Akt/mTORc1 function (hypertrophy), possibly to keep hypertrophy in check [81].

Nitrates and nitrites can serve as food sources for health benefit [82]. NO synthesis in tissue can occur through reduction of nitrate to nitrite, and nitrite can be subsequently reduced to NO as a systemic reservoir for NO production in muscle; thus, a physiological benefit with a provision of nitrite from dietary sources appears likely. However apart from physiological roles, nitrate and nitrite aspects of toxicity related to excess uptake on circulation and muscle remain to be properly assessed [83], in particular with respect to space-related nutritional countermeasures.

Increased inflight oxidative stress has previously been suggested with regard to human spaceflight [84]. To date, antioxidant/anti-inflammatory cocktails have been tested against disuse-induced muscle deconditioning in bed rest [12,42,85] based on nutritional genomics and nutrigenetics to minimize ROS damaging effects during manned spaceflight; such cocktails are currently being assessed for future space travelers and astronauts (in particular, those involving Vitamins A, C, E, and D, selenium, resveratrol, and others) [11]. We previously investigated muscular SNO protein signatures by resistive and vibration exercise in bed rest as an analogue to spaceflight [11], the results of which suggest a high potential for exercise to offset nitrosative stress and damage in chronic disuse conditions on Earth. Exercise-induced changes (endurance, high intensity, resistive) [8] in skeletal muscle antioxidant enzymes have been reported. Likewise, a therapeutic potential of denitrosylation to defend against nitrosative stress in human disease was also recently suggested [9]. For mission success and safety of future spaceflight, disrupted redox homeostasis in LDM astronauts' skeletal muscle, as shown in this work, calls for the development of multimodal inflight countermeasures, with exercise and nutritional supplementation, for improved nitrosative stress management in order to minimize disrupted redox homeostasis of skeletal muscle.

Study limitations exist, including (1) the small number of participants of the study (which is an inherent limitation factor for all space-related research); (2) limited numbers of biopsies, highlighted by the fact that only one ($n = 1$) "non-exercise" SDM crewmember (9 days on ISS) was available for this study, which limits statistical strength; and (3) the fact that regular and rigorous inflight CM exercise protocols (up to 2.5 h daily per week) have already prescribed for LDM astronauts onboard the ISS for many years by the relevant Space Agencies [86,87]. Due to ethical reasons, biopsies from "non-exercise" LDM crew members on ISS (inflight negative controls) or inflight biopsies were not available, for obvious reasons. Nevertheless, small muscle biopsy samples obtained from a unique population of astronauts help to address proof of concept/hypotheses, for instance, with hypotheses centered on microgravity-induced molecular adaptation processes (reflected, for example, by canonical and other muscle-specific biological pathways otherwise being masked by gravity on Earth [39] and with potential benefit for better health management of people on Earth).

5. Conclusions

Aberrant levels of SNO-modified proteins of different subsystems investigated in the slow-type soleus muscle are potential triggers for muscle dysfunctions known from health risk assessments in human spaceflight. The present work showed that sarcolemmal NOS1 dislocation can be at least partly mitigated in actively contracting myofibers of the postural deep calf soleus muscle, suggesting efficacy of the currently available inflight exercise protocol onboard the ISS. However, over-nitrosylation of key functional muscle proteins (structural, metabolic, mitochondria complex) and decreased levels of the two major denitrosylase enzymes, TXNRD1 and GSNOR, are hallmarks of impaired redox homeostasis in astronaut skeletal muscle, independent of the time of microgravity exposure (SDM acute vs. LDM chronic). The complex patterns of under-/over-nitrosylation of specific muscle protein still remain to be analyzed for other muscle groups in order to establish whether peculiar SNO signatures reflect a protective or detrimental mechanism in microgravity adaptation of the astronauts' musculature. This should, however, be considered for the development of next-generation inflight countermeasures targeted to alleviate imbalance in oxidative homeostasis, evidenced by aberrant muscular nitrosoprofiles reported by the present work. Future inflight countermeasures, but also recovery protocols thereafter, should also include nutritional supplementation, for example, by selenoproteins or other antioxidant nutrients to minimize muscular nitrosative stress with impact on mission crew health and performance control in future human deep space exploration missions.

Supplementary Materials: The following supporting information can be downloaded at: https://www.mdpi.com/article/10.3390/antiox13040432/s1, Figure S1: Nitro-DIGE gel images of SOL muscle protein extracts from one short-duration (9 days on ISS, without inflight exercise; acute μG exposure) and four long-duration (>6 month or more on ISS, with routine inflight exercise; chronic μG exposure) mission astronauts (SDM, left panel; LDM, right panel); Table S1: List of identified proteins in Short Duration Mission (SDM) group by NITRO-DIGE and label-free LC-MS/MS analysis; Table S2: List of identified proteins in Long Duration Mission (LDM) group by NITRO-DIGE and label-free LC-MS/MS analysis.

Author Contributions: Conceptualization, D.B., C.G. and M.S.; Muscle biopsies: J.R. and U.L.; Methodology, Data curation, Visualization, Investigation: M.M., M.G., G.T., K.B., S.F., P.V., P.B., D.C. and E.T.; Project administration: M.S. and D.B.; Funding acquisition, D.B., C.G. and M.S.; Writing—review and editing: D.B., C.G. and M.S. All authors have read and agreed to the published version of the manuscript.

Funding: German Federal Ministery of Economy and Energy (BMWi) through the German Aerospace Agency, Deutsches Zentrum für Luft und Raumfahrt e.V. (DLR), Bonn-Oberkassel, Germany (grant# 50WB0521-0821-1121-1421 to D.B.; 50WB2116 to M.S., 50WB2119 to U.L.), Agenzia Spaziale Italiana (ASI) (grant no. 2021-21-U.O ASTRONEMUS to C.G.), and intramural funding from the authors' institutions. We also acknowledge the Open Access Publication Fund of Charité—Universitätsmedizin Berlin.

Institutional Review Board Statement: The study was conducted according to the Declaration of Helsinki (1966), Multilateral Human Research Board (HRMRB) document, including Subject Information handout and Laymen's summary, provided and approved by Institutional Review Boards from National Aeronautics and Space Administration, NASA, (Pro0592, FWA No. 00019876 from 23 June 2015), the Japanese Space Agency, JAXA (Full IRB Approval 2013–2016 from 10 January 2016), the European Space Agency, ESA Medical Board (ESA MB #2022-04-06), and Local Institutional Ethical Board at the Charité—Universitätsmedizin Berlin, Germany (verdict EA4/057/08 from 18 June 2012), before study inclusion.

Informed Consent Statement: Informed consent was obtained from all subjects involved in the study. Participants were allowed to withdraw from the study at any time.

Data Availability Statement: All human datasets analyzed during the current study are not publicly available due to privacy reasons but are available from the corresponding author upon reasonable request.

Acknowledgments: Without the unique passion and commitment of the participating astronauts to science, this work would not have been possible. This study was supported by the European Space Agency (ESA), with the Project Scientists Leonardo Surdo and Giuseppe Correale, and National Aeronautics and Space Administration (NASA). Expert operational support on the ground came from Robert MacGregor, Airbus DS Space Systems Inc., at Johnson's Space Center (JSC) at NASA, Houston, TX, USA, Mark Mouret from the European Astronaut Center (EAC), Cologne, Germany, and by the Mission Crew Surgeons (on site medical supervision) of the Space Agencies involved during the pre- and postflight BDC sessions, which was instrumental to the study. Gudrun Schiffl and Charité Berlin helped with technical laboratory work. Michael Depew and Charité Berlin helped with language editing following manuscript acceptance.

Conflicts of Interest: The authors declare no conflicts of interest. The funders had no role in the design of the study, in the collection, analyses, or interpretation of data, in the writing of the manuscript, or in the decision to publish the results.

Abbreviations

ADH5	alcohol dehydrogenase 5
ALDOA	Fructose-bisphosphate aldolase A
ACO2	aconitate hydratase proteoforms (a, b or c)
ARED	advanced resistive exercise device (on ISS)
BDC	Baseline data collection (on the ground, pre/postflight)
CA3	carbonic anhydrase 3
CEVIS	cycle ergometer with vibration isolation and stabilization (on ISS)
CKM	creatine kinase M-type
CM	countermeasure (physical exercise intervention)
DLR	Deutsches Zentrum für Luft- und Raumfahrt (German Aerospace Agency)
EAC	European Astronaut Center (Cologne)
ESA	European Space Agency
FDR	False discovery rate (statistics)
G	gravity vector (1G Earth/μG Space)
GATD3	glutamine amidotransferase-like class 1 domain-containing protein 3
GOT1	cytoplasmic aspartate aminotransferase
GOT2	mitochondrial aspartate aminotransferase
GSNOR	S-nitrosoglutathione reductase
IDH2	isocitrate dehydrogenase [NADP]
IPG	immobilized pH gradient
ISS	International Space Station
JAXA	Japanese Space Agency

LDM	long-duration mission (>180 days)
MDS	Mice drawer system (habitat module on ISS)
MedOPs	Medical operational team
MDH1	cytoplasmic malate dehydrogenase
MYL3	myosin light chain 3
M.O.M.	mouse-on-mouse (immunocytochemical blocking agent)
NASA	National Aeronautics and Space Administration
Nitro-DIGE	S-Nitrosated proteins identified by 2-D CyDye-Maleimide
NO	nitric oxide
NOS	NO synthase
O.D.	optical density
OGDH	2-oxoglutarate dehydrogenase complex component E2
R+0/1	return day (landing day on Earth)
ROS	reactive oxidative species
RNS	reactive nitrogen species
RONS	reactive oxidative and nitrosative species
SDHA	succinate dehydrogenase [ubiquinone] flavoprotein subunit
SDM	short-duration mission (11 days spaceflight)
T2	treadmill on ISS
TF	serotransferrin
TNNT1	slow skeletal muscle troponin T
TXNRD1	thioredoxin reductase 1
UQCRC1	cytochrome b-c1 complex subunit 1
USOS	United States orbital system
µG	microgravity

References

1. Juhl, O.J.; Buettmann, E.G.; Friedman, M.A.; DeNapoli, R.C.; Hoppock, G.A.; Donahue, H.J. Update on the effects of microgravity on the musculoskeletal system. *NPJ Microgravity* **2021**, *7*, 28. [CrossRef] [PubMed]
2. Lang, T.; Van Loon, J.; Bloomfield, S.; Vico, L.; Chopard, A.; Rittweger, J.; Kyparos, A.; Blottner, D.; Vuori, I.; Gerzer, R.; et al. Towards human exploration of space: The THESEUS review series on muscle and bone research priorities. *NPJ Microgravity* **2017**, *3*, 8. [CrossRef] [PubMed]
3. Bergouignan, A.; Stein, T.P.; Habold, C.; Coxam, V.; O'Gorman, D.; Blanc, S. Towards human exploration of space: The THESEUS review series on nutrition and metabolism research priorities. *NPJ Microgravity* **2016**, *2*, 16029. [CrossRef] [PubMed]
4. Ackermann, M.; van den Bogert, A.J. Predictive simulation of gait at low gravity reveals skipping as the preferred locomotion strategy. *J. Biomech.* **2012**, *45*, 1293–1298. [CrossRef] [PubMed]
5. Antonsen, E.L.; Myers, J.G.; Boley, L.; Arellano, J.; Kerstman, E.; Kadwa, B.; Buckland, D.M.; Van Baalen, M. Estimating medical risk in human spaceflight. *NPJ Microgravity* **2022**, *8*, 8. [CrossRef] [PubMed]
6. Scott, J.P.R.; Weber, T.; Green, D.A. Introduction to the Frontiers Research Topic: Optimization of Exercise Countermeasures for Human Space Flight—Lessons from Terrestrial Physiology and Operational Considerations. *Front. Physiol.* **2019**, *10*, 173. [CrossRef]
7. Indo, H.P.; Majima, H.J.; Terada, M.; Suenaga, S.; Tomita, K.; Yamada, S.; Higashibata, A.; Ishioka, N.; Kanekura, T.; Nonaka, I.; et al. Changes in mitochondrial homeostasis and redox status in astronauts following long stays in space. *Sci. Rep.* **2016**, *6*, 39015. [CrossRef]
8. Powers, S.K.; Goldstein, E.; Schrager, M.; Ji, L.L. Exercise Training and Skeletal Muscle Antioxidant Enzymes: An Update. *Antioxidants* **2022**, *12*, 39. [CrossRef] [PubMed]
9. Yoon, S.; Eom, G.H.; Kang, G. Nitrosative Stress and Human Disease: Therapeutic Potential of Denitrosylation. *Int. J. Mol. Sci.* **2021**, *22*, 9794. [CrossRef]
10. Yfanti, C.; Nielsen, A.R.; Akerstrom, T.; Nielsen, S.; Rose, A.J.; Richter, E.A.; Lykkesfeldt, J.; Fischer, C.P.; Pedersen, B.K. Effect of antioxidant supplementation on insulin sensitivity in response to endurance exercise training. *Am. J. Physiol. Endocrinol. Metab.* **2011**, *300*, E761–E770. [CrossRef]
11. Salanova, M.; Schiffl, G.; Gutsmann, M.; Felsenberg, D.; Furlan, S.; Volpe, P.; Clarke, A.; Blottner, D. Nitrosative stress in human skeletal muscle attenuated by exercise countermeasure after chronic disuse. *Redox Biol.* **2013**, *1*, 514–526. [CrossRef] [PubMed]
12. Arc-Chagnaud, C.; Py, G.; Fovet, T.; Roumanille, R.; Demangel, R.; Pagano, A.F.; Delobel, P.; Blanc, S.; Jasmin, B.J.; Blottner, D.; et al. Evaluation of an Antioxidant and Anti-inflammatory Cocktail Against Human Hypoactivity-Induced Skeletal Muscle Deconditioning. *Front. Physiol.* **2020**, *11*, 71. [CrossRef] [PubMed]
13. Balon, T.W.; Yerneni, K.K. Redox regulation of skeletal muscle glucose transport. *Med. Sci. Sports Exerc.* **2001**, *33*, 382–385. [CrossRef]

14. Nemes, R.; Koltai, E.; Taylor, A.W.; Suzuki, K.; Gyori, F.; Radak, Z. Reactive Oxygen and Nitrogen Species Regulate Key Metabolic, Anabolic, and Catabolic Pathways in Skeletal Muscle. *Antioxidants* **2018**, *7*, 85. [CrossRef] [PubMed]
15. Powers, S.K.; Kavazis, A.N.; McClung, J.M. Oxidative stress and disuse muscle atrophy. *J. Appl. Physiol.* **2007**, *102*, 2389–2397. [CrossRef] [PubMed]
16. Leitner, L.M.; Wilson, R.J.; Yan, Z.; Godecke, A. Reactive Oxygen Species/Nitric Oxide Mediated Inter-Organ Communication in Skeletal Muscle Wasting Diseases. *Antioxid. Redox Signal.* **2017**, *26*, 700–717. [CrossRef] [PubMed]
17. Palomero, J.; Jackson, M.J. Redox regulation in skeletal muscle during contractile activity and aging. *J. Anim. Sci.* **2010**, *88*, 1307–1313. [CrossRef] [PubMed]
18. Bellinger, A.M.; Reiken, S.; Carlson, C.; Mongillo, M.; Liu, X.; Rothman, L.; Matecki, S.; Lacampagne, A.; Marks, A.R. Hypernitrosylated ryanodine receptor calcium release channels are leaky in dystrophic muscle. *Nat. Med.* **2009**, *15*, 325–330. [CrossRef] [PubMed]
19. Salanova, M.; Schiffl, G.; Rittweger, J.; Felsenberg, D.; Blottner, D. Ryanodine receptor type-1 (RyR1) expression and protein S-nitrosylation pattern in human soleus myofibres following bed rest and exercise countermeasure. *Histochem. Cell. Biol.* **2008**, *130*, 105–118. [CrossRef] [PubMed]
20. Bredt, D.S.; Snyder, S.H. Nitric oxide: A physiologic messenger molecule. *Annu. Rev. Biochem.* **1994**, *63*, 175–195. [CrossRef]
21. Kobzik, L.; Reid, M.B.; Bredt, D.S.; Stamler, J.S. Nitric oxide in skeletal muscle. *Nature* **1994**, *372*, 546–548. [CrossRef] [PubMed]
22. Jones, A.M.; Thompson, C.; Wylie, L.J.; Vanhatalo, A. Dietary Nitrate and Physical Performance. *Annu. Rev. Nutr.* **2018**, *38*, 303–328. [CrossRef] [PubMed]
23. Aguiar, A.F.; Vechetti-Junior, I.J.; Souza, R.W.; Piedade, W.P.; Pacagnelli, F.L.; Leopoldo, A.S.; Casonatto, J.; Dal-Pai-Silva, M. Nitric oxide synthase inhibition impairs muscle regrowth following immobilization. *Nitric Oxide* **2017**, *69*, 22–27. [CrossRef]
24. Christova, T.; Grozdanovic, Z.; Gossrau, R. Nitric oxide synthase (NOS) I during postnatal development in rat and mouse skeletal muscle. *Acta Histochem.* **1997**, *99*, 311–324. [CrossRef]
25. Lee, K.H.; Baek, M.Y.; Moon, K.Y.; Song, W.K.; Chung, C.H.; Ha, D.B.; Kang, M.S. Nitric oxide as a messenger molecule for myoblast fusion. *J. Biol. Chem.* **1994**, *269*, 14371–14374. [CrossRef]
26. Tidball, J.G.; Lavergne, E.; Lau, K.S.; Spencer, M.J.; Stull, J.T.; Wehling, M. Mechanical loading regulates NOS expression and activity in developing and adult skeletal muscle. *Am. J. Physiol.* **1998**, *275*, C260–C266. [CrossRef]
27. Godfrey, E.W.; Schwarte, R.C. The role of nitric oxide signaling in the formation of the neuromuscular junction. *J. Neurocytol.* **2003**, *32*, 591–602. [CrossRef]
28. Kasikcioglu, E.; Dinler, M.; Berker, E. Reduced tolerance of exercise in fibromyalgia may be a consequence of impaired microcirculation initiated by deficient action of nitric oxide. *Med. Hypotheses* **2006**, *66*, 950–952. [CrossRef]
29. Vaughn, M.W.; Kuo, L.; Liao, J.C. Effective diffusion distance of nitric oxide in the microcirculation. *Am. J. Physiol.* **1998**, *274*, H1705–H1714. [CrossRef]
30. Filippin, L.I.; Cuevas, M.J.; Lima, E.; Marroni, N.P.; Gonzalez-Gallego, J.; Xavier, R.M. The role of nitric oxide during healing of trauma to the skeletal muscle. *Inflamm. Res.* **2011**, *60*, 347–356. [CrossRef]
31. Anderson, J.E. A role for nitric oxide in muscle repair: Nitric oxide-mediated activation of muscle satellite cells. *Mol. Biol. Cell.* **2000**, *11*, 1859–1874. [CrossRef] [PubMed]
32. Buckwalter, J.B.; Curtis, V.C.; Valic, Z.; Ruble, S.B.; Clifford, P.S. Endogenous vascular remodeling in ischemic skeletal muscle: A role for nitric oxide. *J. Appl. Physiol.* **2003**, *94*, 935–940. [CrossRef]
33. Maxwell, A.J.; Schauble, E.; Bernstein, D.; Cooke, J.P. Limb blood flow during exercise is dependent on nitric oxide. *Circulation* **1998**, *98*, 369–374. [CrossRef]
34. Maddali, S.; Rodeo, S.A.; Barnes, R.; Warren, R.F.; Murrell, G.A. Postexercise increase in nitric oxide in football players with muscle cramps. *Am. J. Sports Med.* **1998**, *26*, 820–824. [CrossRef] [PubMed]
35. Swash, M. Nitric oxide and muscle weakness. *Neurology* **2011**, *76*, 940–941. [CrossRef]
36. Sandona, D.; Desaphy, J.F.; Camerino, G.M.; Bianchini, E.; Ciciliot, S.; Danieli-Betto, D.; Dobrowolny, G.; Furlan, S.; Germinario, E.; Goto, K.; et al. Adaptation of mouse skeletal muscle to long-term microgravity in the MDS mission. *PLoS ONE* **2012**, *7*, e33232. [CrossRef] [PubMed]
37. Rudnick, J.; Puttmann, B.; Tesch, P.A.; Alkner, B.; Schoser, B.G.; Salanova, M.; Kirsch, K.; Gunga, H.C.; Schiffl, G.; Luck, G.; et al. Differential expression of nitric oxide synthases (NOS 1-3) in human skeletal muscle following exercise countermeasure during 12 weeks of bed rest. *FASEB J.* **2004**, *18*, 1228–1230. [CrossRef]
38. Blottner, D.; Moriggi, M.; Trautmann, G.; Hastermann, M.; Capitanio, D.; Torretta, E.; Block, K.; Rittweger, J.; Limper, U.; Gelfi, C.; et al. Space Omics and Tissue Response in Astronaut Skeletal Muscle after Short and Long Duration Missions. *Int. J. Mol. Sci.* **2023**, *24*, 4095. [CrossRef]
39. Jaffrey, S.R.; Snyder, S.H. The biotin switch method for the detection of S-nitrosylated proteins. *Sci. STKE* **2001**, *86*, pl1. [CrossRef]
40. Montagna, C.; Rizza, S.; Cirotti, C.; Maiani, E.; Muscaritoli, M.; Musaro, A.; Carri, M.T.; Ferraro, E.; Cecconi, F.; Filomeni, G. nNOS/GSNOR interaction contributes to skeletal muscle differentiation and homeostasis. *Cell Death Dis.* **2019**, *10*, 354. [CrossRef]
41. Blottner, D.; Capitanio, D.; Trautmann, G.; Furlan, S.; Gambara, G.; Moriggi, M.; Block, K.; Barbacini, P.; Torretta, E.; Py, G.; et al. Nitrosative Redox Homeostasis and Antioxidant Response Defense in Disused Vastus lateralis Muscle in Long-Term Bedrest (Toulouse Cocktail Study). *Antioxidants* **2021**, *10*, 378. [CrossRef]

42. Goodwin, T.J.; Christofidou-Solomidou, M. Oxidative Stress and Space Biology: An Organ-Based Approach. *Int. J. Mol. Sci.* **2018**, *19*, 959. [CrossRef]
43. Berrios, D.C.; Galazka, J.; Grigorev, K.; Gebre, S.; Costes, S.V. NASA GeneLab: Interfaces for the exploration of space omics data *Nucleic Acids Res.* **2021**, *49*, D1515–D1522. [CrossRef]
44. Cope, H.; Willis, C.R.G.; MacKay, M.J.; Rutter, L.A.; Toh, L.S.; Williams, P.M.; Herranz, R.; Borg, J.; Bezdan, D.; Giacomello, S.; et al. Routine omics collection is a golden opportunity for European human research in space and analog environments. *Patterns* **2022**, *3*, 100550. [CrossRef] [PubMed]
45. Rutter, L.; Barker, R.; Bezdan, D.; Cope, H.; Costes, S.V.; Degoricija, L.; Fisch, K.M.; Gabitto, M.I.; Gebre, S.; Giacomello, S.; et al. A New Era for Space Life Science: International Standards for Space Omics Processing. *Patterns* **2020**, *1*, 100148. [CrossRef]
46. Schmidt, M.A.; Goodwin, T.J. Personalized medicine in human space flight: Using Omics based analyses to develop individualized countermeasures that enhance astronaut safety and performance. *Metabolomics* **2013**, *9*, 1134–1156. [CrossRef]
47. Afshinnekoo, E.; Scott, R.T.; MacKay, M.J.; Pariset, E.; Cekanaviciute, E.; Barker, R.; Gilroy, S.; Hassane, D.; Smith, S.M.; Zwart, S.R.; et al. Fundamental Biological Features of Spaceflight: Advancing the Field to Enable Deep-Space Exploration. *Cell* **2020**, *183*, 1162–1184. [CrossRef]
48. Schmidt, M.A.; Meydan, C.; Schmidt, C.M.; Afshinnekoo, E.; Mason, C.E. The NASA Twins Study: The Effect of One Year in Space on Long-Chain Fatty Acid Desaturases and Elongases. *Lifestyle Genom.* **2020**, *13*, 107–121. [CrossRef]
49. Lechado, I.T.A.; Vitadello, M.; Traini, L.; Namuduri, A.V.; Gastaldello, S.; Gorza, L. Sarcolemmal loss of active nNOS (Nos1) is an oxidative stress-dependent, early event driving disuse atrophy. *J. Pathol.* **2018**, *246*, 433–446. [CrossRef]
50. Balke, J.E.; Zhang, L.; Percival, J.M. Neuronal nitric oxide synthase (nNOS) splice variant function: Insights into nitric oxide signaling from skeletal muscle. *Nitric Oxide* **2019**, *82*, 35–47. [CrossRef]
51. Loehr, J.A.; Guilliams, M.E.; Petersen, N.; Hirsch, N.; Kawashima, S.; Ohshima, H. Physical Training for Long-Duration Spaceflight. *Aerosp. Med. Hum. Perform.* **2015**, *86*, A14–A23. [CrossRef]
52. Petersen, N.; Jaekel, P.; Rosenberger, A.; Weber, T.; Scott, J.; Castrucci, F.; Lambrecht, G.; Ploutz-Snyder, L.; Damann, V.; Kozlovskaya, I.; et al. Exercise in space: The European Space Agency approach to in-flight exercise countermeasures for long-duration missions on ISS. *Extrem Physiol. Med.* **2016**, *5*, 9. [CrossRef]
53. Li, J.; Cao, F.; Yin, H.L.; Huang, Z.J.; Lin, Z.T.; Mao, N.; Sun, B.; Wang, G. Ferroptosis: Past, present and future. *Cell Death Dis.* **2020**, *11*, 88. [CrossRef]
54. Khiati, S.; Bonneau, D.; Lenaers, G. Are Your Mitochondria Ready for a Space Odyssey? *Trends Endocrinol. Metab.* **2021**, *32*, 193–195. [CrossRef]
55. Figueiredo-Freitas, C.; Dulce, R.A.; Foster, M.W.; Liang, J.; Yamashita, A.M.; Lima-Rosa, F.L.; Thompson, J.W.; Moseley, M.A.; Hare, J.M.; Nogueira, L.; et al. S-Nitrosylation of Sarcomeric Proteins Depresses Myofilament Ca^{2+} Sensitivity in Intact Cardiomyocytes. *Antioxid. Redox Signal.* **2015**, *23*, 1017–1034. [CrossRef]
56. Horenberg, A.L.; Houghton, A.M.; Pandey, S.; Seshadri, V.; Guilford, W.H. S-nitrosylation of cytoskeletal proteins. *Cytoskeleton* **2019**, *76*, 243–253. [CrossRef]
57. Dowling, P.; Gargan, S.; Swandulla, D.; Ohlendieck, K. Fiber-Type Shifting in Sarcopenia of Old Age: Proteomic Profiling of the Contractile Apparatus of Skeletal Muscles. *Int. J. Mol. Sci.* **2023**, *24*, 2415. [CrossRef]
58. Widrick, J.J.; Knuth, S.T.; Norenberg, K.M.; Romatowski, J.G.; Bain, J.L.; Riley, D.A.; Karhanek, M.; Trappe, S.W.; Trappe, T.A.; Costill, D.L.; et al. Effect of a 17 day spaceflight on contractile properties of human soleus muscle fibres. *J. Physiol.* **1999**, *516 (Pt 3)*, 915–930. [CrossRef]
59. O'Connell, K.; Gannon, J.; Doran, P.; Ohlendieck, K. Proteomic profiling reveals a severely perturbed protein expression pattern in aged skeletal muscle. *Int. J. Mol. Med.* **2007**, *20*, 145–153. [CrossRef]
60. Jiang, L.; Wang, M.; Lin, S.; Jian, R.; Li, X.; Chan, J.; Dong, G.; Fang, H.; Robinson, A.E.; Consortium, G.T.; et al. A Quantitative Proteome Map of the Human Body. *Cell* **2020**, *183*, 269–283.e19. [CrossRef]
61. Dowling, P.; Gargan, S.; Zweyer, M.; Sabir, H.; Swandulla, D.; Ohlendieck, K. Proteomic profiling of carbonic anhydrase CA3 in skeletal muscle. *Expert Rev. Proteom.* **2021**, *18*, 1073–1086. [CrossRef]
62. Staunton, L.; Zweyer, M.; Swandulla, D.; Ohlendieck, K. Mass spectrometry-based proteomic analysis of middle-aged vs. aged vastus lateralis reveals increased levels of carbonic anhydrase isoform 3 in senescent human skeletal muscle. *Int. J. Mol. Med.* **2012**, *30*, 723–733. [CrossRef]
63. Montagna, C.; Di Giacomo, G.; Rizza, S.; Cardaci, S.; Ferraro, E.; Grumati, P.; De Zio, D.; Maiani, E.; Muscoli, C.; Lauro, F.; et al. S-nitrosoglutathione reductase deficiency-induced S-nitrosylation results in neuromuscular dysfunction. *Antioxid. Redox Signal.* **2014**, *21*, 570–587. [CrossRef]
64. Cebula, M.; Schmidt, E.E.; Arner, E.S. TrxR1 as a potent regulator of the Nrf2-Keap1 response system. *Antioxid. Redox Signal.* **2015**, *23*, 823–853. [CrossRef]
65. Maejima, Y.; Kuroda, J.; Matsushima, S.; Ago, T.; Sadoshima, J. Regulation of myocardial growth and death by NADPH oxidase. *J. Mol. Cell. Cardiol.* **2011**, *50*, 408–416. [CrossRef]
66. Sakellariou, G.K.; Vasilaki, A.; Palomero, J.; Kayani, A.; Zibrik, L.; McArdle, A.; Jackson, M.J. Studies of mitochondrial and nonmitochondrial sources implicate nicotinamide adenine dinucleotide phosphate oxidase(s) in the increased skeletal muscle superoxide generation that occurs during contractile activity. *Antioxid. Redox Signal.* **2013**, *18*, 603–621. [CrossRef]

67. Powers, S.K.; Wiggs, M.P.; Duarte, J.A.; Zergeroglu, A.M.; Demirel, H.A. Mitochondrial signaling contributes to disuse muscle atrophy. *Am. J. Physiol. Endocrinol. Metab.* **2012**, *303*, E31–E39. [CrossRef]
68. Boardman, N.T.; Trani, G.; Scalabrin, M.; Romanello, V.; Wust, R.C.I. Intracellular to Interorgan Mitochondrial Communication in Striated Muscle in Health and Disease. *Endocr. Rev.* **2023**, *44*, 668–692. [CrossRef]
69. Da Silveira, W.A.; Fazelinia, H.; Rosenthal, S.B.; Laiakis, E.C.; Kim, M.S.; Meydan, C.; Kidane, Y.; Rathi, K.S.; Smith, S.M.; Stear, B.; et al. Comprehensive Multi-omics Analysis Reveals Mitochondrial Stress as a Central Biological Hub for Spaceflight Impact. *Cell* **2020**, *183*, 1185–1201.e20. [CrossRef]
70. Larsen, F.J.; Schiffer, T.A.; Ortenblad, N.; Zinner, C.; Morales-Alamo, D.; Willis, S.J.; Calbet, J.A.; Holmberg, H.C.; Boushel, R. High-intensity sprint training inhibits mitochondrial respiration through aconitase inactivation. *FASEB J.* **2016**, *30*, 417–427. [CrossRef]
71. Lushchak, O.V.; Piroddi, M.; Galli, F.; Lushchak, V.I. Aconitase post-translational modification as a key in linkage between Krebs cycle, iron homeostasis, redox signaling, and metabolism of reactive oxygen species. *Redox Rep.* **2014**, *19*, 8–15. [CrossRef]
72. Andersson, U.; Leighton, B.; Young, M.E.; Blomstrand, E.; Newsholme, E.A. Inactivation of aconitase and oxoglutarate dehydrogenase in skeletal muscle in vitro by superoxide anions and/or nitric oxide. *Biochem. Biophys. Res. Commun.* **1998**, *249*, 512–516. [CrossRef]
73. Smith, A.J.; Advani, J.; Brock, D.C.; Nellissery, J.; Gumerson, J.; Dong, L.; Aravind, L.; Kennedy, B.; Swaroop, A. GATD3A, a mitochondrial deglycase with evolutionary origins from gammaproteobacteria, restricts the formation of advanced glycation end products. *BMC Biol.* **2022**, *20*, 68. [CrossRef]
74. Blottner, D.; Hastermann, M.; Weber, R.; Lenz, R.; Gambara, G.; Limper, U.; Rittweger, J.; Bosutti, A.; Degens, H.; Salanova, M. Reactive Jumps Preserve Skeletal Muscle Structure, Phenotype, and Myofiber Oxidative Capacity in Bed Rest. *Front. Physiol.* **2019**, *10*, 1527. [CrossRef]
75. Irimia, J.M.; Guerrero, M.; Rodriguez-Miguelez, P.; Cadefau, J.A.; Tesch, P.A.; Cusso, R.; Fernandez-Gonzalo, R. Metabolic adaptations in skeletal muscle after 84 days of bed rest with and without concurrent flywheel resistance exercise. *J. Appl. Physiol. (1985)* **2017**, *122*, 96–103. [CrossRef]
76. Edgerton, V.R.; Zhou, M.Y.; Ohira, Y.; Klitgaard, H.; Jiang, B.; Bell, G.; Harris, B.; Saltin, B.; Gollnick, P.D.; Roy, R.R.; et al. Human fiber size and enzymatic properties after 5 and 11 days of spaceflight. *J. Appl. Physiol.* **1995**, *78*, 1733–1739. [CrossRef]
77. Jackson, M.J.; Pollock, N.; Staunton, C.; Jones, S.; McArdle, A. Redox Control of Signalling Responses to Contractile Activity and Ageing in Skeletal Muscle. *Cells* **2022**, *11*, 1698. [CrossRef]
78. Powers, S.K.; Deminice, R.; Ozdemir, M.; Yoshihara, T.; Bomkamp, M.P.; Hyatt, H. Exercise-induced oxidative stress: Friend or foe? *J. Sport Health Sci.* **2020**, *9*, 415–425. [CrossRef]
79. Bellinger, A.M.; Reiken, S.; Dura, M.; Murphy, P.W.; Deng, S.X.; Landry, D.W.; Nieman, D.; Lehnart, S.E.; Samaru, M.; LaCampagne, A.; et al. Remodeling of ryanodine receptor complex causes "leaky" channels: A molecular mechanism for decreased exercise capacity. *Proc. Natl. Acad. Sci. USA* **2008**, *105*, 2198–2202. [CrossRef]
80. Suhr, F.; Gehlert, S.; Grau, M.; Bloch, W. Skeletal muscle function during exercise-fine-tuning of diverse subsystems by nitric oxide. *Int. J. Mol. Sci.* **2013**, *14*, 7109–7139. [CrossRef]
81. Hord, N.G.; Tang, Y.; Bryan, N.S. Food sources of nitrates and nitrites: The physiologic context for potential health benefits. *Am. J. Clin. Nutr.* **2009**, *90*, 1–10. [CrossRef] [PubMed]
82. Butler, A.R.; Feelisch, M. Therapeutic uses of inorganic nitrite and nitrate: From the past to the future. *Circulation* **2008**, *117*, 2151–2159. [CrossRef] [PubMed]
83. Stein, T.P. Space flight and oxidative stress. *Nutrition* **2002**, *18*, 867–871. [CrossRef] [PubMed]
84. Barbacini, P.; Blottner, D.; Capitanio, D.; Trautmann, G.; Block, K.; Torretta, E.; Moriggi, M.; Salanova, M.; Gelfi, C. Effects of Omega-3 and Antioxidant Cocktail Supplement on Prolonged Bed Rest: Results from Serum Proteome and Sphingolipids Analysis. *Cells* **2022**, *11*, 2120. [CrossRef] [PubMed]
85. Gomez, X.; Sanon, S.; Zambrano, K.; Asquel, S.; Bassantes, M.; Morales, J.E.; Otanez, G.; Pomaquero, C.; Villarroel, S.; Zurita, A.; et al. Key points for the development of antioxidant cocktails to prevent cellular stress and damage caused by reactive oxygen species (ROS) during manned space missions. *NPJ Microgravity* **2021**, *7*, 35. [CrossRef] [PubMed]
86. Hackney, K.J.; Scott, J.M.; Hanson, A.M.; English, K.L.; Downs, M.E.; Ploutz-Snyder, L.L. The Astronaut-Athlete: Optimizing Human Performance in Space. *J. Strength Cond. Res.* **2015**, *29*, 3531–3545. [CrossRef]
87. Petersen, N.; Lambrecht, G.; Scott, J.; Hirsch, N.; Stokes, M.; Mester, J. Postflight reconditioning for European Astronauts—A case report of recovery after six months in space. *Musculoskelet Sci. Pract.* **2017**, *27* (Suppl. 1), S23–S31. [CrossRef]

antioxidants

MDPI

Article

Zearalenone Exposure Damages Skeletal Muscle Through Oxidative Stress and Is Alleviated by Glutathione, Nicotinamide Mononucleotide, and Melatonin

Dandan Li [1,2,3,†], Wei Fu [1,2,3,†], Jiyue Zhang [1,2,3], Yaqiu Lin [1,2,3], Xianrong Xiong [1,2,3], Jian Li [1,2,3] and Yan Xiong [1,2,3,*]

[1] Key Laboratory of Qinghai-Tibetan Plateau Animal Genetic Resource Reservation and Utilization, Ministry of Education, Southwest Minzu University, Chengdu 610041, China; lidandan@mnu.cn (D.L.); fuwei@swun.edu.cn (W.F.); zhangjiyue0416@163.com (J.Z.); yaqlin@163.com (Y.L.); xiongxianrong@163.com (X.X.); lijian@swun.edu.cn (J.L.)
[2] College of Animal & Veterinary Sciences, Southwest Minzu University, Chengdu 610041, China
[3] Key Laboratory of Animal Science of National Ethnic Affairs Commission of China, Southwest Minzu University, Chengdu 610041, China
* Correspondence: xiongyan0910@126.com
† These authors contributed equally to this work.

Abstract: Zearalenone (ZEN), a mycotoxin, is toxic to skeletal muscle, and the solution to alleviate its damage remains unknown. Here, we analyzed the toxic effect of ZEN on muscle and the mitigation of antioxidants (GSH, NMN, and melatonin) for this toxicity. The results showed that 0.02 mmol/L ZEN inhibited myoblast viability and myogenic differentiation, accompanied by reducing *Type I* and *Type IIA* and increasing *Type IIX* myofibers. Antioxidants (NMN with 0.5 mmol/L, GSH with 1 mmol/L, and melatonin with 1 mmol/L) rescued these phenotypes. Mice that were delivered 3 mg/kg body weight (BW)/day of ZEN by gavage for 35 days exhibited a similar trend of muscle fiber composition, but the gavage of antioxidants (NMN with 500 mg/kg BW/day, GSH with 300 mg/kg BW/day, and melatonin with 100 mg/kg BW/day) abolished this phenotype. Mechanistically, ZEN treatment increased ROS production, resulting in oxidative stress, mitochondrial dysfunction, and, subsequently, myofiber changes. Additionally, ZEN indirectly contributed to its damage, decreasing the abundance of *Lactobacillus* at the genus level and increasing *Streptococcus* sp. at the species level, which was associated with lactic acid production. Antioxidants partially rescued this microbiota composition. This study explores ZEN toxicity effects and alleviation of antioxidants, which provides new insights and attenuation solutions for ZEN damage to skeletal muscle. However, the underlying molecular mechanism of ZEN and antioxidants in the skeletal muscle still needs to be explored.

Keywords: zearalenone; toxicity; skeletal muscle; antioxidant; muscle fiber

1. Introduction

Mycotoxin contamination is widespread in the world and has become a global concern, which can lead to significant economic losses in agriculture and seriously threaten the safety of food and the health of humans [1]. ZEN, also known as F-2 toxin, is a kind of mycotoxin that has been classified as a Group 3 carcinogen by the International Agency for Research on Cancer (IARC), primarily produced by *Fusarium graminearum* and *Fusarium culmorum* in the moldy corn, sorghum, wheat, and their products of cultivation area [2]. Extensive research reported that animals were exposed to ZEN, leading to reproductive toxicity in the sow [3] and immunotoxicity or cytotoxicity in various tissues of other animals, including the liver [4] and kidney [5]. The skeletal muscle is the largest locomotor organ, protein reservoir, and primary glucose metabolism site, accounting for ~40% of individual body weight [6]. Myofibers are the basic structural units of skeletal muscle, which are roughly divided into fast (*Type IIA*, *Type IIB*, and *Type IIX*) and slow (*Type I*) muscle fibers, respectively, with glycolytic and oxidative metabolic characteristics [7]. For humans, the composition and transformation of myofiber types are linked to individual normal physiological function and exercise capacity [8], while a higher proportion of slow-twitch fibers would improve meat quality for the animals [9]. Currently, the specific damage caused by ZEN exposure to skeletal muscle and muscle fiber composition is largely unknown.

Recently, several works provided evidence for high concentrations of ZEN detected in the muscle tissues when the animals are exposed to ZEN [10,11], which could be carried over in animal food products to further threaten human health. For instance, the highest concentrations of ZEN were observed in muscle tissue (12.033 ng/g) among the measured tissues of clinically healthy female wild boars fed naturally contaminated with ZEN at 50 mg/kg BW/day [10]. Consistently, ZEN residues are found in the leg muscle (0.069 μg/kg) of Hy-Line Brown laying hens fed a basal diet containing 60% moldy corn [11]. However, the deleterious effects of ZEN on skeletal muscle remain largely unknown. Damiano et al. found that ZEN natural exposure changed the morphology and myofiber diameter slightly in the muscles of wild boars from Southern Italy [12]. In addition, sows fed with ZEN-contaminated diets (2.77 mg/kg BW) from 35 to 70 days of gestation seem to be particularly sensitive and even decrease myofiber number and diameter in the muscle tissue of newborns and weaning piglets [13]. Thus, these studies suggested that animals exposed to ZEN would have damaged skeletal muscles. However, it was difficult to identify the direct pathological changes of skeletal muscle mediated by ZEN because of the complicated effects of the coexistence of multiple mycotoxins in moldy cereal crops and their products.

Previous studies have shown that oxidative stress is one of the main pathways through which ZEN toxicity causes impaired cell viability, mitochondrial dysfunction, etc. [14,15]. Therefore, antioxidant therapy may be a potential solution to alleviate the toxicity of ZEN to muscles. Nicotinamide mononucleotide (NMN), as another potential antioxidant, could increase oxidized nicotinamide adenine dinucleotide (NAD^+) levels and rescue mitochondrial dysfunction in C2C12 cells and levator ani muscle (LAM) damaged by mechanical stress (MS) [16]. Glutathione (GSH) is a functional dietary supplement with detoxification [17] and antioxidant [18] functions, etc. Melatonin is a broad-spectrum free radical scavenger with strong antioxidant ability, which can directly or indirectly regulate the expression of cellular antioxidant enzymes, thereby enhancing the antioxidant capacity of cells [19]. Thus, GSH, NMN, and melatonin were selected as the candidate antioxidants to alleviate the damage to muscles mediated by ZEN.

Herein, we aimed to explore the toxicity of ZEN on skeletal muscle by measuring cell viability, myogenesis, and muscle fiber composition. Then, the underlying mechanism was revealed by measurement of ROS levels, mitochondrial function, and gut microbial composition. Moreover, we explored whether three antioxidants (GSH, NMN, and melatonin) could alleviate the muscle damage caused by ZEN in vitro and in vivo.

2. Materials and Methods

2.1. Cell Culture

C2C12 myoblast cell line (ATCC, Rockville, MD, USA) was cultured in a Growth medium consisting of Dulbecco's modified Eagle medium (DMEM) (Gibco, C11995500BT, Shanghai, China), 10% Fetal bovine serum (FBS) (AlphaCell, 100061, Guangdong, China), and 1% penicillin-streptomycin (PS) (Biosharp, BL505A, Hefei, China). The C2C12 cells with a confluence of 80–90% were induced myogenic differentiation by a differentiation medium (DM) containing 2% horse serum (Solarbio, S9050, Beijing, China) and 1% PS. All the cell plates were maintained in a cell incubator at a constant temperature and humidity (37 °C, 5% CO_2), and the medium was changed every two days. For the drug treatment, the stock solution of ZEN (HPLC $\geq$ 99.0%, Biopike, Waltham, MA, USA), NMN (HPLC $\geq$ 95.0%, Sigma-Aldrich, St. Louis, MO, USA), GSH (HPLC $\geq$ 98.0%, Sigma-Aldrich), and melatonin (HPLC: 99.95%, Selleck, Houston, TX, USA) were dissolved in DMSO and stored at −20 °C in the dark, which was used at a work solution concentration of ZEN with 0.02 mmol/L [20], NMN with 0.5 mmol/L [16], GSH with 1 mmol/L [21,22], and melatonin with 1 mmol/L [21,23], respectively. The Control (0.1% DMSO), ZEN, ZEN + NMN, ZEN + GSH, and ZEN + Melatonin treatment groups were designed. The drug is added to the C2C12 cells along with the DM, and collected 4 days after cell differentiation.

2.2. Animals and Study Approval

C57BL/6J female mice (8 weeks old) were purchased from the Chengdu Dossay Laboratory Animal Center. The mice were kept in a room with a temperature of 25 $\pm$ 1 °C, a humidity of 55 $\pm$ 5%, and a 12-h light cycle (light at 8:00 a.m.–8:00 p.m., dark at 8:00 p.m.–8:00 a.m.). Thirty female mice were randomly divided into 5 groups, including the Control (0.2 mL normal saline per day), ZEN (ZEN: 3 mg/kg BW/day [24]), ZEN + NMN (ZEN: 3 mg/kg BW/day + NMN: 500 mg/kg BW/day [24]), ZEN + GSH (ZEN: 3 mg/kg BW/day + GSH: 300 mg/kg BW/day [21,25]), ZEN + Melatonin groups (ZEN: 3 mg/kg BW/day + Melatonin: 100 mg/kg BW/day [21,26]). According to the drug solubility, ZEN, NMN, GSH, and melatonin were weighed and dissolved in DMSO and stored at −20 °C in the dark. Normal saline was added 30 min before gavage for dissolution, and each mouse was gavaged with 0.2 mL per day. Daily weight and food intake were recorded. After 35 days of gavage, the mice were sacrificed by cervical dislocation. The gastrocnemius (GAS), soleus (SOL), tibialis anterior (TA), and extensor digitorum longus (EDL) muscles of the hind legs in mice were collected and frozen with liquid nitrogen. All the tissues were stored in an ultra-low temperature freezer at −80 °C.

2.3. Cell Counting Kit 8 (CCK-8) Assay

C2C12 cells were cultured in 96-well plates, and six biological replicates were set up for each treatment group. For the CCK-8 assay, the cells were washed 2–3 times in PBS, added 100 μL of growth medium and 10 μL of CCK-8 (AbMole, M4839, Shanghai, China), and incubated in an incubator (37 °C) for 1 to 4 h sequentially. The full-band microplate reader (Perkin Elmer) measured the absorbance (450 nm) after 48 h of culture. The formula is cell viability = [A(dosing) − A(blank)]/[A(control) − A(blank)] × 100%. A (dosing): absorbance of cells, CCK-8 solution, and drug solution pores; A (control): absorbance of cells and CCK-8 solution pores; A (blank): absorbance of culture medium and CCK-8 solution pores.

2.4. Total RNA Extraction and Quantitative Real-Time PCR (RT-qPCR)

Collected C2C12 cells for 4 days of differentiation and leg muscles of mice after 35 days of gavage. The TRiZOL (Vazyme, R401-01, Nanjing, China) reagent was used to extract total RNA. The quality and concentration of RNA were detected by Nanodrop ND-1000 (Thermo Fisher, Waltham, MA, USA), and the OD_{260}/OD_{280} with 1.8–2.0 was used to synthesize cDNA according to the manufacturer's instructions (Prime-Script RT Master Mix, Takara, Dalian, China). Real-time quantitative PCR analysis was performed using a SYBR real-time PCR mixture (BioTeke, Beijing, China). The primer sequences for RT-qPCR are listed in Table 1. Each sample was repeated thrice, and the $2^{-\Delta\Delta Ct}$ method was used to calculate the relative mRNA levels.

Table 1. RT-qPCR primer sequences.

Gene	Forward Primer (5′–3′)	Reverse Primer (5′–3′)	Length (bp)	Gene Accession Number
MYHC	GCGAATCGAGGCTCAGAACAA	GTAGTTCCGCCTTCGGTCTTG	138	XM_017314318.3
MYOG	GAGACATCCCCCTATTTCTACCA	GCTCAGTCCGCTCATAGCC	106	NM_031189.2
MYF5	AAGGCTCCTGTATCCCCTCAC	AAGGCTCCTGTATCCCCTCAC	213	NM_175686.3
RPL7	TGGTTTAGGAGAGTAAGGTTGCT	TGGTTTAGGAGAGTAAGGTTGCT	351	XM_006529040.3
MYOD1	CCACTCCGGGACATAGACTTG	AAAAGCGCAGGTCTGGTGAG	109	NM_010866.2
CCND1	TAGGCCCTCAGCCTCACTC	CCACCCCTGGGATTGGTTTA	338	XM_006529043.2
CCNB1	CTTGCAGTGAGTGACGTAGAC	CCAGTTGTCGGAGATAAGCATAG	94	NM_172301.3
CDK2	CAAAGCCAAGCACGTAGAGAC	TGCACCACATATTGACTGTCC	141	NM_053180.2
MYH7	GAATGGCAAGACGGTGACTGTG	GAAGCGTAGCGCTCCTTGAG	233	gi∣1698894
MYH2	ATCAACCAGCAGCTGGACACCA	TCCAGCACGAACATGTGGTGGT	249	gi∣5360745
MYH4	ACAGACTAAAGTGAAAGCCTACAA	CACATTTTGTGATTTCTCCTGTCAC	257	gi∣5360749
MYH1	CCAATGAAACCAAGACTCCTGG	TGCTATCGATGAACTGTCCCTC	234	gi∣5360747

2.5. Immunofluorescence Staining

The cells were fixed with 4% paraformaldehyde (Biosharp, BL539A, Hefei, China) and then washed with PBS three times. A total of 100 mM glycine was incubated for 10 min at room temperature and washed 3 times with PBS. A blocking buffer (2% BSA, 0.1% sodium azide, 0.2% Triton X-100, and 5% goat serum) was added to incubate cells for 45 min at room temperature. Added Primary antibody (MYHC, DSHB Hybridoma Product MF 20) for immunized cells overnight at 4 °C. After 3 washes of PBS, fluorescent secondary antibody, including Alexa FluorTM 568 Goat anti-rabbit IgG (H + L) (Invitrogen, A11011, Carlsbad, CA, USA) and Goat anti-mouse IgG H&L/FITC (Bioss, bs-0296G-FITC, Beijing, China), was incubated at room temperature for 1 h. The confocal microscope (Zeiss, LSM800, Oberkochen, Baden-Württemberg, Germany) was used to capture images. The fluorescence intensity (total fluorescence intensity divided by area), myotube fusion index (the ratio of the number of nuclei in the myotube to the total number), and total myonuclei from randomly selected areas of 3 biological replicates were analyzed by ImageJ software V1.8.0.112 [27].

2.6. Mito-Tracker Staining

The nucleus was stained with DAPI, and mitochondrial morphology was marked by Mito-Tracker Red staining (C1049B, Beyotime, Shanghai, China) according to the manufacturer's instructions. Then, the cells were washed 3 times with PBS. Subsequently, the images were captured by a confocal microscope, and the red fluorescence density per area unit was measured by ImageJ software V1.8.0.112 [27]. There were 3 biological replicates in each group.

2.7. Reactive Oxygen Species (ROS) Evaluation by 2′,7′-Dichlorodihydrofluorescein Diacetate (DCFH-DA) Staining

The plate was seeded at a cell density of 3×10^5 cells. Then, the cells were incubated with 10 μmol/L DCFH-DA (D6883; Sigma-Aldrich, St. Louis, MO, USA) in serum-free DMEM at 37 °C for 20 min. Cells were washed 3 times with serum-free DMEM to remove DCFH-DA that had not entered the cells adequately. The confocal microscope was used to capture images, and the ImageJ software V1.8.0.112 [27] was used to count fluorescence intensity.

2.8. Hematoxylin-Eosin (H&E) Staining

The GAS tissue was fixed in 4% paraformaldehyde for 24 h and dehydrated for 16 h. After routine embedding with paraffin, the tissue was cut into sections (5 μm). Then, the sections were placed in an oven (60 °C) for 1 h and were respectively dewaxed with xylene and ethanol and then washed with running water. The sections were further stained with hematoxylin for 5 min and placed under running water to remove residual color. Differentiated cells with 5% acetic acid for 10 s and washed with running water, then stained with 0.1% eosin for 1 min. After dehydration with alcohol, xylene is used to clear the tissue. Finally, the sections were sealed with neutral gelatin after drying. The confocal microscope was used to capture images, and ImageJ software V1.8.0.112 [27] was used to count the cross-sectional area.

2.9. Transmission Electron Microscopy (TEM)

The collected skeletal muscle was cut into less than $1 \times 1 \times 1$ mm^3 cubes, added 3% glutaraldehyde fixative solution within 1 min, rinsed in PBS, and then applied 1% osmium acid fixative solution for 2 h (pH = 7.3) sequentially. The samples were rinsed for 30 min with PBS and then dehydrated with acetone. Epoxy resin Epon 812 embedded tissue. An ultra-thin microtome machine was used to prepare 60–90 nm sections, and the sections were placed on a copper mesh with a supporting film. At room temperature, stained with uranyl acetate for 15 min. Then, the sections were stained with citrate for 2 min. A JEM-1400FLASH transmission electron microscope (JEOL Ltd., Tokyo, Japan) was used to capture images of copper meshes.

2.10. 16S rDNA Sequencing

The E.Z.N.A. Soil DNA Kit (Omega Bio-tek, Inc., Norcross, GA, USA) was used to extract genomic DNA from 15 mouse intestinal contents, and DNA quality and concentration were detected by Nanodrop 2000 (ThermoFisher Scientific, Inc., Waltham, MA, USA). Universal primers 338F (5′-ACTCCTACGGGAGGCAGCAG-3′) and 806R (5′-GGACTACHVGGGTWTCTAAT-3′) amplify the V3–V4 region of the bacterial 16S rRNA gene. A total of 8 bp barcode sequences were added to the 5′ ends of the upstream and downstream primers to distinguish different samples. Finally, universal primers with barcode sequences were synthesized and amplified on an ABI 9700 PCR instrument (Applied Biosystems, Inc., Foster City, CA, USA), and the size of the amplified target band was detected by 1% agarose gel electrophoresis. The Agencourt AMPure XP (Beckman Coulter, Inc., Pasadena, CA, USA) Nucleic Acid Purification Kit was used to automate the purification of PCR products, and the Caliper LabChip GX Touch HT (PerkinElmer, Inc., Waltham, MA, USA) measured the PCR product concentration and fragment size. The PCR products were then library constructed using the NEB Next Ultra II DNA Library Prep Kit (New England Biolabs, Inc., Ipswich, MA, USA) and purified with the Agencourt AMPure XP (Beckman Coulter, Inc., Pasadena, CA, USA) Nucleic Acid Purification Kit, Nanodrop

2000 (ThermoFisher Scientific, Inc., Waltham, MA, USA), Agilent 2100 Bioanalyzer (Agilent Technologies, Inc., Santa Clara, CA, USA) for library fragment size, and ABI StepOne-Plus Real-Time PCR System (Applied Biosystems, Inc., Foster City, CA, USA) accurately quantified library concentration. Finally, the libraries were sequenced on the Illumina Miseq/Novaseq 6000 (Illumina, Inc., San Diego, CA, USA) platform with a sequencing strategy of PE250/PE300.

2.11. Statistics Analysis

The high-throughput 16S rDNA gene sequencing data were analyzed using the Qiime platform at Allwegene Company (Beijing, China). Each experiment was repeated thrice, and 3–6 biological replicates were set for each experimental group. All significant analyses were performed using SPSS 18.0 and GraphPad Prism 8.0 software. The data were presented as Mean $\pm$ SD and analyzed by One-way ANOVA; all differences were considered to be statistically significant at $p < 0.05$.

3. Results

3.1. ZEN Inhibits the Proliferation of Myoblast and Antioxidant Treatments Alleviate This Inhibitory Effect

The molecular formula of ZEN is $C_{18}H_{22}O_5$ (Figure 1A), whose treatment concentration exceeded 0.02 mmol/L and significantly decreased the cell viability of C2C12 cells (Figure 1B), and this concentration was used in the further treatment of this study. Further, the CCK-8 assay showed a ~60% decrease in cell activity after ZEN treatment compared with the control group (Figure 1C). Compared with the ZEN treatment group, the combined treatment of antioxidants (ZEN + NMN, ZEN + GSH, and ZEN + Melatonin) all partially rescued cell viability and maintained cell morphology (Figure 1C,D). RT-qPCR detected the mRNA levels of proliferation-related genes, and data showed that cyclin d1 (*CCND1*) significantly decreased to 50% (Figure 1E) and cyclin b1(*CCNB1*) significantly decreased to 60% (Figure 1G) in the ZEN treatment compared with the control group. In contrast, the mRNA levels of *CCND1*, cyclin-dependent kinase 2 (*CDK2*), and *CCNB1* in the ZEN + NMN, ZEN + GSH, and ZEN + Melatonin groups were increased compared to those of ZEN treatment (Figure 1E–G). In addition, compared with the ZEN + NMN or ZEN + Melatonin groups, the ZEN + GSH group significantly promoted the expression of *CCNB1* (Figure 1G). Thus, these data indicated that ZEN inhibited the proliferation of myoblasts, and the inhibitory effect was alleviated after the combined treatment with NMN, GSH, and Melatonin to different extents, of which GSH had a better alleviation effect.

3.2. ZEN Suppresses the Differentiation of Myoblast and Antioxidant Treatment Rescues This Suppression Effect

Next, the effect of ZEN on myogenic differentiation was measured, which is involved in mononuclear myoblasts fusing to form multinucleated myotubes [28]. The ZEN treatment significantly decreased myotube formation compared with that of the control (Figure 2A), indicated by the reduction of MYHC fluorescence intensity (Figure 2B, $p < 0.05$), myotube fusion index (Figure 2C, $p < 0.05$), and the proportion of >4 myonuclei myofiber (Figure 2D, $p < 0.05$). Interestingly, the inhibition of myotube formation caused by ZEN was rescued by the combination with the treatment of NMN, GSH, and melatonin (Figure 2B–D). At the mRNA level, ZEN treatment suppressed mRNA levels of myogenic markers, myosin heavy chain (*MYHC*) with a ~90% significant decrease (Figure 2E), myogenic differentiation 1 (*MYOD1*) with a ~80% significant decrease (Figure 2F), myogenin (*MYOG*) with a ~30% decrease (Figure 2G), and myogenic factor 5 (*MYF5*) with a ~75% significantly decrease (Figure 2H) compared with the control group, respectively. Compared with the ZEN group, the combined treatment of ZEN and antioxidants significantly up-regulated the expression levels of *MYHC*, *MYOD1*, *MYF5*, and *MYOG*, except for *MYOG* of the ZEN + NMN group (Figure 2E–H). Among the ZEN + GSH, ZEN + NMN, and ZEN + Melatonin groups, ZEN + Melatonin treatment significantly up-regulated most of the myogenic genes (Figure 2E–H). These results indicated that ZEN exposure inhibits myogenesis, which was rescued by antioxidant treatments (NMN, GSH, and Melatonin) to different extents.

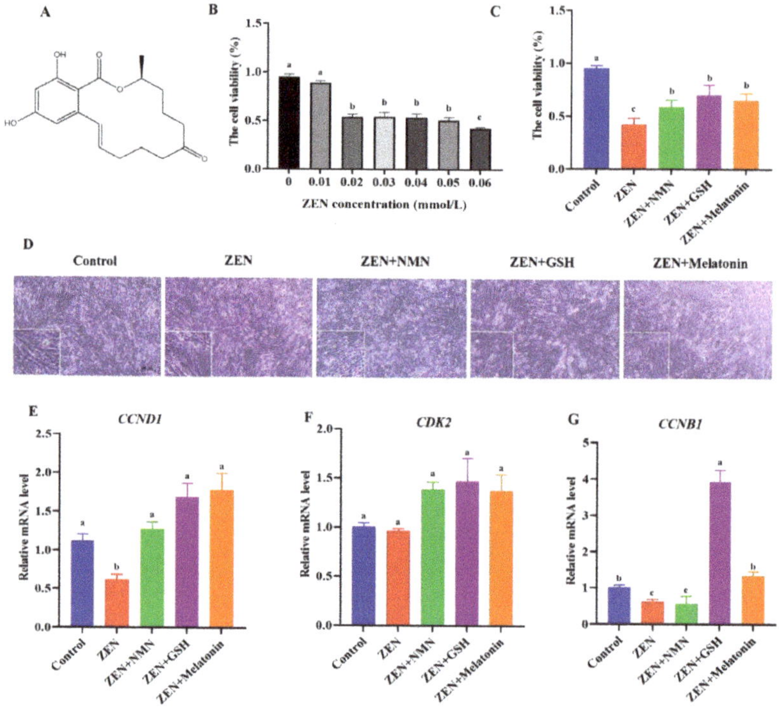

Figure 1. ZEN inhibits the cellular activity of myoblasts, and antioxidant treatment rescues this inhibition. (**A**) The chemical structural formula of ZEN, (**B**) CCK-8 assay detected the proliferation activity of myoblast treated by ZEN at concentrations of 0.01, 0.02, 0.03, 0.04, 0.05, and 0.06 mmol/L, respectively, (**C**,**D**) the cell viability (**C**) and cell morphology (**D**) (scale bar: 100 μm) were observed in Control, ZEN (0.02 mmol/L), ZEN + NMN (NMN with 0.5 mmol/L), ZEN + GSH (GSH with 1 mmol/L), and ZEN + Melatonin (melatonin with 1 mmol/L) groups, (**E**–**G**) the mRNA level of proliferation-related genes, including *CCND1* (**E**), *CDK2* (**F**), and *CCNB1* (**G**) in different treatment groups. Different letters (a–c) indicate significant differences ($p < 0.05$), and the same letters indicate no difference, $n = 3$.

3.3. ZEN Induces Slow-to-Fast Myofiber Shift and Antioxidant Treatment Rescues This Phenotype

Next, whether ZEN treatment induced the muscle fiber shift remains unknown. Oxidative myofibers are enriched in mitochondria, which can resist fatigue from long-term endurance exercise, while glycolytic myofibers rely more on glycolytic enzymes to produce energy quickly for rapid bursts of resistance movement [7]. Myofiber identification by RT-qPCR showed that the mRNA levels of oxidative myofiber *Type I* (*MYH7*) (Figure 3A) and oxidative myofiber *Type IIA* (*MYH2*) (Figure 3D) were significantly decreased in the ZEN treatment group compared with the control group. However, ZEN treatment increased the mRNA level of glycolytic myofiber *Type IIX* (*MYH1*), with ~2-fold significant changes compared to the control (Figure 3C, $p < 0.05$). For the antioxidant treatment, the gene expression of *Type I* and *Type IIA* in the ZEN + NMN, ZEN + GSH, and ZEN + Melatonin groups all showed a rescued effect compared with the ZEN group (Figure 3A,D). At the same time, the mRNA level of *Type IIX* significantly decreased to ~20%, ~4%, and ~10% of the control, respectively (Figure 3C). Furthermore, the ZEN + Melatonin group significantly increased the mRNA levels of *Type IIA* compared with the ZEN + NMN and ZEN + GSH groups (Figure 3D). Thus, it is concluded that ZEN affected the level of muscle fiber type-related genes, and antioxidant (NMN, GSH, and melatonin) treatments alleviated this change. Specifically, the rescued effect of melatonin was better compared with NMN and GSH.

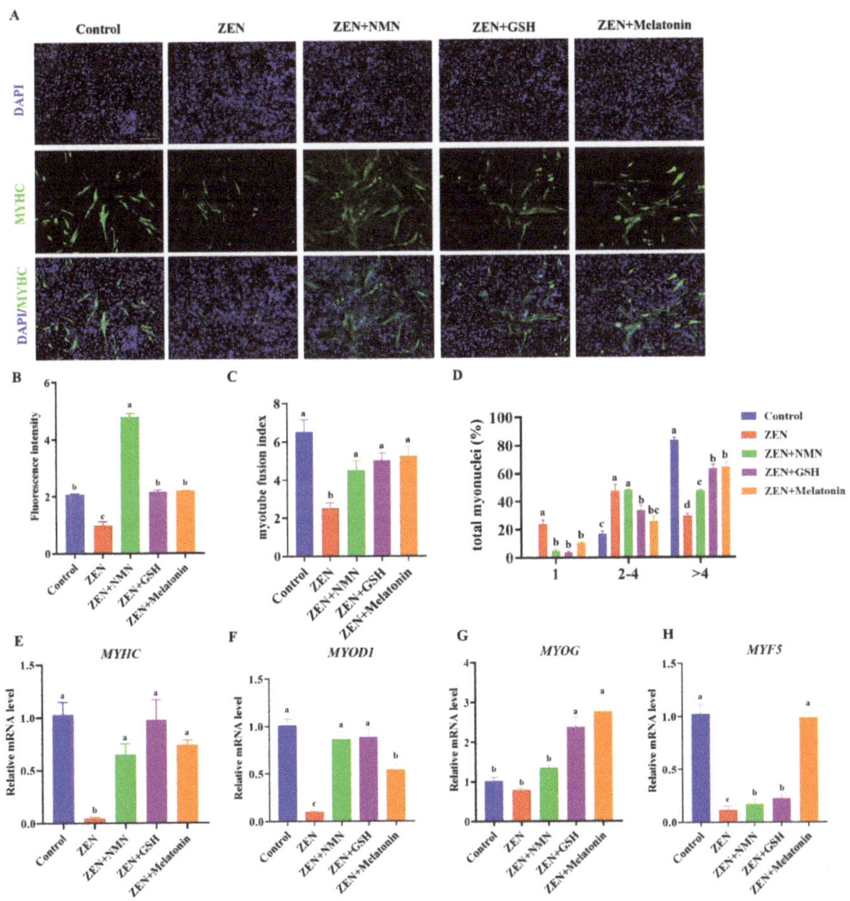

Figure 2. ZEN suppresses myogenic differentiation, and antioxidant treatment rescues this phenotype. (**A**) The images of MYHC immunofluorescence staining in the Control, ZEN, ZEN + NMN, ZEN + GSH, and ZEN + Melatonin groups, scale bar: 100 μm, (**B–D**) the count analysis of MYHC immunofluorescence intensity (**B**), the myotube fusion index (**C**), and the proportion distribution of different myo-nucleus (**D**), (**E–H**) the mRNA level of myogenic-related genes, including *MYHC* (**E**), *MYOD1* (**F**), *MYOG* (**G**), and *MYF5* (**H**) in the Control, ZEN, ZEN + NMN, ZEN + GSH, and ZEN + Melatonin groups. Different letters (a–d) indicate significant differences ($p < 0.05$), and the same letters indicate no difference, $n = 3$.

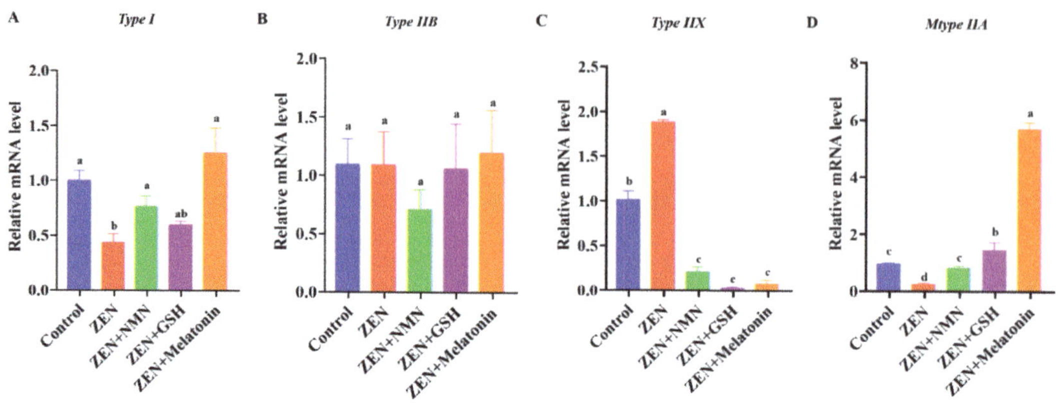

Figure 3. ZEN affected muscle fiber type-related gene levels, and antioxidant treatment rescued the myofiber change. (**A**–**D**) The mRNA levels of myofiber types of *Type I* (**A**), *Type IIB* (**B**), *Type IIX* (**C**), and *Type IIA* (**D**) in the aforementioned groups. Different letters (a–d) indicate significant differences ($p < 0.05$), and the same letters indicate no difference, $n = 3$.

3.4. ZEN Exposure Decreases Skeletal Muscle Myogenic-Related Gene Levels and Changes Myofiber Composition In Vivo

To explore the ZEN effect on skeletal muscle and screen an antioxidant to protect from ZEN damage in vivo, this mycotoxin and the antioxidants were delivered by gavage (Figure 4A). The results showed that the body weight (Figure 4B), feed intake (Figure 4C), skeletal muscle weight (Figure 4D), and muscle morphology (Figure 4E) were not significantly changed among the Control, ZEN, ZEN + NMN, ZEN + GSH, and ZEN + Melatonin groups.

Next, the H&E staining and counting analysis found that the average area and diameter of myofibers in the ZEN group were significantly smaller than the control group (Figure 5A,B). Compared with the ZEN group, the average area and diameter of myofibers in the ZEN + Melatonin group were upward, and the average area of myofibers in the ZEN + NMN group was also increased, but the diameter of myofibers in the ZEN + NMN group was significantly increased (Figure 5B). In addition, the results of TEM showed that the ZEN group muscle had obvious lesions such as mitochondrial swelling, z-line irregularity, unclear light and dark bands, and myolysis compared with the control group (Figure 5C). However, the ZEN + NMN, ZEN + GSH, and ZEN + Melatonin groups all reduced lesions in comparison with ZEN treatment (Figure 5C). The expression levels of *MYF5* (Figure 5D) and *MYOG* (Figure 5F) were significantly lower in the ZEN group than those of the control, and *MYHC* (Figure 5G) was down-regulated. However, compared with the ZEN group, the mRNA levels of *MYF5*, *MYOD1*, *MYOG*, and *MYHC* were all partly or totally rescued in the ZEN + Melatonin, ZEN + GSH, and ZEN + NMN groups (Figure 5D–G). Among the three antioxidant treatment groups (ZEN + NMN, ZEN + GSH, and ZEN + Melatonin), the ZEN + Melatonin group had fewer lesions caused by ZEN (Figure 5C). From the above results, ZEN exposure damaged muscle in vivo, and antioxidants partly or totally rescued the deleterious effects, with melatonin having a more pronounced effect.

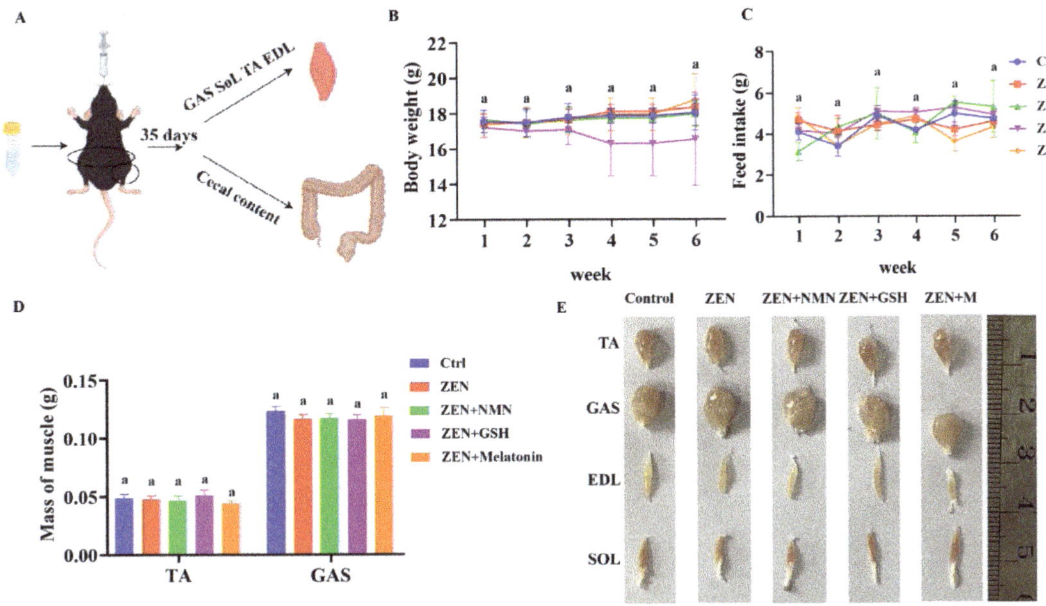

Figure 4. The mass of skeletal muscle is influenced by ZEN combined with antioxidant treatment. (**A**) A schematic diagram of gavage ZEN and antioxidant in mice was drawn by "Figdraw 2.0" software, (**B**,**C**) the body weight (**B**) and feed intake (**C**) of mice treated by ZEN and combined with antioxidants (NMN, GSH, and melatonin) for 6 weeks, (**D**,**E**) the mass (**D**) and images (**E**) of muscles (TA, tibialis anterior; GAS, gastrocnemius muscle; EDL, extensor digitorum longus; SOL, soleus) in the Control, ZEN, ZEN + NMN, ZEN + GSH, and ZEN + Melatonin groups. The same letter (a) indicates no difference, $n = 6$.

Furthermore, the mRNA levels of *Type I* (Figure 5H) and *Type IIA* (Figure 5I) were significantly lower, and that of *Type IIX* (Figure 5K) was significantly higher in the ZEN group than the Control group in vivo, which was consistent with the trend in vitro (Figure 3A–D). In the ZEN + NMN group, the mRNA levels of oxidative myofiber, including *Type I* (Figure 5H) and *Type IIA* (Figure 5I), were significantly higher, and myofiber *Type IIX* (Figure 5K) was significantly lower than in the ZEN group. In the ZEN + GSH group, *Type I* (Figure 5H), *Type IIA* (Figure 5I), and *Type IIX* (Figure 5K) all increased significantly compared with the ZEN group. Specifically, the mRNA expression of *Type I* ($p < 0.05$) and *Type IIA* ($p < 0.05$) in the ZEN + Melatonin group completely rescued the myofiber shift phenotype caused by ZEN (Figure 5H,I). Moreover, the ZEN + Melatonin group had significantly higher mRNA levels of *Type IIA* than the ZEN + GSH and ZEN + NMN groups (Figure 5I). Collectively, ZEN exposure affected muscle fiber composition, which was rescued by NMN, GSH, and melatonin in vivo, and melatonin had a better-rescued effect on ZEN damage.

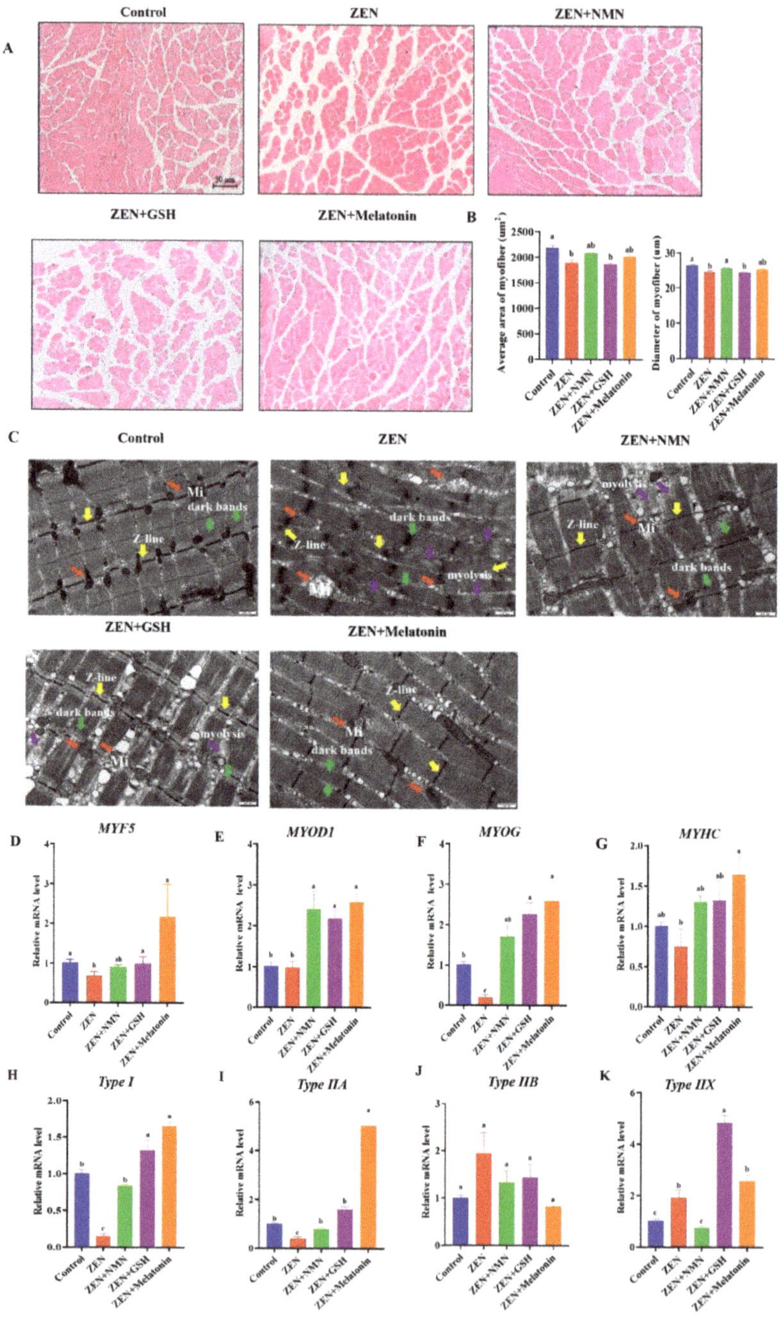

Figure 5. ZEN affected muscle fiber composition, and antioxidant treatment rescued the myofiber change in vivo. (**A**,**B**) The H and E staining (**A**) (scale bar: 50 μm) and average fiber area analysis (**B**) of GAS from Control, ZEN, ZEN + NMN, ZEN + GSH, and ZEN + Melatonin treated mice, (**C**) the

ultrastructure of skeletal muscle observed by TEM, scale bar: 500 nm, (Red arrow: mitochondrial (Mi); Yellow arrow: Z-line; Green arrow: dark bands; Purple arrow: myolysis), (**D–K**) the mRNA level of myogenic genes (*MYF5* (**D**), *MYOD1* (**E**), *MYOG* (**F**), and *MYHC* (**G**)) and fiber type-specific genes (*Type I* (**H**), *Type IIA* (**I**), *Type IIB* (**J**), and *Type IIX* (**K**)) in the aforementioned groups. Different letters (a–c) indicate significant differences ($p < 0.05$), and the same letters indicate no difference, $n = 6$.

3.5. ZEN Induces Slow-to-Fast Myofiber Shift Through Oxidative Stress and Mitochondrial Dysfunction

ROS levels and mitochondrial content were detected to explore the mechanism of action of ZEN on myofibers further. The data showed that the fluorescence intensity of ROS levels in the ZEN group significantly increased compared with the control group, while the ROS levels significantly reduced after the treatment of NMN, GSH, and melatonin (Figure 6A,B). Thus, Mito-Tracker staining results showed that the mitochondrial content of the ZEN group was significantly reduced compared with the Control group, while the combined treatment of NMN, GSH, and melatonin in the ZEN group significantly restored the mitochondrial staining signal (Figure 6C,D). Further, the ZEN + GSH group significantly decreased the fluorescence intensity of ROS levels compared with the ZEN + NMN and ZEN + Melatonin groups (Figure 6B). The ZEN + NMN group had the strongest mitochondria staining signal compared with the ZEN + GSH and ZEN + Melatonin groups (Figure 6D). Therefore, ZEN exposure resulted in excessive ROS accumulation, mitochondrial dysfunction, and subsequent changes in myofiber composition. Antioxidants (NMN, GSH, and melatonin) alleviated these symptoms. Among the three antioxidants, NMN has a better ability to increase mitochondrial content and reduce oxidative stress.

3.6. ZEN Combined with Antioxidant Treatment Indirectly Affects Slow-to-Fast Myofiber Shift by Altering the Proportion of the Gut Microbiota

Previously, research found that the gut-muscle axis indirectly affects skeletal muscle mass by microbiota [29]. Thus, we performed 16S rDNA sequencing to explore the indirect effects of ZEN and antioxidants on skeletal muscle through gut microbiota composition. As shown in the Venn diagram, there were a total of 14,888 operational taxonomic units (OTUs) in these five groups (Figure 7A). Of which, 1515 OTUs overlapped in these five groups, and 97, 104, 229, 170, and 121 specific OTUs respectively existed in the control, ZEN, ZEN + NMN, ZEN + GSH, and ZEN + Melatonin groups (Figure 7A). The chao1 index was higher in the four treated groups than in the control group (Figure 7B). In the principal component analysis (PCA) of the first principal factor (19.58%) and the second principal factor (16.92%), these five groups of intestinal samples were divided into five obvious clusters, indicating that they had significant differences in intestinal microbial structure (Figure 7C). It was reported that feeding *Lactobacillus paracasei PS23* (LPPS23) and *Lactobacillus casei Shirota* (LcS) attenuates age-related sarcopenia in SAMP8 mice via the gut-muscle axis [30]. At the genus level, ZEN (8.88%) decreased the abundance of *Lactobacillus* compared with the control (16.62%) (Figure 7D and Table S1). Compared with the ZEN group, the addition of NMN (22.51%, $p < 0.05$), GSH (26.79%, $p < 0.05$), and melatonin (10.68%) increased the abundance of *Lactobacillus*, respectively (Figure 7D and Table S1). Furthermore, *Streptococcus* can produce lactic acid [31], and *Streptococcus pneumoniae* can affect glycolytic activity [32] and may be associated with fast-twitch fiber biogenesis [33]. Among the top 20 bacteria at the species level, the relative abundance of *Streptococcus* sp. in the ZEN group (0.24%) was significantly increased compared with the control group (0.02%). However, this kind of bacteria in the ZEN + NMN group (0.16%), ZEN + GSH group (0.06%, $p < 0.05$), and ZEN + Melatonin group were decreased compared with the ZEN group (Figure 7E and Table S2). These results suggested that the levels of the dominant genus of the gut microbiota in mice exposed to ZEN were altered, which might affect myofiber types.

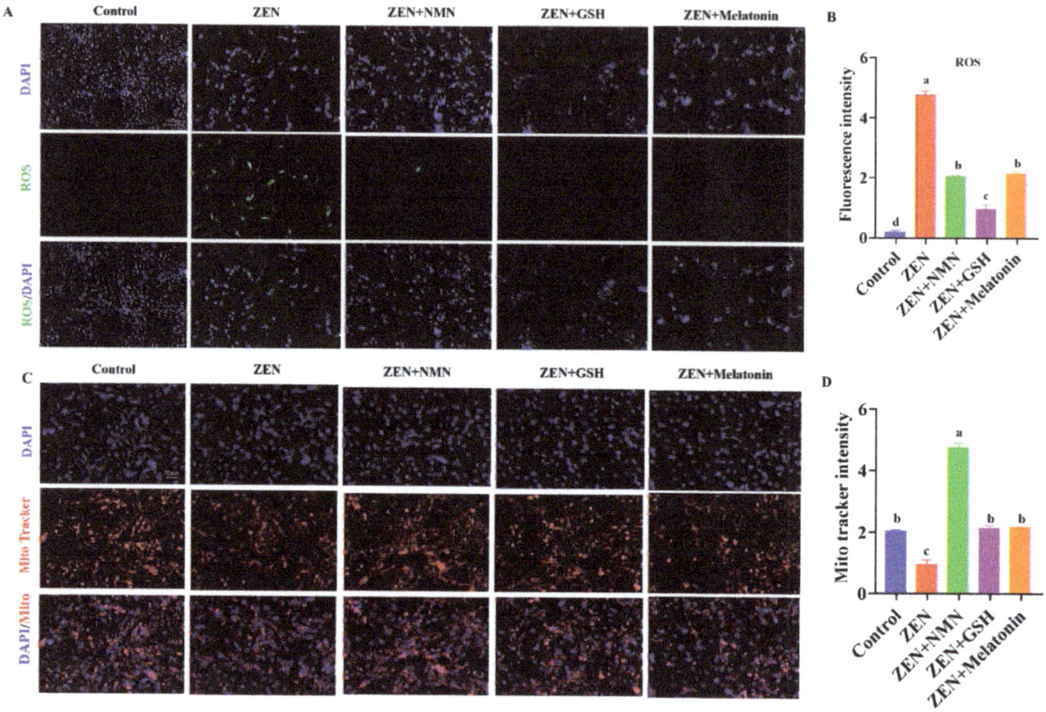

Figure 6. ZEN combined antioxidants affect muscle fiber transformation by mitochondria content and oxidative stress. (**A**,**B**) The ROS probe (**A**) (scale bar: 100 μm) and count analysis (**B**) of myoblasts in the aforementioned groups, (**C**,**D**) the images (**C**) (scale bar: 50 μm) and count analysis (**D**) of mito-tracker staining in myoblasts treated by ZEN or ZEN combined with antioxidant (NMN, GSH, and melatonin). Different letters (a–d) indicate significant differences ($p < 0.05$), and the same letters indicate no difference, $n = 3$.

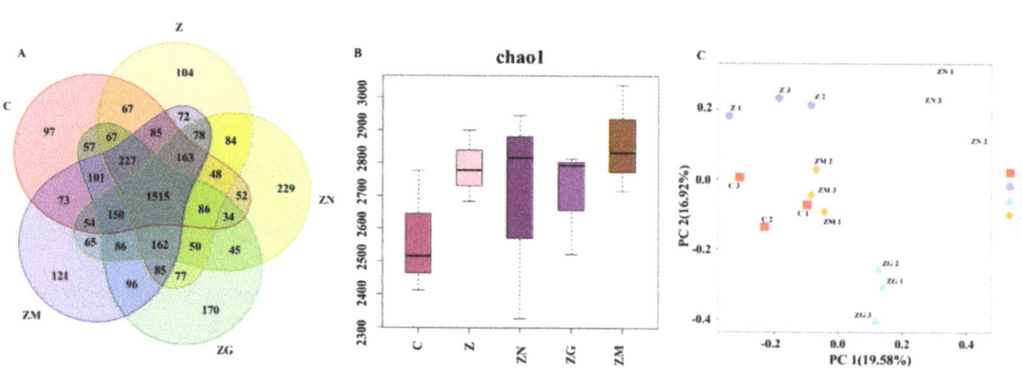

Figure 7. *Cont.*

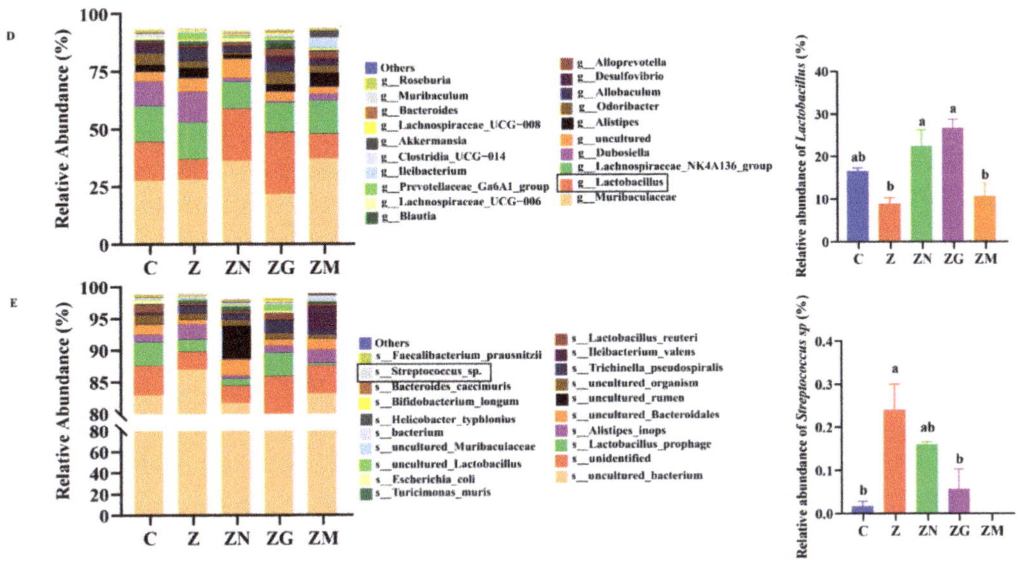

Figure 7. ZEN combined with antioxidants affects myofiber types through changes in the proportion of gut microbes. (**A**) Venn diagram of microbial OTUs in five groups (Z, ZEN group; ZN, ZEN + NMN group; ZG, ZEN + GSH group; ZM, ZEN + Melatonin group; C, Control group) of mice, (**B**) microbial richness (Chao1 index), (**C**) PCA plot of mice in five groups, (**D**,**E**) differences in the genus level (**D**) and species level (**E**) of gut microflora in the five groups of C57BL/6J mice. Different letters (a and b) indicate significant differences ($p < 0.05$), and the same letters indicate no difference, $n = 3$.

4. Discussion

ZEN contamination has become a significant global threat to animal and human health. Some evidence indicates that ZEN-contaminated food fed to livestock and poultry (wild boars [10] and Hy-Line Brown laying hens [11], etc.) could be detected in their residues in muscle tissue. However, the exact damage of ZEN on skeletal muscle remains largely unknown. In this study, ZEN exposure in vitro decreased the mRNA levels of the proliferative marker genes (*CCND1*, *CCNB1*, and *CDK2*) and the myogenic marker genes (*MYHC*, *MYOD1*, *MYOG*, and *MYF5*) in myoblasts, but GSH, melatonin, and NMN reversed this effect. Similar to these results, studies have shown that the addition of 10 μmol/L and 30 μmol/L of ZEN to donkey granulosa cells (dGCs) and porcine GCs cultured for 72 h in vitro significantly reduced mRNA expression of the *CDK2* and *CCNB1* genes [34,35]. Furthermore, the literature showed that the expression of *MYOG* and *MYOD1* genes in zebrafish larvae exposed to 4 μmol/L ZEN was significantly decreased, and their development was defective [36]. Gao et al. found that Yorkshire sow feeding of ZEN-contaminated food reduced the expression of *MYF5* and *MYOD1* in offspring [13]. These results indicated that ZEN exposure inhibits the proliferation and myogenic differentiation of myoblasts.

Oxidative stress refers to the stimulation of organisms by harmful factors, resulting in the excessive production of ROS, which disrupts the balance of the antioxidant system, impairs cell function, and ultimately leads to apoptosis or necrosis [37,38]. We found that ZEN increased ROS levels, which causes oxidative stress and mitochondrial damage. Studies have shown that ZEN can induce ROS production and cause oxidative stress at different dosages [37,39]. Cao et al. found that 25 μmol/L ZEN increased ROS production, decreased antioxidant enzyme activity, and induced oxidative damage in Sertoli cells in piglets [40]. The toxicity of ZEN to cardiac cells (H9c2) is achieved by increasing ROS levels, loss of mitochondrial transmembrane potential ($\Delta\Psi m$), and activation of caspase [41]. Furthermore, 10 μmol/L ZEN could induce oxidative stress in porcine embryos, increase reactive oxygen species (ROS) levels, and cause mitochondrial dysfunction to affect development, and melatonin could rescue the reproductive toxicity of ZEN to embryos [42]. Our results also showed that ZEN elevated ROS levels, decreased mitochondrial content, and caused mitochondrial swelling, but antioxidants attenuated the pathological symptoms. For the potential molecular mechanism of ZEN regulation on oxidative stress, Long et al. found that the sertoli cell TM4 of mice exposed to ZEN decreased the mRNA and protein levels of NF-E2-related factor-2 (Nrf2) and decreased the mRNA and protein levels of heme oxygenase 1 (HO-1), quinone oxidoreductase 1 (NQO1), γ-glutamylcysteine synthetase (γ-GCS), and glutathione peroxidase (GPX) in the downstream part of the Nrf2 signal pathway [20], which suggested that ZEN may inhibit the Nrf2/ARE signaling pathway, thus caused oxidative damage.

In this study, we found that ZEN produced oxidative stress, then reduced the average area and diameter of muscle fibers and exhibited the pathology of myolysis. Consistently, Yorkshire sows ingested ZEN-contaminated food, accompanied by pale sarcoplasm and optically empty vacuoles in the muscles, and even led to a decrease in myofiber diameter for the offspring [13]. Recently, Wang et al. found that zebrafish larvae treated with ZEN had a phenotype of reduced trunk myofibrils, reduced density, nuclei lysis, sarcoplasmic reticulum swelling, reduced distance of motion, and reduced mean velocity [36]. Except for these symptoms, ZEN exposure induced slow myofiber changes into fast myofiber in our work, but the addition of NMN, GSH, and melatonin restored this phenomenon. It was known that slow-twitch fibers' proportion was positively correlated to meat quality, including color, tenderness, water-holding capacity, etc. [43]. The slow-to-fast twitch fiber shift might be explained by slow-twitch fibers being more susceptible to oxidative stress than fast-twitch fibers, which was verified in the prepubertal gilts' slow-twitch fibers under ZEN exposure, increasing malondialdehyde (MDA) concentration and decreasing GPX-1 protein abundance and activity, but not in the fast-twitch myofibers [44]. In contrast, mice increased the mRNA and protein levels of Nrf2 and increased the protein levels of NQO1 and HO-1, thereby enhancing antioxidant capacity and inducing the fast-to-slow twitch fibers shift [45]. In this current work, we preliminarily explore the damaging effect of ZEN on muscle fibers caused by oxidative stress (increasing ROS levels and decreasing mitochondrial count), but the pathways (signaling pathways or cellular processes) through which ZEN affects have not yet been explored. It will be the focus of our future work.

Generally, the metabolism of ZEN and its derivatives in animals is through the liver and gastrointestinal transformation pathways and is finally excreted through feces and urine [46]. However, several works showed that ZEN and its metabolites were detected in the muscle tissues of pigs (hybrids of *Deutsches Edelschwein* and *Pietrain*) [46] and wild boar (*Sus scrofa*) [47] fed with ZEN-contaminated food. Specifically, the mean concentrations of ZEN in the liver, muscle, and kidney samples of wild boar (*Sus scrofa*) were 1.71 ng/g, 1.49 ng/g, and 0.65 ng/g, respectively, and the mean α-zearalenol values were 0.65 ng/g, 0.66 ng/g, and 0.77 ng/g, respectively. In pigs (hybrids of *Deutsches Edelschwein* and *Pietrain*) muscle tissue, zeranol with concentrations of up to 13.3 μg/kg along with α-zearalenol (up to 14.5 μg/kg) and traces of ZEN and taleranol were detected. This suggested that ZEN and its metabolites were able to accumulate in muscle tissue, and muscle was also a toxicity target tissue of ZEN. Currently, our work has found that ZEN direct treatment inhibited myogenic differentiation and changed the myofiber compositions in vitro, which indicated that ZEN at least partly causes these effects. However, we still cannot exclude the effect of its metabolites contributing to these phenotypes, and their contributions will be explored in a further study.

Beyond the direct effect of mycotoxin (such as ZEN) on skeletal muscle, mycotoxin also disrupts the intestinal flora [48], which influences bacterial community diversity, involving muscle function and development through the gut-muscle axis and leading to impairing bodily functions and ultimately reducing the quality of life [49]. Studies have shown that Lactobacillus strains (*Lactobacillus fermentum DR9*, *Lactobacillus paracasei OFS 0291*, and *L. helveticus OFS 1515*) protected the skeletal muscle from aging through antioxidant effects, of which *Lactobacillus fermentum DR9* had the strongest regulatory role [50]. We found that the results of 16S rDNA sequencing of cecal contents in this study showed that ZEN treatment decreased the relative abundance of *Lactobacillus* at the genus level, and ZEN + NMN and ZEN + GSH treatment groups rescued its abundance. In addition, at the species level, the relative abundance of *Streptococcus* sp. in the ZEN group was higher than that of the Control group, which was reported to produce lactic acid [51]. The limitation was that the exact roles of gut microbiota alteration induced by ZEN and antioxidant regulation on myofiber composition need to be further studied. Studies have shown that the small intestine of Merino Wethers can absorb lactic acid, which then enters the blood [52]. Furthermore, plasma lactate can affect muscle metabolic performance and muscle mass [53]. Compared with slow-twitch fibers, fast-twitch fibers facilitate the production of lactic acid through the glycolytic pathway [32,33]. However, whether lactic acid produced by *Streptococcus* sp. affects the composition of muscle fibers would be the focus of our future work.

In this study, ZEN exposure increases the level of ROS and damages mitochondria to cause oxidative stress, subsequently affecting the composition of myofibers. Interestingly, antioxidant gavage (NMN, GSH, and melatonin) all exhibited a protective effect on ZEN-damaged skeletal muscle to different extents. It was reported that NMN oral administration prevents doxorubicin-induced cardiotoxicity and skeletal muscle loss in mice by defending against oxidative stress, indicating by enhancing NAD^+ turnover and decreasing the accumulation of 4-hydroxy-2-nonenal (4-HNE), a biologically active aldehyde of lipid peroxidation [54]. Consistently, NMN rescued the ZEN-induced side effect of skeletal muscle also through the alleviation of oxidative stress, manifested by the reduction of ROS level and increasing mitochondrial content. However, the potential molecular mechanism of NMN against oxidative stress in skeletal muscle induced by ZEN needs further exploration.

GSH is an endogenous antioxidant, and the salvage mechanism of exogenous GSH addition to damaged skeletal muscle is still unclear. The content of GSH in mice skeletal muscle stem cells (MuSCs) changed with age, and a high content of endogenous GSH increased the expression of *NRF2* by increasing the total transcript levels of microsomal glutathione S-transferase 1 (Mgst1), as an endoplasmic reticulum glutathione transferase and peroxidase, increased mitochondrial turnover to decrease mitochondrial ROS and promote myogenesis [55]. Similarly, the exogenous addition of GSH rescued the proliferation of myoblasts in ZEN-injured skeletal muscle in our study, but whether GSH rescued damage to skeletal muscle induced by ZEN through the NRF2 signaling pathway needs further study. For melatonin, 50 mg/kg BW partly alleviated the muscle damage induced by carbon tetrachloride (CCl_4) [56]. In our work, myolysis pathology was not observed in the ZEN+ Melatonin group, which was also associated with the promotion of fast-to-slow myofiber shift. However, the molecular mechanism of melatonin rescue ZEN inducing myolysis and the shift in slow-to-fast twitch fibers still need to be further explored.

5. Conclusions

Summarily, we found that ZEN inhibits myoblast proliferation and differentiation in vitro, and antioxidants (NMN, GSH, and melatonin) rescue these effects. In addition, ZEN exposure induced slow-to-fast myofiber shift in vivo and in vitro, which was restored by antioxidant treatments. Pathologically, ZEN led to ROS production and oxidative stress, and subsequently, pathological symptoms occurred, including mitochondrial dysfunction, myolysis, etc., which were combated by antioxidant treatment. Moreover, ZEN and antioxidants might indirectly affect muscle fibers via a reduced relative abundance of *Lactobacillus* and an increased relative abundance of *Streptococcus* sp., respectively (Figure 8). Collectively, this study explores the pathology of ZEN on muscle and the salvage roles of NMN, GSH, and melatonin for ZEN toxicity, which would provide new insights and promising attenuation solutions for ZEN-damaging skeletal muscle. However, the underlying molecular mechanism of ZEN and the 3 antioxidants (NMN, GSH, and melatonin) in the skeletal muscle still needs to be explored.

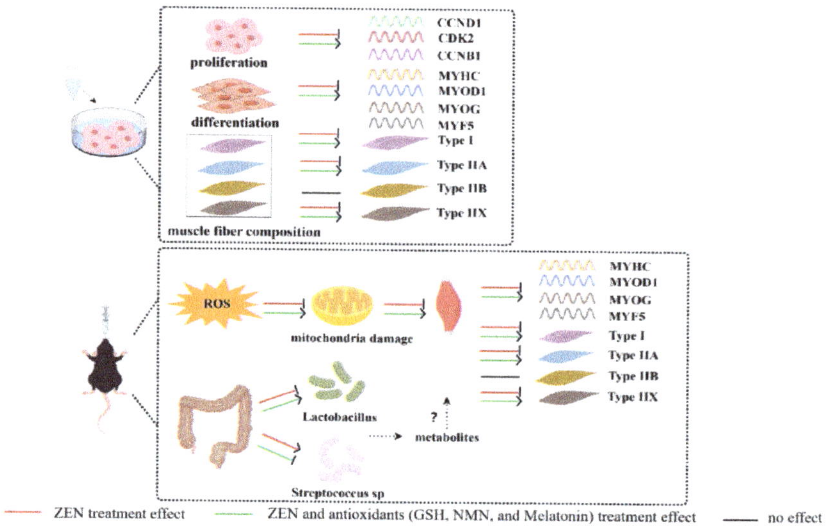

Figure 8. A schematic diagram of ZEN damage to skeletal muscle in vitro and in vivo and salvage of 3 antioxidants (NMN, GSH, and melatonin) was drawn by "Figdraw 2.0" software.

Supplementary Materials: The following supporting information can be downloaded at: https://www.mdpi.com/article/10.3390/antiox14050528/s1.

Author Contributions: Y.X.: conceptualization, methodology, resources, and writing—review and editing; J.Z., W.F. and D.L.: software, validation, formal analysis, investigation, visualization, data curation, and writing—original draft preparation, Y.L., X.X. and J.L.: supervision, project administration, funding acquisition. All authors have read and agreed to the published version of the manuscript.

Funding: This research was funded by the National Natural Sciences Foundation of China (Grant No. 31902154), the Natural Science Foundation of Sichuan Province (Grant No. 2023NSFSC0233), the Fundamental Research Funds for the Central Universities of Southwest Minzu University (Grant No. ZYN2023095&ZYN2024084), and the Sichuan Province Science and Technology Plan (SSTP) (Grant No. 2024ZHXC0003).

Institutional Review Board Statement: All experimental procedures were approved by the Animals Care and Ethics Committee of Southwest Minzu University (SMU-CAVS-220601001).

Informed Consent Statement: Not applicable.

Data Availability Statement: The data that support the findings of this study are available from the corresponding author upon reasonable request. The raw data supporting the findings of this study are available from the corresponding authors upon reasonable request. The 16S rDNA sequencing raw data during this study were submitted to the National Center of Biotechnology Information (No. PRJNA1173076).

Acknowledgments: We are grateful for the help and understanding of all the laboratory staff during the experiment.

Conflicts of Interest: The authors declare no conflict of interest.

Abbreviations

CCND1	Cyclin d1
CCNB1	Cyclin b1
CDK2	Cyclin-dependent kinase 2
MYHC	Myosin heavy chain
MYOD1	Myogenic differentiation 1
MYOG	Myogenin
MYF5	Myogenic factor 5
TA	Tibialis anterior
GAS	Gastrocnemius muscle
EDL	Extensor digitorum longus
SOL	Soleus
DAPI	4′,6-diamidino-2-phenylindole
ROS	Reactive oxygen species
DCFH-DA	2′,7′-dichlorodihydrofluorescein diacetate
TEM	Transmission electron microscopy

References

1. Li, Z.; Ma, T.; Liu, Y.; Liu, W.; Zhao, X.; Zhang, G.; Wang, J.; Zhang, Y. Screening and Mechanism Study of Three Antagonistic Drugs, Oxysophoridine, Rutin, and Phellodendrine, against Zearalenone-Induced Reproductive Toxicity in Ovine Oocytes. *Antioxidants* **2024**, *13*, 752. [CrossRef] [PubMed]
2. Yan, J.; Kong, L.; Zhang, X.; Yu, M.; Zhu, K.; Zhao, A.; Shi, D.; Sun, Y.; Wang, J.; Shen, W.; et al. Maternal Zearalenone Exposure Affects Gut Microbiota and Follicular Development in Suckled Offspring. *J. Agric. Food Chem.* **2022**, *70*, 15570–15582. [CrossRef] [PubMed]
3. Zhou, J.; Zhao, L.; Huang, S.; Liu, Q.; Ao, X.; Lei, Y.; Ji, C.; Ma, Q. Zearalenone toxicosis on reproduction as estrogen receptor selective modulator and alleviation of zearalenone biodegradative agent in pregnant sows. *J. Anim. Sci. Biotechnol.* **2022**, *13*, 36. [CrossRef] [PubMed]
4. Bao, L.; Huang, Y.; Gu, F.; Liu, W.; Guo, Y.; Chen, H.; Wang, K.; Wu, Z.; Li, J. Zearalenone induces liver injury in mice through ferroptosis pathway. *Sci. Total. Environ.* **2024**, *952*, 175875. [CrossRef]
5. Xu, J.; Li, S.; Jiang, L.; Gao, X.; Liu, W.; Zhu, X.; Huang, W.; Zhao, H.; Wei, Z.; Wang, K.; et al. Baicalin protects against zearalenone-induced chicks liver and kidney injury by inhibiting expression of oxidative stress, inflammatory cytokines and caspase signaling pathway. *Int. Immunopharmacol.* **2021**, *100*, 108097. [CrossRef]
6. Shirakawa, T.; Miyawaki, A.; Kawamoto, T.; Kokabu, S. Natural Compounds Attenuate Denervation-Induced Skeletal Muscle Atrophy. *Int. J. Mol. Sci.* **2021**, *22*, 8310. [CrossRef]
7. Campbell, W.G.; Gordon, S.E.; Carlson, C.J.; Pattison, J.S.; Hamilton, M.T.; Booth, F.W. Differential global gene expression in red and white skeletal muscle. *Am. J. Physiol. Cell Physiol.* **2001**, *280*, C763–C768. [CrossRef]
8. Ciciliot, S.; Rossi, A.C.; Dyar, K.A.; Blaauw, B.; Schiaffino, S. Muscle type and fiber type specificity in muscle wasting. *Int. J. Biochem. Cell Biol.* **2013**, *45*, 2191–2199. [CrossRef]
9. Huo, W.; Weng, K.; Gu, T.; Zhang, Y.; Zhang, Y.; Chen, G.; Xu, Q. Effect of muscle fiber characteristics on meat quality in fast- and slow-growing ducks. *Poult. Sci.* **2021**, *100*, 101264. [CrossRef]

10. Gajecka, M.; Slawuta, P.; Nicpon, J.; Kolacz, R.; Kielbowicz, Z.; Zielonka, L.; Dabrowski, M.; Szweda, W.; Gajecki, M.; Nicpon, J. Zearalenone and its metabolites in the tissues of female wild boars exposed per os to mycotoxins. *Toxicon* **2016**, *114*, 1–12. [CrossRef]
11. Zhu, F.; Zhu, L.; Xu, J.; Wang, Y.; Wang, Y. Effects of moldy corn on the performance, antioxidant capacity, immune function, metabolism and residues of mycotoxins in eggs, muscle, and edible viscera of laying hens. *Poult. Sci.* **2023**, *102*, 102502. [CrossRef] [PubMed]
12. Damiano, S.; Longobardi, C.; Ferrara, G.; Piscopo, N.; Riccio, L.; Russo, V.; Meucci, V.; De Marchi, L.; Esposito, L.; Florio, S.; et al. Oxidative Status and Histological Evaluation of Wild Boars' Tissues Positive for Zearalenone Contamination in the Campania Region, Southern Italy. *Antioxidants* **2023**, *12*, 1748. [CrossRef] [PubMed]
13. Gao, R.; Meng, Q.; Li, J.; Liu, M.; Zhang, Y.; Bi, C.; Shan, A. Modified halloysite nanotubes reduce the toxic effects of zearalenone in gestating sows on growth and muscle development of their offsprings. *J. Anim. Sci. Biotechnol.* **2016**, *7*, 14. [CrossRef] [PubMed]
14. Bouaziz, C.; Sharaf El Dein, O.; El Golli, E.; Abid-Essefi, S.; Brenner, C.; Lemaire, C.; Bacha, H. Different apoptotic pathways induced by zearalenone, T-2 toxin and ochratoxin A in human hepatoma cells. *Toxicology* **2008**, *254*, 19–28. [CrossRef]
15. Kouadio, J.H.; Mobio, T.A.; Baudrimont, I.; Moukha, S.; Dano, S.D.; Creppy, E.E. Comparative study of cytotoxicity and oxidative stress induced by deoxynivalenol, zearalenone or fumonisin B1 in human intestinal cell line Caco-2. *Toxicology* **2005**, *213*, 56–65. [CrossRef]
16. Huang, G.; He, Y.; Hong, L.; Zhou, M.; Zuo, X.; Zhao, Z. Restoration of NAD(+) homeostasis protects C2C12 myoblasts and mouse levator ani muscle from mechanical stress-induced damage. *Anim. Cells Syst.* **2022**, *26*, 192–202. [CrossRef]
17. Morel, G.; Bonnet, P.; Cossec, B.; Morel, S.; Cour, C.; Lambert, A.M.; Roure, M.B.; Brondeau, M.T. The role of glutathione and cysteine conjugates in the nephrotoxicity of o-xylene in rats. *Arch. Toxicol.* **1998**, *72*, 553–558. [CrossRef]
18. Lana, J.V.; Rios, A.; Takeyama, R.; Santos, N.; Pires, L.; Santos, G.S.; Rodrigues, I.J.; Jeyaraman, M.; Purita, J.; Lana, J.F. Nebulized Glutathione as a Key Antioxidant for the Treatment of Oxidative Stress in Neurodegenerative Conditions. *Nutrients* **2024**, *16*, 2476. [CrossRef]
19. Gunata, M.; Parlakpinar, H.; Acet, H.A. Melatonin: A review of its potential functions and effects on neurological diseases. *Rev. Neurol.* **2020**, *176*, 148–165. [CrossRef]
20. Long, M.; Yang, S.H.; Shi, W.; Li, P.; Guo, Y.; Guo, J.; He, J.B.; Zhang, Y. Protective effect of proanthocyanidin on mice Sertoli cell apoptosis induced by zearalenone via the Nrf2/ARE signalling pathway. *Environ. Sci. Pollut. Res. Int.* **2017**, *24*, 26724–26733. [CrossRef]
21. Fu, W.; Dai, C.; Ma, Z.; Li, Q.; Lan, D.; Sun, C.; Wu, X.; Li, J.; Wang, S. Enhanced glutathione production protects against zearalenone-induced oxidative stress and ferroptosis in female reproductive system. *Food Chem. Toxicol.* **2024**, *185*, 114462. [CrossRef] [PubMed]
22. Jani, N.; Ziogas, J.; Angus, J.A.; Wright, C.E. Exogenous glutathione is essential in the testing of antioxidant capacity using radical-induced haemolysis. *J. Pharmacol. Toxicol. Methods* **2012**, *65*, 142–146. [CrossRef]
23. Chen, B.; You, W.; Shan, T. Myomaker, and Myomixer-Myomerger-Minion modulate the efficiency of skeletal muscle development with melatonin supplementation through Wnt/beta-catenin pathway. *Exp. Cell Res.* **2019**, *385*, 111705. [CrossRef]
24. Dai, C.; Hou, M.; Yang, X.; Wang, Z.; Sun, C.; Wu, X.; Wang, S. Increased NAD(+) levels protect female mouse reproductive system against zearalenone-impaired glycolysis, lipid metabolism, antioxidant capacity and inflammation. *Reprod. Toxicol.* **2024**, *124*, 108530. [CrossRef]
25. Han, B.H.; Lee, H.K.; Jang, S.H.; Tai, A.L.; Jang, Y.J.; Yoon, J.J.; Kim, H.Y.; Lee, H.S.; Lee, Y.J.; Kang, D.G. Effect of Geumgwe-Sinkihwan on Renal Dysfunction in Ischemia/Reperfusion-Induced Acute Renal Failure Mice. *Nutrients* **2021**, *13*, 3859. [CrossRef]
26. Khan, S.; Adhikari, J.S.; Rizvi, M.A.; Chaudhury, N.K. Radioprotective potential of melatonin against (6)(0)Co gamma-ray-induced testicular injury in male C57BL/6 mice. *J. Biomed. Sci.* **2015**, *22*, 61. [CrossRef]
27. Jensen, E.C. Quantitative analysis of histological staining and fluorescence using ImageJ. *Anat. Rec.* **2013**, *296*, 378–381. [CrossRef]
28. Arneson-Wissink, P.C.; Doles, J.D. Quantification of Muscle Stem Cell Differentiation Using Live-Cell Imaging and Eccentricity Measures. *Methods. Mol. Biol.* **2022**, *2429*, 455–471.
29. Lv, W.Q.; Lin, X.; Shen, H.; Liu, H.M.; Qiu, X.; Li, B.Y.; Shen, W.D.; Ge, C.L.; Lv, F.Y.; Shen, J.; et al. Human gut microbiome impacts skeletal muscle mass via gut microbial synthesis of the short-chain fatty acid butyrate among healthy menopausal women. *J. Cachexia Sarcopenia Muscle* **2021**, *12*, 1860–1870. [CrossRef]
30. Chen, L.H.; Chang, S.S.; Chang, H.Y.; Wu, C.H.; Pan, C.H.; Chang, C.C.; Chan, C.H.; Huang, H.Y. Probiotic supplementation attenuates age-related sarcopenia via the gut-muscle axis in SAMP8 mice. *J. Cachexia Sarcopenia Muscle* **2022**, *13*, 515–531. [CrossRef]

31. Misra, S.; Raghuwanshi, S.; Gupta, P.; Saxena, R.K. Examine growth inhibition pattern and lactic acid production in Streptococcus mutans using different concentrations of xylitol produced from Candida tropicalis by fermentation. *Anaerobe* **2012**, *18*, 273–279. [CrossRef]
32. Friman, G.; Ilback, N.G.; Beisel, W.R. Effects of Streptococcus pneumoniae, Salmonella typhimurium and Francisella tularensis infections on oxidative, glycolytic and lysosomal enzyme activity in red and white skeletal muscle in the rat. *Scand. J. Infect. Dis.* **1984**, *16*, 111–119. [CrossRef]
33. Hansen, A.K.; Clausen, T.; Nielsen, O.B. Effects of lactic acid and catecholamines on contractility in fast-twitch muscles exposed to hyperkalemia. *Am. J. Physiol. Cell Physiol.* **2005**, *289*, C104–C112. [CrossRef]
34. Zhang, G.L.; Song, J.L.; Ji, C.L.; Feng, Y.L.; Yu, J.; Nyachoti, C.M.; Yang, G.S. Zearalenone Exposure Enhanced the Expression of Tumorigenesis Genes in Donkey Granulosa Cells via the PTEN/PI3K/AKT Signaling Pathway. *Front. Genet.* **2018**, *9*, 293. [CrossRef]
35. Zhang, G.L.; Song, J.L.; Zhou, Y.; Zhang, R.Q.; Cheng, S.F.; Sun, X.F.; Qin, G.Q.; Shen, W.; Li, L. Differentiation of sow and mouse ovarian granulosa cells exposed to zearalenone in vitro using RNA-seq gene expression. *Toxicol. Appl. Pharmacol.* **2018**, *350*, 78–90. [CrossRef]
36. Wang, Y.; Li, S.; Cheng, Z.; Zhang, Z.; Xu, Y.; Zhang, H.; Xu, T.; Chen, J.; Yin, D.; Yan, W.; et al. Caspase-8 dependent apoptosis contributes to dyskinesia caused by muscle defects and neurotoxicity in zebrafish exposed to zearalenone. *Food Chem. Toxicol.* **2024**, *186*, 114516. [CrossRef]
37. Tatay, E.; Font, G.; Ruiz, M.J. Cytotoxic effects of zearalenone and its metabolites and antioxidant cell defense in CHO-K1 cells. *Food Chem. Toxicol.* **2016**, *96*, 43–49. [CrossRef]
38. Feng, Y.Q.; Zhao, A.H.; Wang, J.J.; Tian, Y.; Yan, Z.H.; Dri, M.; Shen, W.; De Felici, M.; Li, L. Oxidative stress as a plausible mechanism for zearalenone to induce genome toxicity. *Gene* **2022**, *829*, 146511. [CrossRef]
39. Salem, I.B.; Boussabbeh, M.; Neffati, F.; Najjar, M.F.; Abid-Essefi, S.; Bacha, H. Zearalenone-induced changes in biochemical parameters, oxidative stress and apoptosis in cardiac tissue: Protective role of crocin. *Hum. Exp. Toxicol.* **2016**, *35*, 623–634. [CrossRef]
40. Cao, L.; Zhao, J.; Ma, L.; Chen, J.; Xu, J.; Rahman, S.U.; Feng, S.; Li, Y.; Wu, J.; Wang, X. Lycopene attenuates zearalenone-induced oxidative damage of piglet sertoli cells through the nuclear factor erythroid-2 related factor 2 signaling pathway. *Ecotoxicol. Environ. Saf.* **2021**, *225*, 112737. [CrossRef]
41. Ben Salem, I.; Boussabbeh, M.; Pires Da Silva, J.; Guilbert, A.; Bacha, H.; Abid-Essefi, S.; Lemaire, C. SIRT1 protects cardiac cells against apoptosis induced by zearalenone or its metabolites alpha- and beta-zearalenol through an autophagy-dependent pathway. *Toxicol. Appl. Pharmacol.* **2017**, *314*, 82–90. [CrossRef]
42. Xu, Y.; Zhang, K.H.; Sun, M.H.; Lan, M.; Wan, X.; Zhang, Y.; Sun, S.C. Protective Effects of Melatonin Against Zearalenone Toxicity on Porcine Embryos in vitro. *Front. Pharmacol.* **2019**, *10*, 327. [CrossRef]
43. Senoner, T.; Dichtl, W. Oxidative Stress in Cardiovascular Diseases: Still a Therapeutic Target? *Nutrients* **2019**, *11*, 2090. [CrossRef]
44. Rudolph, T.E.; Roach, C.M.; Baumgard, L.H.; Ross, J.W.; Keating, A.F.; Selsby, J.T. The impact of Zearalenone on heat-stressed skeletal muscle in pigs. *J. Anim. Sci.* **2022**, *100*, skac215. [CrossRef]
45. Huang, S.; Sun, H.; Lin, D.; Huang, X.; Chen, R.; Li, M.; Huang, J.; Guo, F. Camellia oil exhibits anti-fatigue property by modulating antioxidant capacity, muscle fiber, and gut microbial composition in mice. *J. Food. Sci.* **2024**, *89*, 2465–2481. [CrossRef]
46. Zollner, P.; Jodlbauer, J.; Kleinova, M.; Kahlbacher, H.; Kuhn, T.; Hochsteiner, W.; Lindner, W. Concentration levels of zearalenone and its metabolites in urine, muscle tissue, and liver samples of pigs fed with mycotoxin-contaminated oats. *J. Agric. Food Chem.* **2002**, *50*, 2494–2501. [CrossRef]
47. Longobardi, C.; Damiano, S.; Ferrara, G.; Montagnaro, S.; Meucci, V.; Intorre, L.; Bacci, S.; Esposito, L.; Piscopo, N.; Rubino, A.; et al. Zearalenone (ZEN) and Its Metabolite Levels in Tissues of Wild Boar (*Sus scrofa*) from Southern Italy: A Pilot Study. *Toxins* **2023**, *15*, 56. [CrossRef]
48. Gao, Y.; Meng, L.; Liu, H.; Wang, J.; Zheng, N. The Compromised Intestinal Barrier Induced by Mycotoxins. *Toxins* **2020**, *12*, 619. [CrossRef]
49. Grosicki, G.J.; Fielding, R.A.; Lustgarten, M.S. Gut Microbiota Contribute to Age-Related Changes in Skeletal Muscle Size, Composition, and Function: Biological Basis for a Gut-Muscle Axis. *Calcif. Tissue. Int.* **2018**, *102*, 433–442. [CrossRef]
50. Hor, Y.Y.; Ooi, C.H.; Lew, L.C.; Jaafar, M.H.; Lau, A.S.; Lee, B.K.; Azlan, A.; Choi, S.B.; Azzam, G.; Liong, M.T. The molecular mechanisms of probiotic strains in improving ageing bone and muscle of d-galactose-induced ageing rats. *J. Appl. Microbiol.* **2021**, *130*, 1307–1322. [CrossRef]
51. Kupp, L.I.; Rosen, S.; Beck, F.M. Effect of anti-oxidants on growth and lactic acid production by Streptococcus mutans. *J. Dent. Res.* **1985**, *64*, 1016–1018. [CrossRef]

52. Ding, Z.; Xu, Y. Lactic acid is absorbed from the small intestine of sheep. *J. Exp. Zool. A Comp. Exp. Biol.* **2003**, *295*, 29–36. [CrossRef]
53. Giron, M.; Thomas, M.; Dardevet, D.; Chassard, C.; Savary-Auzeloux, I. Gut microbes and muscle function: Can probiotics make our muscles stronger? *J. Cachexia Sarcopenia Muscle* **2022**, *13*, 1460–1476. [CrossRef]
54. Margier, M.; Kuehnemann, C.; Hulo, N.; Morales, J.; Ashok Kumaar, P.V.; Cros, C.; Cannelle, H.; Charmetant, J.; Verdin, E.; Canault, M.; et al. Nicotinamide Mononucleotide Administration Prevents Doxorubicin-Induced Cardiotoxicity and Loss in Physical Activity in Mice. *Cells* **2022**, *12*, 108. [CrossRef]
55. Benjamin, D.I.; Brett, J.O.; Both, P.; Benjamin, J.S.; Ishak, H.L.; Kang, J.; Kim, S.; Chung, M.; Arjona, M.; Nutter, C.W.; et al. Multiomics reveals glutathione metabolism as a driver of bimodality during stem cell aging. *Cell. Metab.* **2023**, *35*, 472–486.e476. [CrossRef]
56. Sokolovic, D.T.; Lilic, L.; Milenkovic, V.; Stefanovic, R.; Ilic, T.P.; Mekic, B.; Ilic, I.; Stojanovic, N.M.; Ilic, I.R. Effects of melatonin on oxidative stress parameters and pathohistological changes in rat skeletal muscle tissue following carbon tetrachloride application. *Saudi. Pharm. J.* **2018**, *26*, 1044–1050. [CrossRef]

antioxidants

MDPI

Review

Antioxidant Defense and Redox Signaling in Elite Soccer Players: Insights into Muscle Function, Recovery, and Training Adaptations

Qing Meng [1,2] and Chun-Hsien Su [3,4,*]

1 School of Physical Education, Huaqiao University, Xiamen 361021, China; mq@hqu.edu.cn
2 Sport and Health Research Center, Huaqiao University, Xiamen 361021, China
3 Department of Exercise and Health Promotion, Chinese Culture University, Taipei 111369, Taiwan
4 College of Kinesiology and Health, Chinese Culture University, Taipei 111369, Taiwan
* Correspondence: chsu@ulive.pccu.edu.tw; Tel.: +886-975159678

Abstract: Elite soccer places significant neuromuscular and metabolic stress on athletes, leading to elevated production of reactive oxygen and nitrogen species (RONS), particularly in skeletal muscle, where intense contractile activity and increased oxygen flux drive oxidative processes. These reactive species play a dual role in skeletal muscle, supporting adaptive signaling at controlled levels while causing oxidative damage when poorly regulated. This paper presents an integrated synthesis of current knowledge on redox biology in elite soccer players, focusing on the origins and regulation of RONS, the functions of enzymatic and non-enzymatic antioxidant systems, and how both RONS and antioxidant responses influence muscle performance, fatigue, recovery, and long-term physiological adaptation. Drawing on studies conducted between 2000 and 2025, the discussion underscores the seasonal fluctuations in oxidative stress, individual variability in redox responses, and the potential adverse effects of unsystematic antioxidant supplementation. The analysis also emphasizes the value of using biomarker-guided, periodized antioxidant interventions tailored to training demands. Future directions include longitudinal tracking and the use of AI-assisted monitoring to enable personalized strategies for maintaining redox balance and optimizing performance in elite sport.

Keywords: mitochondrial biogenesis; exercise-induced stress; reactive species regulation; training periodization; injury prevention; biomarker monitoring

1. Introduction

Redox regulation is essential for muscle adaptation and performance in elite soccer, where players are regularly exposed to intense physical and metabolic stress. Reactive oxygen and nitrogen species (RONS), generated during exercise, influence muscle remodeling, fatigue, and recovery [1,2]. While moderate levels of RONS support physiological adaptation, excessive accumulation can disrupt cellular balance and contribute to oxidative stress and tissue damage [3,4]. The ability to maintain redox homeostasis through endogenous antioxidant systems is therefore critical to sustaining performance across training cycles and competitive demands [5].

This review integrates current knowledge on redox biology and antioxidant defense in elite soccer players. It begins with a foundational overview of the physiological stressors encountered in the sport and the key antioxidant systems in skeletal muscle. Subsequent sections explore molecular signaling mechanisms, redox biomarkers, and practical strategies for recovery and performance optimization.

1.1. Physiological Demands of Elite Soccer and Oxidative Load

The high-intensity, intermittent nature of elite soccer involves repeated sprints, rapid directional changes, jumping, and decelerative actions. These movement patterns result in elevated mitochondrial oxygen consumption and increased production of RONS in skeletal muscle, especially during eccentric contractions and transient ischemia–reperfusion events [6–10]. If not properly regulated, this oxidative load can impair calcium handling, promote inflammation, and delay muscle recovery [9,11].

The physiological strain is magnified during congested match schedules, where recovery periods are shortened and redox imbalances can accumulate [12,13]. Biomarkers such as creatine kinase and C-reactive protein have been shown to rise in response to such stress, reflecting muscle damage and systemic inflammation [10,14,15]. Managing oxidative stress through individualized recovery strategies, nutritional support, and training load adjustment is essential to prevent fatigue-related decline and reduce injury risk [11].

1.2. Antioxidant Systems in Skeletal Muscle

Skeletal muscle maintains redox balance through a coordinated antioxidant defense system. This includes enzymatic antioxidants such as superoxide dismutase (SOD), catalase (CAT), and glutathione peroxidase, which work to neutralize reactive species produced during exercise [9,10]. In addition, non-enzymatic antioxidants such as glutathione, vitamin C, and vitamin E support this system by stabilizing reactive intermediates and protecting cellular components from oxidative damage [16,17].

In elite athletes, regular training stimulates adaptations in the antioxidant system, improving the muscle's ability to handle oxidative stress and enhancing recovery capacity [4,5]. However, these benefits can be diminished when antioxidant capacity is exceeded or when exogenous supplements are used inappropriately, potentially interfering with essential redox signaling pathways [18–22]. For this reason, antioxidant interventions should be carefully matched to the athlete's training load and physiological condition in order to maintain adaptive benefits while avoiding disruption of beneficial molecular responses [9,23].

2. Literature Selection and Scope

This work utilizes a narrative approach to explore antioxidant defense mechanisms and redox signaling pathways in the context of elite soccer performance. The aim is to present an integrative perspective on skeletal muscle oxidative stress, adaptive responses, and their relevance to training, recovery, and performance optimization in high-level athletes. The discussion emphasizes conceptual synthesis and the practical application of redox biology to support effective performance management strategies.

The relevant literature was identified through a focused search of peer-reviewed journal articles published between 2000 and 2025 across databases including PubMed, Scopus, Web of Science, and EBSCO. Search terms were combinations of the following keywords: "reactive oxygen species", "reactive nitrogen species", "oxidative stress", "antioxidants", "skeletal muscle", "soccer performance", "redox signaling", "exercise-induced adaptation", and "training recovery". Articles were selected based on their relevance to three core themes: (1) the physiological production and regulatory role of reactive species during exercise; (2) antioxidant defense systems in skeletal muscle; and (3) training and recovery strategies in elite soccer players that influence redox balance.

Both original research and review papers were considered to ensure comprehensive coverage of current knowledge. Studies involving human participants, exceptionally trained or elite athletic populations, were prioritized. Research with direct implications for soccer-specific performance, rather than general fitness or clinical populations, was emphasized. Rather than following a strict systematic review protocol, this narrative review allows for thematic exploration across key domains, including mitochondrial adaptations, antioxidant enzyme regulation, oxidative stress biomarkers, periodization, and nutritional interventions. This flexible approach facilitates a critical discussion of emerging evidence, identifies knowledge gaps, and highlights practical considerations for sports scientists and performance practitioners.

3. Redox Biology and Antioxidant Defenses in Skeletal Muscle

3.1. Sources and Types of Reactive Species During Exercise

During intense physical activity, particularly under the high-intensity, intermittent demands of elite soccer, the generation of reactive oxygen and nitrogen species (RONS) increases markedly. These include reactive oxygen species (ROS), such as superoxide anion ($O_2\bullet^-$), hydrogen peroxide (H_2O_2), and hydroxyl radical (•OH), as well as reactive nitrogen species like nitric oxide (NO•) and its downstream product peroxynitrite ($ONOO^-$), each varying in reactivity, half-life, and cellular target [24,25].

The mitochondria are the primary intracellular source of ROS during aerobic metabolism. Electron leakage at complexes I and III of the electron transport chain leads to partial reduction of oxygen and formation of superoxide, particularly under high ATP turnover in active muscle fibers [7]. Other enzymatic systems also contribute—NADPH oxidases (NOX2/NOX4) are activated by mechanical stress and inflammatory cues [26]; xanthine oxidase becomes active during ischemia–reperfusion events; and lipoxygenases and cyclooxygenases contribute additional ROS under inflammatory conditions [25].

A key component of RNS production is nitric oxide (NO), which plays an essential role in skeletal muscle vasodilation, mitochondrial respiration regulation, and redox signaling. NO is synthesized from L-arginine by the enzyme nitric oxide synthase (NOS), which exists in three isoforms: neuronal NOS (nNOS), endothelial NOS (eNOS), and inducible NOS (iNOS) [24]. In skeletal muscle, nNOS is the dominant isoform and is localized near the sarcolemma, where it regulates local blood flow, glucose uptake, and contractile efficiency [25]. eNOS, primarily expressed in vascular endothelium, contributes to perfusion regulation via shear-stress–mediated NO release [26]. Under physiological conditions, nNOS and eNOS generate NO in a controlled manner, supporting adaptive cellular processes.

However, under oxidative stress or cofactor depletion—such as reduced tetrahydrobiopterin availability—NOS may become uncoupled, producing superoxide instead of NO [27]. Furthermore, NO can rapidly react with superoxide to form peroxynitrite ($ONOO^-$), a potent oxidant that can damage proteins, lipids, and nucleic acids. While NO plays a signaling role at physiological levels, its dysregulation contributes to nitrosative stress and impairs muscle function when homeostasis is lost [28].

The extent and location of RONS generation are influenced by contraction type, oxygen availability, and fiber type composition. Fast-twitch (type II) fibers, which dominate sprinting and explosive efforts, are particularly prone to oxidative stress due to high metabolic activity and relatively low antioxidant capacity [28,29]. These fibers are heavily recruited in elite soccer, especially during repeated efforts under fatigue.

In addition, cycles of restricted blood flow and reoxygenation—as seen during intermittent play—mimic ischemia–reperfusion conditions, further elevating ROS and RNS levels [30]. This dynamic redox environment reinforces the need for well-coordinated antioxidant defenses and recovery strategies tailored to the demands of elite soccer competition.

3.2. Antioxidant Defense Systems in Skeletal Muscle

To counteract exercise-induced RONS, skeletal muscle relies on a well-organized antioxidant defense system composed of enzymatic and non-enzymatic components. These systems maintain redox homeostasis, protect muscle integrity, and facilitate recovery and adaptation [31].

The primary enzymatic antioxidants include superoxide dismutase (SOD), CAT, and glutathione peroxidase (GPx). SOD converts superoxide into hydrogen peroxide, which is then broken down by CAT or reduced by GPx using glutathione as a substrate. These enzymes are compartmentalized within the cell: SOD1 is cytosolic; SOD2 is mitochondrial; and CAT resides in peroxisomes [31–33]. Their expression is regulated by redox-sensitive transcription factors, particularly Nuclear Factor Erythroid 2–Related Factor 2 (Nrf2), which is activated under oxidative stress to enhance antioxidant gene expression [31].

Complementing the enzymatic defenses are non-enzymatic antioxidants such as reduced glutathione (GSH), vitamin C, and vitamin E [34–36]. GSH is a direct scavenger and a cofactor in enzymatic detoxification [34]. Vitamin C exerts its antioxidant effects primarily in aqueous compartments, where it neutralizes free radicals and facilitates the regeneration of oxidized vitamin E to its active, reduced form, thereby preserving lipid membrane stability and extending the functional lifespan of vitamin E molecules [35]. In turn, vitamin E contributes to the stabilization of lipid membranes by preventing lipid peroxidation [37]. Additionally, dietary polyphenols such as quercetin, curcumin, and resveratrol play a role by modulating oxidative signaling pathways and reducing inflammation [38].

In elite athletes, these antioxidant systems are not static but dynamically adapt to training load, recovery status, and nutritional intake. As seen in training, regular exposure to moderate oxidative stress can upregulate these systems, enhancing resilience and performance [4]. However, disruptions in this balance, through overtraining, poor recovery, or inappropriate supplementation, can compromise adaptation [39]. Thus, understanding and supporting these endogenous defenses are critical for optimizing athletic output and long-term physiological health.

3.3. Dual Role of RONS: Signaling and Damage

Reactive oxygen and nitrogen species (RONS) exert a concentration-dependent influence on skeletal muscle, functioning as critical signaling molecules and potential sources of cellular damage. At physiological levels, RONS are indispensable for initiating and regulating adaptations to exercise [7]. They activate several redox-sensitive signaling pathways, including AMP-activated protein kinase (AMPK), mitogen-activated protein kinases (MAPKs), and Nrf2 [7,40]. These pathways mediate key processes such as mitochondrial biogenesis, antioxidant enzyme upregulation, glucose uptake, and muscle fiber remodeling [41,42].

Nrf2, in particular, serves as a central regulatory node in oxidative stress responses. Upon activation by moderate RONS levels, Nrf2 translocates to the nucleus [43]. It binds to antioxidant response elements (AREs), promoting the transcription of genes encoding detoxifying and antioxidant enzymes like SOD and GPx [43,44]. Similarly, AMPK and MAPKs are activated under increased oxidative load and energy stress, facilitating metabolic flexibility and enhancing cellular resilience [45]. Through these mechanisms, RONS are necessary stimuli that drive the adaptations required for improved muscle performance and recovery [43].

However, when the production of RONS exceeds the buffering capacity of the antioxidant defense system, oxidative stress occurs. This pathological condition is marked by structural and functional impairments. Elevated levels of RONS can trigger lipid peroxidation, protein carbonylation, and DNA damage, ultimately compromising cellular integrity and interfering with essential physiological functions [46]. In muscle cells, oxidative modifications to calcium-handling proteins, such as ryanodine receptors and Sarcoplasmic/Endoplasmic Reticulum Calcium ATPase (SERCA) pumps, interfere with excitation–contraction coupling [28,47]. This impairs calcium homeostasis, reducing contractile efficiency and delaying recovery [46].

Such oxidative disruptions are particularly detrimental in elite soccer players, who often face high-frequency neuromuscular demands and condensed recovery periods. Persistent oxidative stress diminishes short-term performance and contributes to chronic fatigue, muscle soreness, systemic inflammation, and elevated injury risk [2,48]. Importantly, this damage is not simply a byproduct of overexertion; it also reflects a failure of the redox system to adapt to cumulative training and match stress [49].

Therefore, maintaining a precise balance in RONS levels is critical enough to stimulate beneficial adaptations, but not so much as to trigger deleterious effects [50]. This delicate interplay emphasizes the need for targeted strategies that support redox homeostasis through well-regulated training loads, recovery protocols, and nutrition [48]. Recognizing the dual nature of RONS as both signal and stressor provides the foundation for developing individualized approaches to optimize performance while safeguarding long-term muscular health. A summarized comparison of reactive species sources, antioxidant defenses, and their dual role in muscle physiology is presented in Table 1.

Table 1. Integrated overview of redox species, antioxidant defenses, and physiological effects in skeletal muscle during exercise.

Mechanism	Primary Sources and Examples	Key Molecular Targets/ Components	Effects at Physiological Levels	Effects when Excessive	Practical Implications for Elite Soccer Players	References
Reactive Species Generation	Mitochondria (Complex I and III), NADPH oxidase (NOX2/4), xanthine oxidase, uncoupled NOS	Superoxide ($O_2\bullet^-$), H_2O_2, •OH, NO•, $ONOO^-$	Signal transduction, mitochondrial biogenesis, and glucose uptake	Oxidative stress, lipid/protein/DNA damage, impaired calcium handling	Requires modulation through recovery protocols and load management	[7,24–28,30]
Antioxidant Defense—Enzymatic	SOD (cytosolic/mitochondrial), CAT, GPx	Detoxification of superoxide and hydrogen peroxide	Maintains redox homeostasis, supports adaptation	Inadequate during overtraining or insufficient recovery	Endogenous enzyme expression can be enhanced through training hormesis	[31–33,43]
Antioxidant Defense—Non-Enzymatic	Glutathione (GSH), vitamins C and E, polyphenols (quercetin, resveratrol)	Neutralization of free radicals, regeneration of enzymatic antioxidants	Protects membranes and proteins, supports recovery	Risk of adaptation blunting if over-supplemented	Timing and dosing of supplements must align with a training phase	[34–38]
Signaling Pathways	RONS-mediated activation of AMPK, Nrf2, MAPKs	Antioxidant gene transcription, mitochondrial biogenesis	Promotes metabolic flexibility, muscle remodeling	Suppressed adaptation if RONS is excessively neutralized	Controlled RONS exposure is essential for long-term conditioning	[7,40–45]
Muscle Fiber Susceptibility	Fast-twitch (Type II) fibers in sprinting, cutting	Lower intrinsic antioxidant levels	Greater capacity for power but higher vulnerability to oxidative damage	Increased injury risk, especially under congested schedules	Customized antioxidant support is needed based on positional demands	[28,29,49]

Note: for the full terms of abbreviations used in this table, please refer to the Abbreviations Section at the end of the manuscript.

4. Redox Homeostasis and Exercise in Elite Soccer Players

4.1. Exercise Demands and RONS Generation in Soccer

Elite soccer is defined by high-intensity, intermittent efforts that require a combination of aerobic endurance, anaerobic bursts, and rapid neuromuscular responses. Players frequently perform sprints, jumps, tackles, and directional changes, which place considerable metabolic and mechanical stress on the body. These repeated efforts elevate mitochondrial respiration, increasing electron leakage at complexes I and III of the electron transport chain and promoting superoxide formation [51,52]. This oxidative load intensifies as match intensity increases, with soccer matches often involving over 150 high-intensity actions [52].

In addition to mitochondrial sources, nonmitochondrial enzymes such as NADPH oxidase (NOX), xanthine oxidase, and nitric oxide synthase isoforms including nNOS and eNOS are activated in response to mechanical loading, inflammation, and transient hypoxia, all of which are frequently encountered during competitive play [53]. Eccentric muscle contractions, including those involved in braking, cutting, and deceleration, further increase RONS production by causing localized muscle damage and triggering inflammatory responses [54].

Additionally, the intermittent nature of play leads to repeated cycles of restricted blood flow and reoxygenation, mimicking ischemia–reperfusion conditions that favor oxidative reactions. These physiological stressors are exacerbated during congested match schedules, limiting recovery time and overwhelming antioxidant defenses [55]. If not properly regulated, the resulting oxidative imbalance may disrupt cellular signaling, impair calcium handling, and initiate catabolic processes. These effects can ultimately reduce athletic performance and increase the risk of injury [15].

Understanding the specific contexts in which RONS production is elevated, such as during sprints, directional changes, or periods of high match load, can inform the design of targeted training, recovery, and antioxidant strategies that preserve redox balance without suppressing necessary adaptive signals [56].

4.2. Acute and Chronic Oxidative Stress Responses

Acute bouts of high-intensity exercise transiently elevate RONS production, primarily in skeletal muscle. This short-term oxidative stress activates redox-sensitive signaling pathways such as AMP-activated protein kinase (AMPK), mitogen-activated protein kinases (MAPKs), and Nuclear Factor Erythroid 2–Related Factor 2 (Nrf2) [40,57]. These pathways promote mitochondrial biogenesis, antioxidant enzyme expression, and tissue repair mechanisms, supporting improved recovery and muscular resilience [40].

In contrast, chronic oxidative stress arises from repeated intense efforts without adequate recovery. Persistent RONS exposure may exceed the muscle's adaptive capacity, leading to lipid peroxidation, protein carbonylation, DNA damage, and mitochondrial dysfunction [58]. In elite soccer players, this can impair calcium handling, neuromuscular function, and elevate inflammation, ultimately contributing to performance decline and increased injury risk [21].

These opposing outcomes reflect the principle of hormesis, where moderate oxidative stress promotes beneficial adaptations, while excessive stress is detrimental [59]. Well-trained athletes often exhibit enhanced antioxidant defenses, including increased expression of SOD, GPx, and CAT, and elevated levels of glutathione and vitamin C [4]. However, individual responses vary depending on genetic background, training load, nutrition, and recovery strategies.

Given this variability, redox management in elite soccer cannot follow a one-size-fits-all approach. Instead, recovery and nutritional strategies should be personalized based on regularly monitoring redox biomarkers, training load, and individual responsiveness [3], thereby reducing the risk of oxidative maladaptation while promoting consistent performance. To better illustrate the complex interplay between RONS production, antioxidant defense, redox signaling, and athletic performance, a schematic overview is provided in Figure 1. This diagram integrates the molecular pathways discussed and highlights their implications for fatigue, recovery, and adaptation in elite soccer players.

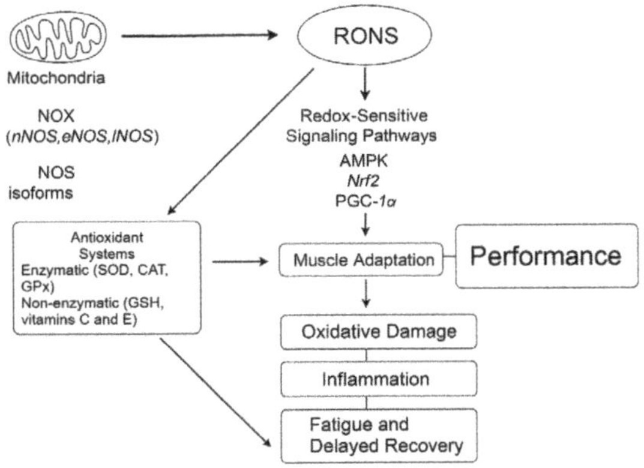

Figure 1. Redox regulation in elite soccer performance. During high-intensity exercise, reactive oxygen and nitrogen species (RONS) are produced by mitochondria, NADPH oxidases (NOX), and nitric oxide synthase (NOS) isoforms. Moderate levels of RONS activate redox-sensitive signaling pathways, such as AMPK, Nrf2, and PGC-1α, promoting muscle adaptation and performance. Antioxidant systems—including enzymatic (SOD, CAT, GPx) and non-enzymatic (GSH, vitamins C and E) components—counteract oxidative overload. Imbalance between RONS production and antioxidant capacity leads to oxidative damage, inflammation, and fatigue, ultimately impairing recovery and performance.

4.3. Biomarker Monitoring and Practical Applications

Effectively managing redox homeostasis in elite athletes requires objective tools to assess oxidative stress and antioxidant status. Biomarkers offer a non-invasive and quantifiable way to monitor the balance between RONS production and antioxidant capacity, providing insights into an athlete's physiological state and recovery needs.

Commonly used oxidative stress biomarkers include malondialdehyde (MDA) and F_2-isoprostanes for lipid peroxidation, protein carbonyls for protein oxidation, and 8-hydroxy-2′-deoxyguanosine (8-OHdG) for oxidative DNA damage [60–62]. These can be measured through blood, urine, or saliva, making them practical for laboratory and field settings. Complementing these are markers of antioxidant defense, such as total antioxidant capacity (TAC) and activities of enzymes like SOD, CAT, and GPx.

When incorporated into athlete monitoring systems, these biomarkers help identify redox imbalance before it results in noticeable fatigue or performance drops [63,64]. This allows for proactive adjustments in training loads, recovery protocols, and dietary support. Advances in portable diagnostics have also improved the feasibility of frequent sampling, enabling real-time feedback [65].

It is important to interpret biomarker data within the appropriate physiological and training context. Isolated measurements may be misleading; therefore, longitudinal tracking that considers training phases, performance indicators, and subjective recovery assessments is essential for accurate evaluation [66]. Personalized profiles can guide evidence-based decisions on antioxidant use and recovery interventions, reducing injury risk and supporting long-term adaptation [64].

Redox biomarker monitoring, combined with other physiological and psychological indicators (e.g., HRV, neuromuscular function, perceived exertion), provides a comprehensive framework for optimizing training and recovery in elite soccer [63,67]. It marks a key step toward precision performance management tailored to the unique oxidative demands of each athlete. A synthesized overview of exercise-induced redox stress, adaptation mechanisms, and biomarker applications in elite soccer is presented in Table 2.

Table 2. Overview of redox stress, adaptive responses, and biomarker-guided strategies in elite soccer.

Mechanism or Concept	Description and Key Features	Effects/Outcomes	Implications for Elite Soccer Players	References
RONS Generation in Match Play	High-intensity actions, ischemia–reperfusion cycles, eccentric contractions	Elevated oxidative load, disrupted calcium handling, and inflammation	Needs monitoring during congested schedules; impacts recovery and injury risk	[15,51–55]
Acute Oxidative Stress	Transient RONS production post-exercise; activates signaling pathways (e.g., AMPK, Nrf2)	Stimulates mitochondrial biogenesis, antioxidant enzyme expression	Essential for adaptation; should not be overly suppressed	[40,57]
Chronic Oxidative Stress	Repeated exposure due to overtraining, poor nutrition, and insufficient rest	Lipid/protein/DNA damage, mitochondrial dysfunction, fatigue	Leads to maladaptation and injury risk if unmanaged	[21,58]
Hormesis Principle	Moderate stress promotes adaptation; excessive stress causes damage	Enhances redox resilience vs. elevates injury and inflammation	Guides training load and antioxidant strategy design	[4,59]
Individual Variability	Influenced by genetics, age, training, and nutrition	Determines antioxidant capacity and stress response	Necessitates personalized recovery and supplementation plans	[3,4,21]
Oxidative Stress Biomarkers	MDA, 8-OHdG, F_2-isoprostanes, protein carbonyls	Reflect lipid, DNA, and protein oxidation status	Enables early detection of redox imbalance	[60–62]
Antioxidant Defense Biomarkers	Total antioxidant capacity, SOD, CAT, GPx activity	Assess redox buffering capacity and recovery state	Support real-time monitoring and intervention tailoring	[63–65]
Practical Monitoring Applications	Longitudinal biomarker tracking + HRV, RPE, performance data	Informs precision training, recovery, and nutrition	Enhances resilience and minimizes maladaptation risks	[63,66,67]

5. Redox Signaling and Muscle Function in Soccer Performance

5.1. Redox Regulation of Neuromuscular Function and Energy Metabolism

RONS exert significant control over neuromuscular function by modulating the molecular processes that govern excitation–contraction (E–C) coupling and cellular energy metabolism. In skeletal muscle, E–C coupling involves the precise sequence of electrical stimulation, calcium release, actin–myosin interaction, and subsequent calcium reuptake [68,69]. Moderate concentrations of RONS, particularly hydrogen peroxide (H_2O_2) and nitric oxide (NO•), can enhance this process by reversibly oxidizing thiol groups on proteins such as ryanodine receptors (RyR1) and SERCA pumps, thereby facilitating calcium handling and improving contractile performance [68,70].

This redox-sensitive modulation is especially relevant during submaximal or repetitive efforts common in elite soccer, where efficient calcium cycling supports sustained muscular output [71]. However, when RONS accumulate excessively, they induce irreversible oxidative modifications, such as carbonylation or nitrosylation, of contractile proteins, impairing calcium flux and reducing muscle fiber responsiveness [72,73]. This deterioration in E–C coupling contributes to decreased force production, particularly in later stages of match play or under cumulative fatigue [73].

In addition to regulating contractile function, RONS influence skeletal muscle energy metabolism by activating AMPK, a key energy sensor that responds to increased AMP:ATP ratios and oxidative cues [74]. Once activated, AMPK promotes adaptive processes including glucose uptake via GLUT4 translocation, fatty acid oxidation, and mitochondrial biogenesis [75]. These effects are primarily mediated through upregulation of peroxisome proliferator-activated receptor gamma coactivator 1-alpha (PGC-1α), a master regulator of oxidative metabolism and mitochondrial function [75].

The interaction between RONS, AMPK, and PGC-1α enhances metabolic flexibility, allowing soccer players to transition efficiently between energy systems during variable-intensity play [40]. Redox-sensitive pathways involving sirtuins (e.g., SIRT1) and FOXO transcription factors support mitochondrial quality control and protect against oxidative injury [41]. However, sustained oxidative overload can interfere with these signaling pathways, impair the mitochondrial membrane potential, and decrease ATP synthesis. These effects collectively compromise endurance capacity and contribute to increased fatigue.

Therefore, maintaining RONS within a functional range is crucial for preserving neuromuscular performance and metabolic resilience. Training, recovery, and nutritional strategies must be designed with an understanding of redox dynamics to support performance and adaptation without tipping into oxidative dysfunction.

5.2. Redox Contribution to Fatigue, Recovery, and Muscle Damage

In elite soccer, repeated high-intensity efforts with limited recovery elevate RONS production, contributing to fatigue and muscle stress [13,73]. While moderate RONS levels promote beneficial adaptations, excessive accumulation disrupts cellular functions essential for endurance and recovery [76]. A key mechanism involves impairment of excitation–contraction coupling through oxidative modification of calcium-handling proteins such as ryanodine receptor 1 (RyR1) and sarcoplasmic/endoplasmic reticulum Ca^{2+}-ATPase (SERCA), leading to reduced calcium availability and slower reuptake [73]. Concurrently, oxidation of metabolic enzymes like glyceraldehyde-3-phosphate dehydrogenase and creatine kinase impairs ATP regeneration, compromising energy supply during prolonged activity [73].

These effects are pronounced in fast-twitch fibers, which are preferentially recruited during sprinting and possess lower antioxidant capacity [2]. Additionally, eccentric contractions cause structural muscle damage that initiates an inflammatory response, further amplifying RONS production through immune cell infiltration [77]. The resulting oxidative stress prolongs inflammation, delays repair processes, and contributes to delayed-onset muscle soreness (DOMS) and increased injury risk when recovery is inadequate [78].

Despite these risks, RONS are vital for initiating the repair and remodeling processes necessary for long-term adaptation [79]. They activate transcription factors such as NF-κB and AP-1, which upregulate cytokines and growth factors involved in muscle regeneration, satellite cell activation, and extracellular matrix remodeling [72]. Consequently, completely suppressing RONS, such as through indiscriminate antioxidant use, may interfere with natural recovery processes and hinder adaptation.

Effective recovery management, therefore, requires a nuanced understanding of redox physiology. Interventions to limit excessive oxidative stress (e.g., antioxidant support, cryotherapy) should be timed strategically to avoid blunting redox-sensitive adaptation pathways. By combining redox biomarker monitoring with neuromuscular assessments and subjective fatigue ratings, practitioners can better support muscle recovery and maintain high-level performance throughout the competitive season. A summary of redox signaling pathways and their influence on muscle function, fatigue, and recovery in elite soccer is presented in Table 3.

Table 3. Summary of redox signaling effects on neuromuscular function, fatigue, and muscle recovery in elite soccer.

Mechanism or Concept	Description and Key Features	Effects/Outcomes	Implications for Elite Soccer Players	References
Excitation–Contraction (E–C) Coupling Modulation	RONS modulate RyR1 and SERCA through reversible oxidation	Enhances calcium handling and contractile performance	Supports sustained submaximal output; vulnerable to oxidative damage in fatigue	[68–71]
Excessive RONS and E–C Disruption	Irreversible protein oxidation impairs calcium flux	Reduces force production and muscle responsiveness	Leads to late-match fatigue, especially under high load	[72,73]
AMPK–PGC-1α Pathway Activation	AMPK is activated by oxidative and energetic stress	Promotes glucose uptake, FAO, mitochondrial biogenesis	Enhances energy efficiency and endurance performance	[40,74,75]

Table 3. *Cont.*

Mechanism or Concept	Description and Key Features	Effects/Outcomes	Implications for Elite Soccer Players	References
Redox and Mitochondrial Regulation	RONS-sensitive transcription (e.g., SIRT1, FOXO) affects mitochondrial health	Improves metabolic flexibility and quality control	Disrupted under chronic oxidative overload, reducing ATP	[40,41]
Fast-Twitch Fiber Susceptibility	Type II fibers with low antioxidant capacity under high load	Accumulate oxidative damage, reduce contraction efficiency	Require targeted recovery and antioxidant strategies	[2,73]
RONS Role in Muscle Damage and DOMS	Secondary oxidative stress from immune activation post-exercise	Lipid and protein oxidation, prolonged inflammation	Increases soreness and injury risk under poor recovery	[13,73,77,78]
RONS in Muscle Repair and Remodeling	Activate NF-κB and AP-1, stimulating cytokines and growth factors	Support satellite cell activation and muscle regeneration	Necessary for adaptation; indiscriminate antioxidant use may impair the process	[72,79]
Recovery Management Strategy	Combines redox biomarker monitoring with neuromuscular and fatigue data	Enables precise intervention timing (e.g., antioxidant, cryotherapy)	Prevents maladaptation and maintains performance	[2,76,79]

6. Antioxidant Defense and Recovery Strategies

6.1. Endogenous Adaptation Versus Exogenous Antioxidant Supplementation

Skeletal muscle has a well-developed endogenous antioxidant system that adjusts in response to oxidative stress induced by training. This adaptive response is governed by the principle of hormesis, in which exposure to moderate oxidative stress activates protective cellular mechanisms, thereby enhancing resilience and overall performance capacity. Central to this process is the transcription factor Nrf2, which, upon activation by RONS, translocates to the nucleus and promotes the expression of antioxidant enzymes such as SOD, CAT, and GPx, as well as enzymes involved in glutathione synthesis [2,80,81].

These adaptive responses enhance mitochondrial efficiency, reinforce redox buffering capacity, and assist in the regulation of inflammation. Together, these effects are essential for maintaining high-level athletic performance. Importantly, repeated exposure to manageable oxidative stress during training triggers the expression of PGC-1α, a key coactivator of mitochondrial biogenesis and oxidative metabolism [82]. This leads to improved endurance, delayed onset of fatigue, and greater efficiency in energy utilization, all of which are essential attributes for elite soccer players.

On the other hand, exogenous antioxidant supplementation introduces a layer of complexity. Widely used compounds such as vitamin C, E, and N-acetylcysteine (NAC) are intended to reduce oxidative damage and inflammation [22,83]. At the same time, polyphenols like quercetin, curcumin, and resveratrol have additional signaling effects [84]. While these supplements can reduce muscle soreness and improve subjective recovery in certain contexts, their indiscriminate or chronic use may interfere with the very signaling pathways that mediate beneficial training adaptations [84].

Studies have shown that high doses of antioxidants can blunt the activation of redox-sensitive kinases such as AMPK and p38 MAPK, reducing PGC-1α expression and impairing mitochondrial and metabolic adaptations [22,83,85]. These findings challenge the notion that oxidative stress is always harmful, highlighting its essential role in driving physiological improvements.

Thus, while antioxidant supplementation may be useful in specific contexts, such as injury recovery, high-altitude exposure, or during periods of extreme match congestion, it should not be applied universally. A periodized and personalized approach, aligned with training goals, recovery needs, and individual redox profiles, is increasingly regarded as the most effective strategy to optimize both adaptation and performance in elite soccer.

6.2. Nutritional and Physiological Recovery Interventions

In elite soccer, recovery strategies are crucial to counteract the neuromuscular and oxidative strain from frequent high-intensity activity. Both nutritional antioxidants and physiological methods are used to support tissue repair, reduce inflammation, and maintain performance continuity, though their effectiveness depends heavily on timing, context, and dosage.

From a nutritional perspective, several antioxidants are commonly used to buffer exercise-induced oxidative stress. Vitamin C operates in the aqueous phase to scavenge radicals and regenerate other antioxidants, while vitamin E protects lipid membranes from peroxidation [86]. NAC serves as a precursor for glutathione synthesis and has shown promise in preserving muscle function under high oxidative loads [87]. Polyphenols like quercetin, resveratrol, and curcumin offer dual benefits: direct antioxidant activity and modulation of redox-sensitive signaling pathways such as NF-κB and Nrf2 [88].

These compounds can be delivered through diet or supplements, but their absorption, efficacy, and interaction with training stress must be carefully considered. For instance, while acute supplementation may reduce inflammation and muscle damage post-exercise [89], chronic high-dose intake may suppress signaling pathways required for adaptation [90]. This underscores the need to align antioxidant intake with the training cycle, applying them more liberally during high-stress periods and sparingly during adaptation-focused blocks [26,91].

Physiological interventions such as cold-water immersion and cryotherapy are also widely used to manage inflammation and oxidative stress [92]. By limiting neutrophil activity and oxidative bursts following muscle damage, these techniques can aid short-term recovery. However, excessive or untimely use may blunt anabolic signaling and delay long-term adaptation [93].

Sleep is another vital component of redox regulation, as it influences hormonal rhythms and endogenous antioxidant activity, including melatonin production [94]. Sleep deprivation can impair redox signaling and reduce antioxidant capacity, emphasizing the need for consistent sleep hygiene as a non-negotiable element of recovery [95]. Additionally, low-intensity active recovery and massage can improve circulation, promote metabolic byproduct clearance, and reduce oxidative load [96].

Taken together, these interventions form a multifaceted recovery toolkit. Their success depends on individual responsiveness, the training phase, and the broader performance context. When strategically implemented, they can accelerate recovery, preserve adaptations, and extend the competitive longevity of elite soccer players.

6.3. Strategic Application and the Adaptation-Blunting Debate

Antioxidant strategies in elite sport must be applied with precision, as inappropriate timing or excessive dosing may impair training-induced adaptations. The adaptation-blunting hypothesis suggests that high-dose supplementation can suppress redox-sensitive pathways, such as AMPK, PGC-1α, and Nrf2, disrupting mitochondrial biogenesis and metabolic remodeling [26,97]. To preserve the benefits of oxidative signaling, antioxidant use should be aligned with specific training phases and individual recovery needs.

In elite soccer, where training focus shifts throughout the year, from capacity-building in the preseason to performance preservation during congested match periods, antioxidant strategies must be periodized [9]. During phases of high training load or competition density, moderate antioxidant use may help manage cumulative fatigue and protect against excessive oxidative stress [98]. However, during adaptation-intensive blocks, minimizing supplementation may preserve the oxidative signals necessary for physiological gains.

Individual variability further complicates blanket recommendations. Genetic polymorphisms in redox-related enzymes (e.g., SOD2, GSTP1), baseline antioxidant status, dietary habits, and positional demands all influence an athlete's oxidative profile and responsiveness to interventions [1,99]. For example, midfielders and wide players typically engage in more high-intensity running than defenders or goalkeepers, potentially experiencing greater oxidative stress.

To tailor antioxidant support, objective data are essential. Biomarkers such as MDA, protein carbonyls, 8-OHdG, and enzyme activities (e.g., SOD, GPx) can provide insights into redox status when interpreted alongside performance and recovery metrics [26,100]. This enables practitioners to detect maladaptation early and adjust interventions accordingly.

Ultimately, the future application of antioxidants in elite sport is expected to focus on precision recovery. This approach involves the use of biomarker data, training load metrics, and athlete feedback to guide decisions regarding the timing, dosage, and method of intervention [101,102]. By doing so, it ensures that performance is supported without interfering with the molecular pathways essential for long-term development and physiological resilience. A comparative summary of antioxidant strategies, adaptive responses, and recovery interventions in elite soccer is presented in Table 4.

Table 4. Summary of endogenous and exogenous antioxidant strategies and recovery applications in elite soccer.

Mechanism or Concept	Description and Key Features	Effects/Outcomes	Implications for Elite Soccer Players	References
Endogenous Antioxidant Adaptation	Nrf2 activation induces antioxidant enzymes (SOD, CAT, GPx) in response to RONS	Enhances redox resilience and mitochondrial efficiency	Promotes natural adaptation; training must provide a moderate oxidative stimulus	[2,80,81]
PGC–1α–Mediated Adaptation	Triggered by oxidative and energetic stress; supports mitochondrial biogenesis and metabolism	Increases endurance, delays fatigue, and improves energy utilization	Critical for long-term adaptation and performance enhancement	[82]
Exogenous Supplementation—Vitamins C, E, NAC	Reduces oxidative stress and inflammation but may blunt signaling if misused	Can impair AMPK and PGC-1α activation, reducing adaptation	Use cautiously during injury or overload periods, not chronically	[22,83,85]
Polyphenol-Based Supplementation	Quercetin, curcumin, resveratrol have antioxidant and signaling effects	Aid inflammation control and recovery; modulate NF-κB and Nrf2	Apply selectively based on training intensity and goals	[84,88,90]
Physiological Interventions	Cryotherapy, cold-water immersion, sleep, massage, and active recovery	Support acute recovery and tissue repair; reduce inflammatory oxidative bursts	Effective when timed appropriately; avoid overuse to preserve adaptations	[92–96]
Adaptation-Blunting Hypothesis	High antioxidant doses suppress redox-sensitive pathways (AMPK, PGC-1α, Nrf2)	May limit mitochondrial biogenesis and oxidative capacity	Highlights the importance of precise timing and dosing	[26,83,97]
Training Periodization of Antioxidants	Strategy varies between adaptation-focused and performance-focused phases	Balances oxidative signaling and damage control	Periodized planning enhances recovery and preserves adaptation	[9,98]
Individual Variability and Precision Recovery	Redox response is influenced by genetics, diet, training, and playing position	Biomarkers (MDA, 8-OHdG, GPx) enable targeted intervention	Supports personalized antioxidant and recovery protocols	[1,26,99–102]

7. Periodization, Redox Adaptation, and Long-Term Conditioning

7.1. Seasonal Variation and Oxidative Stress Profiles

Throughout the annual training and competition cycle, elite soccer players are exposed to varying levels of physiological stress that impact their oxidative status. The different phases of the season, including the preseason, in-season, and off-season, each present unique metabolic, neuromuscular, and recovery demands that collectively influence redox balance [103,104]. Understanding how oxidative stress fluctuates across these periods is essential for designing effective training and recovery programs that support long-term conditioning and resilience.

During the preseason, training loads intensify as players rebuild fitness, strength, and coordination. This phase typically includes unfamiliar or high-volume eccentric exercises and increased total workload, all of which elevate RONS production [2,80]. Mitochondrial respiration, NADPH oxidase activity, and inflammatory responses from muscle microtrauma contribute to elevated oxidative stress, as reflected by increased biomarkers such as MDA and protein carbonyls [105]. However, when managed properly, this oxidative challenge acts as a beneficial stimulus, activating redox-sensitive pathways like Nrf2 and promoting antioxidant enzyme expression and mitochondrial biogenesis [106]. These adaptations lay the physiological foundation for enduring performance throughout the season.

The in-season phase for elite soccer players is characterized by a focus on maintaining performance, preventing injuries, and ensuring efficient recovery between matches. This period is marked by frequent high-intensity games, travel, and limited rest, which collectively contribute to increased oxidative stress. For players with high playing minutes and limited rotation, the endogenous antioxidant capacity may be insufficient to counteract the repeated exposure to RONS. This persistent oxidative overload can lead to reduced recovery quality, elevated muscle damage markers, and a higher incidence of soft tissue injuries. Therefore, in-season management often requires targeted nutritional support, individualized recovery interventions, and close monitoring of oxidative stress to prevent maladaptation. Effective nutritional strategies for post-match recovery emphasize the timely intake of approximately 1.2 g/kg/h of carbohydrates and 40 g of protein within 20 min to restore glycogen and stimulate muscle protein synthesis. Daily recommendations include 6–10 g/kg of carbohydrates and over 1.5 g/kg of protein, ideally in divided doses [107]. Supplementation with vitamin D, omega-3 fatty acids, creatine, and antioxidants may further support recovery, with compounds like curcumin and bromelain showing potential benefits pending further research [108]. Monitoring oxidative stress through markers such as creatine kinase (CK) and C-reactive protein helps assess muscle damage and inflammation, as CK typically peaks 24–48 h post-match and normalizes within 72 h [9,14]. Antioxidant enzymes, including glutathione reductase and CAT, exhibit seasonal variation, suggesting adaptive buffering of oxidative stress, despite stable overall marker levels [2].

The off-season is a critical phase for redox recalibration in elite soccer players, providing an opportunity for physiological regeneration, reduction in systemic inflammation, and restoration of antioxidant capacity. Decreased training volume lowers oxidative stress, allowing replenishment of key antioxidants such as glutathione and vitamins C and E [26,109,110]. This period also supports improved sleep and circadian rhythm regulation, which enhances recovery and immune function [111,112]. Regular in-season oxidative stress may enhance endogenous defenses, which can be recalibrated during the off-season to restore redox balance [4]. Although antioxidant supplementation may aid recovery, its effectiveness depends on exercise intensity and individual requirements [26,110]. Low-level conditioning should be maintained to preserve mitochondrial function and prevent detraining [4,113], while the shift toward a more reducing environment supports long-term performance and health [113]. This phase also allows for dietary reassessment to better prepare for oxidative demands in the upcoming preseason [111,114].

By recognizing the seasonal rhythm of oxidative stress, practitioners can better align antioxidant strategies with physiological demands, ensuring recovery and adaptation are both supported throughout the training year.

7.2. Periodized Antioxidant Strategies and Individual Redox Profiling

Antioxidant strategies in elite soccer should be periodized to match the oxidative demands of specific training phases. High-dose supplementation during adaptation-focused periods, such as preseason or strength blocks, may blunt training-induced adaptations by interfering with redox-sensitive pathways like AMPK, PGC-1α, and Nrf2, which are essential for mitochondrial biogenesis and metabolic remodeling [97,115]. Endogenous antioxidant systems, strengthened through controlled oxidative stress, play a key role in promoting adaptation and performance by facilitating ROS-mediated signaling and reducing inflammation [26]. In contrast, during high-intensity competition phases, short-term antioxidant support may help mitigate fatigue and muscle damage, particularly when appropriately timed and dosed [26,116]. Although some evidence suggests that antioxidants can attenuate mitochondrial and antioxidant adaptations, the redundancy in skeletal muscle signaling pathways indicates that training may offset these effects in certain contexts [97,115].

In elite soccer players, individualized redox management is essential due to variability in oxidative stress susceptibility, influenced by genetic polymorphisms in antioxidant enzymes such as SOD2, GPX1, and CAT [50,117]. These genetic differences, along with training history, position-specific demands, and dietary patterns, affect oxidative stress responses and recovery needs [2,4,118]. Personalized nutritional interventions, including vitamin C and antioxidant-rich diets, can support redox balance and reduce markers like 8-OHdG and F2α-isoprostane [117,118]. Biomarker profiling, which includes measures such as MDA, 8-OHdG, protein carbonyls, and total antioxidant capacity, along with data from GPS tracking, HRV, and subjective fatigue assessments, provides the basis for developing individualized antioxidant strategies [118,119]. Applying these interventions across training cycles helps regulate oxidative stress, enhances performance, and reduces the risk of overtraining [4,119]. A seasonal and personalized overview of redox adaptation and antioxidant strategy in elite soccer is presented in Table 5.

Table 5. Seasonal variation, periodized antioxidant strategy, and personalized redox profiling in elite soccer.

Mechanism or Concept	Description and Key Features	Effects/Outcomes	Implications for Elite Soccer Players	References
Preseason Redox Challenge	High training volume and eccentric exercises increase RONS via mitochondrial respiration, NOX activity, and inflammation	Activates Nrf2, enhances antioxidant enzyme expression, and promotes mitochondrial biogenesis	Forms an adaptive foundation for in-season performance; must be monitored for overload	[2,80,103–106]
In-Season Oxidative Load	Frequent high-intensity matches, travel, and limited rest elevate oxidative stress, CK, and inflammation	Can impair recovery, increase muscle damage, and raise injury risk	Demands targeted nutrition (CHO, protein, vitamin D), biomarker monitoring, and rotation strategies	[9,14,107,108]
Off-Season Redox Recalibration	Reduced load allows antioxidant restoration and inflammation resolution; improved sleep supports redox reset	Enhances glutathione, vitamin C/E levels, immune function, and mitochondrial maintenance	Ideal for physiological regeneration and dietary reassessment	[4,26,109–114]
Periodized Antioxidant Strategy	High-dose antioxidants during adaptation phases may suppress AMPK, PGC-1α, and Nrf2 pathways	May blunt training adaptations and mitochondrial gains	Strategy should match training cycle—minimal during adaptation blocks, moderate during congestion	[26,97,115,116]
Redox Genetic Variability	Individual differences in antioxidant enzyme genes (SOD2, GPX1, CAT) modulate oxidative response	Affects susceptibility to oxidative stress and efficacy of supplementation	Highlights the need for genetically informed nutrition and recovery protocols	[2,4,50,117]
Biomarker-Guided Personalization	Redox markers (MDA, 8-OHdG, protein carbonyls, TAC) with GPS, HRV, and RPE data provide a recovery profile	Supports individualized training, antioxidant timing, and stress management	Reduces overtraining risk and optimizes performance through precision conditioning	[4,117–119]

8. Future Research Directions

Future investigations should prioritize longitudinal studies that capture redox fluctuations across an entire competitive season in elite soccer players. Current research predominantly focuses on acute responses to isolated training or matches, offering limited insight into cumulative oxidative stress and its implications for injury risk, fatigue, and adaptation. Monitoring oxidative biomarkers alongside performance and recovery metrics during preseason, in-season, and off-season phases would provide a more integrated understanding of redox dynamics and their role in athlete resilience and performance sustainability.

Advancements in omics technologies, including metabolomics and transcriptomics, provide new opportunities to explore redox-sensitive pathways involved in muscle adaptation. These approaches can assist in detecting early molecular changes that are associated with maladaptation or overtraining. Additionally, future studies should aim to develop individualized antioxidant strategies based on athletes' genetic profiles, training demands, and biomarker data. Integrating these insights into AI-assisted monitoring systems could enable real-time decision-making in training and recovery planning, facilitating precision-based redox management tailored to each athlete's physiological needs.

9. Conclusions

RONS play a central role in skeletal muscle physiology, particularly under the physical demands of elite soccer. At moderate levels, RONS serve as important signaling molecules that stimulate beneficial adaptations. These include mitochondrial biogenesis, induction of endogenous antioxidant enzymes, and muscle remodeling processes that collectively enhance endurance, recovery, and overall performance.

However, when RONS production exceeds the buffering capacity of antioxidant defenses due to excessive training loads, poor recovery, or suboptimal nutrition, the result can be maladaptive. Excess RONS may disrupt redox homeostasis, impair calcium signaling, damage mitochondrial function, and accelerate muscle fatigue and inflammation. Over time, this can lead to performance decline and a greater risk of soft tissue injuries.

Recognizing this dual nature of RONS, future strategies should avoid uniform antioxidant use and instead prioritize personalized, periodized approaches. These may include monitoring redox biomarkers, adapting antioxidant support to the athlete's physiological state, and integrating advancements in digital performance tracking and nutrition science. Such individualized interventions can help maintain a balance that supports adaptive signaling while preventing oxidative damage, thereby promoting long-term athletic resilience and peak performance in elite soccer players.

Author Contributions: Q.M. and C.-H.S. contributed to this expert review. All authors have read and agreed to the published version of the manuscript.

Funding: This research received funding from the Research Project on Education and Teaching Reform of HUAQIAO University, No. HQJG202114.

Institutional Review Board Statement: Not applicable.

Informed Consent Statement: Not applicable.

Data Availability Statement: Not applicable.

Conflicts of Interest: The authors declare no conflicts of interest.

Abbreviations

The following abbreviations are used in this manuscript:

RONS	Reactive Oxygen and Nitrogen Species
E–C coupling	Excitation–Contraction Coupling
ROS	Reactive Oxygen Species
NADPH	Nicotinamide Adenine Dinucleotide Phosphate
NOX	NADPH Oxidase
NOS	Nitric Oxide Synthase
SOD	Superoxide Dismutase
GPx	Glutathione Peroxidase
CAT	Catalase
GSH	Reduced Glutathione
AMPK	AMP-Activated Protein Kinase
PGC-1α	Peroxisome Proliferator-Activated Receptor Gamma Coactivator 1-alpha
Nrf2	Nuclear Factor Erythroid 2–Related Factor 2
NF-κB	Nuclear Factor Kappa-Light-Chain-Enhancer of Activated B Cells
TAC	Total Antioxidant Capacity
8-OHdG	8-Hydroxy-2′-Deoxyguanosine
DOMS	Delayed-Onset Muscle Soreness
CK	Creatine Kinase
HRV	Heart Rate Variability
SERCA	Sarcoplasmic/Endoplasmic Reticulum Calcium ATPase
MAPK	Mitogen-Activated Protein Kinase
FOXO	Forkhead Box O Transcription Factor
SIRT1	Sirtuin 1
CHO	Carbohydrates
NAC	N-Acetylcysteine
FAO	Fatty Acid Oxidation
GPS	Global Positioning System (Athlete Tracking)
RPE	Rating of Perceived Exertion

References

1. Rago, V.; Leo, I.; Pizzuto, F.; Costa, J.; Angeicchio, G.; Tafuri, D.; Raiola, G. Variation of Oxidative Stress of Elite Football Players during Pre-Season and In-Season. *J. Phys. Educ. Sport* **2016**, *16*, 326–329. [CrossRef]
2. Radák, Z.; Powers, S.; Jackson, M.; Ji, L. Reactive Oxygen Species Promote Endurance Exercise-Induced Adaptations in Skeletal Muscles. *J. Sport Health Sci.* **2024**, *13*, 780–792. [CrossRef]
3. Toledo, I.C. Clinical, Cellular and Molecular Approaches to Oxidative Stress in Athletes' Bodies: A Systematic and Integrative Review. *Int. J. Nutrology* **2023**, *16*, e23106. [CrossRef]
4. Ruzicic, R.; Jakovljevic, V.; Djordjevic, D. Oxidative Stress in Training, Overtraining and Detraining: From Experimental to Applied Research. *Serb. J. Exp. Clin. Res.* **2016**, *17*, 343–348. [CrossRef]
5. Gomez-Cabrera, M.C.; Carretero, A.; Millan-Domingo, F.; Garcia-Dominguez, E.; Correas, Á.G.; Olaso-Gonzalez, G.; Viña, J. Redox-Related Biomarkers in Physical Exercise. *Redox Biol.* **2021**, *42*, 101956. [CrossRef] [PubMed]
6. Zulfahmidah, Z.; Safei, I. The Role of Reactive Oxygen Species in Muscle: Beneficial/Harmful. *Green Med. J.* **2022**, *4*, 84–92. [CrossRef]
7. Chen, M.; Li, Y.; Deng, S.-L.; Zhao, Y.; Lian, Z.; Yu, K. Mitochondrial Function and Reactive Oxygen/Nitrogen Species in Skeletal Muscle. *Front. Cell Dev. Biol.* **2022**, *10*, 826981. [CrossRef]
8. Zhang, X.; Yan, X.; Davison, G.; Zhou, Y.; Baker, J. Redox Signaling and Skeletal Muscle Adaptation during Aerobic Exercise. *iScience* **2024**, *27*, 109643. [CrossRef]
9. Owen, A.L. Biochemical Response Comparisons of a Competitive Microcycle vs. Congested Fixture Periods in Elite Level European Champions League Soccer Players. *J. Complement. Med. Alt. Healthc.* **2019**, *10*, 555778. [CrossRef]

10. Murach, K.; Dyar, K.; Smith, J.; Zierath, J. Exercise Metabolism and Adaptation in Skeletal Muscle. *Nat. Rev. Mol. Cell Biol.* **2023**, *24*, 607–632. [CrossRef]
11. Mohr, M.; Vigh-Larsen, J.F.; Krustrup, P. Muscle Glycogen in Elite Soccer—A Perspective on the Implication for Performance, Fatigue, and Recovery. *Front. Sports Act. Living* **2022**, *4*, 876534. [CrossRef] [PubMed]
12. Zouhal, H.; Granacher, U.; Laher, I.; Abderrahman, A.; Hage, R.; Saeidi, A.; Saidi, K.; Hackney, A.; Bideau, B. Immune Inflammation Markers and Physical Fitness during a Congested Match Play Period in Elite Male Soccer Players. *Sci. Rep.* **2024**, *14*, 81225. [CrossRef]
13. Liu, G. Extended Match Time Exacerbates Fatigue and Impacts Physiological Responses in Male Soccer Players. *Med. Sci. Sports Exerc.* **2022**, *55*, 80–92. [CrossRef]
14. Saidi, K.; Zouhal, H.; Boullosa, D.; Dupont, G.; Hackney, A.C.; Bideau, B.; Granacher, U.; Ben Abderrahman, A. Biochemical Markers and Wellness Status During a Congested Match Play Period in Elite Soccer Players. *Int. J. Sports Physiol. Perform.* **2022**, *17*, 605–620. [CrossRef]
15. Saidi, K.; Zouhal, H.; Rhibi, F.; Tijani, J.M.; Boullosa, D.; Chebbi, A.; Hackney, A.C.; Granacher, U.; Bideau, B.; Ben Abderrahman, A. Effects of a Six-Week Period of Congested Match Play on Plasma Volume Variations, Hematological Parameters, Training Workload and Physical Fitness in Elite Soccer Players. *PLoS ONE* **2019**, *14*, e0219692. [CrossRef]
16. Mankowski, R.T.; Anton, S.D.; Buford, T.W.; Leeuwenburgh, C. Dietary Antioxidants as Modifiers of Physiologic Adaptations to Exercise. *Med. Sci. Sports Exerc.* **2015**, *47*, 1857–1868. [CrossRef]
17. Bhol, N.; Samal, R.; Duttaroy, A.; Jena, A. Cellular Red-Ox System in Health and Disease: The Latest Update. *Biomed. Pharmacother.* **2023**, *162*, 114606. [CrossRef]
18. Kang, I.; Ha, J.; Fu, M.; Choe, W.; Yoon, K. Crosstalk Between Antioxidants and Adipogenesis: Mechanistic Pathways and Their Roles in Metabolic Health. *Antioxidants* **2025**, *14*, 203. [CrossRef]
19. Kawamura, T.; Muraoka, I. Exercise-Induced Oxidative Stress and the Effects of Antioxidant Intake from a Physiological Viewpoint. *Antioxidants* **2018**, *7*, 119. [CrossRef]
20. Supruniuk, E.; Górski, J.; Chabowski, A. Endogenous and Exogenous Antioxidants in Skeletal Muscle Fatigue Development during Exercise. *Antioxidants* **2023**, *12*, 501. [CrossRef]
21. Vargas-Mendoza, N.; Angeles-Valencia, M.; Morales-González, Á.; Madrigal-Santillán, E.; Morales-Martínez, M.; Madrigal-Bujaidar, E.; Álvarez-González, I.; Gutiérrez-Salinas, J.; Esquivel-Chirino, C.; Chamorro-Cevallos, G.; et al. Oxidative Stress, Mitochondrial Function and Adaptation to Exercise: New Perspectives in Nutrition. *Life* **2021**, *11*, 1269. [CrossRef] [PubMed]
22. McLeay, Y.; Stannard, S.R.; Houltham, S.D.; Starck, C. Dietary Thiols in Exercise: Oxidative Stress Defence, Exercise Performance, and Adaptation. *J. Int. Soc. Sports Nutr.* **2017**, *14*, 12. [CrossRef] [PubMed]
23. Fransson, D.; Nielsen, T.S.; Olsson, K.; Christensson, T.; Bradley, P.S.; Fatouros, I.G.; Krustrup, P.; Nordsborg, N.B.; Mohr, M. Skeletal Muscle and Performance Adaptations to High-Intensity Training in Elite Male Soccer Players: Speed Endurance Runs versus Small-Sided Game Training. *Eur. J. Appl. Physiol.* **2018**, *118*, 111–121. [CrossRef]
24. Dutka, T.L.; Mollica, J.P.; Lamb, G.D. Differential Effects of Peroxynitrite on Contractile Protein Properties in Fast- and Slow-Twitch Skeletal Muscle Fibers of Rat. *J. Appl. Physiol.* **2011**, *110*, 705–716. [CrossRef]
25. da Silva, A.N.; Freitas-Lima, L.C. The Association between Physical Exercise and Reactive Oxygen Species (ROS) Production. *J. Sports Med. Doping Stud.* **2015**, *5*, 1000152. [CrossRef]
26. Clemente-Suárez, V.J.; Bustamante-Sánchez, Á.; Mielgo-Ayuso, J.; Martínez-Guardado, I.; Martín-Rodríguez, A.; Tornero-Aguilera, J.F. Antioxidants and Sports Performance. *Nutrients* **2023**, *15*, 2371. [CrossRef] [PubMed]
27. Yavari, A.; Javadi, M.; Mirmiran, P.; Bahadoran, Z. Exercise-Induced Oxidative Stress and Dietary Antioxidants. *Asian J. Sports Med.* **2015**, *6*, e24898. [CrossRef]
28. Townsend, J.; Hoffman, J.; Gonzalez, A.; Pinzone, A. Supplementation with Nitric Oxide Precursors for Strength Performance: A Review of the Current Literature. *Nutrients* **2023**, *15*, 660. [CrossRef]
29. Mobley, C.; Mueller, B.; Judd, R.; Roberts, M.; Kavazis, A. Nitric Oxide in Exercise Physiology: Past and Present Perspectives. *Front. Physiol.* **2025**, *15*, 1504978. [CrossRef]
30. Aldous, J.W.F.; Chrismas, B.C.R.; Akubat, I.; Dascombe, B.J.; Abt, G.; Taylor, L. Hot and Hypoxic Environments Inhibit Simulated Soccer Performance and Exacerbate Performance Decrements When Combined. *Front. Physiol.* **2015**, *6*, 421. [CrossRef]
31. Powers, S.K.; Goldstein, E.R.; Schrager, M.A.; Ji, L.L. Exercise Training and Skeletal Muscle Antioxidant Enzymes: An Update. *Antioxidants* **2022**, *12*, 39. [CrossRef] [PubMed]
32. Kozakowska, M.; Pietraszek-Gremplewicz, K.; Jozkowicz, A.; Dulak, J. The Role of Oxidative Stress in Skeletal Muscle Injury and Regeneration: Focus on Antioxidant Enzymes. *J. Muscle Res. Cell Motil.* **2015**, *36*, 377–393. [CrossRef]
33. Bjelakovic, L.; Kocic, G.; Radovanović, D.; Antić, V.; Bjelakovic, B.; Antić, Z. Antioxidants and Their Importance during Muscular Exercise: A Review. *Facta Univ. Ser. Med. Biol.* **2017**, *19*, 48–56. [CrossRef]

34. Da Silva, T.; Meierjohann, S.; Mosteo, L.; Goding, C.; Santos, A.; Kreß, J.; Hufnagel, A.; Jessen, C.; Angeli, J.; Schmitz, W. The Integrated Stress Response Effector ATF4 Is an Obligatory Metabolic Activator of NRF2. *Cell Rep.* **2023**, *42*, 112724. [CrossRef]
35. Lee, H.; Choi, J. MLL1 Histone Methyltransferase and UTX Histone Demethylase Functionally Cooperate to Regulate the Expression of NRF2 in Response to ROS-Induced Oxidative Stress. *Free Radic. Biol. Med.* **2024**, *217*, 34–59. [CrossRef]
36. Gutowicz, M. Antioxidant and Detoxyactive Mechanisms in Central Nervous System. *Postep. Hig. Med. Dosw.* **2020**, *74*, 1–11. [CrossRef]
37. Baktiyani, S.C.W. The Changes of H_2O_2 Level and Glutathione/Glutathione Dioxide Ratio with the Administration of N-Acetylcysteine, Vitamin C, and Vitamin E towards In Vitro Eclampsia Human Umbilical Vein Endothelial Cell Model. *Indones. J. Obstet. Gynecol.* **2011**, *34*, 64–68.
38. Arulselvan, P.; Fard, M.T.; Tan, W.S.; Gothai, S.; Fakurazi, S.; Norhaizan, M.E.; Kumar, S.S. Role of Antioxidants and Natural Products in Inflammation. *Oxid. Med. Cell. Longev.* **2016**, *2016*, 5276130. [CrossRef]
39. Maskhulia, L.; Kakhabrishvili, Z.; Akhalkatsi, V.; Tskhvediani, N.; Kapetivadze, I. Evaluation of the Effect of Oxidative Stress on Dynamics of Cardiac Biomarkers in Georgian Elite Athletes with Overtraining Syndrome. *Eur. J. Prev. Cardiol.* **2023**, *30* (Suppl. S1), zwad125.126. [CrossRef]
40. Bouviere, J.; Fortunato, R.S.; Dupuy, C.; Werneck-de-Castro, J.P.; Carvalho, D.P.; Louzada, R.A.N. Exercise-Stimulated ROS Sensitive Signaling Pathways in Skeletal Muscle. *Antioxidants* **2021**, *10*, 537. [CrossRef]
41. Nemes, R.; Koltai, E.; Taylor, A.W.; Suzuki, K.; Gyori, F.; Radak, Z. Reactive Oxygen and Nitrogen Species Regulate Key Metabolic, Anabolic, and Catabolic Pathways in Skeletal Muscle. *Antioxidants* **2018**, *7*, 85. [CrossRef]
42. Llanos, P.; Palomero, J. Reactive Oxygen and Nitrogen Species (RONS) and Cytokines—Myokines Involved in Glucose Uptake and Insulin Resistance in Skeletal Muscle. *Cells* **2022**, *11*, 4008. [CrossRef] [PubMed]
43. Arif, I.S.; Kamal, Y.M.; Raoof, I.B. Nrf2 as a Modulator of Oxidative Stress. *Al Mustansiriyah J. Pharm. Sci.* **2022**, *21*, 17–23. [CrossRef]
44. Haghdoost, S.; Emami, S.; Cesário, R.; Godoy, P.; Raftari, M.; Hammad, M.; Salma, R. Roles of Oxidative Stress and Nrf2 Signaling in Pathogenic and Non-Pathogenic Cells: A Possible General Mechanism of Resistance to Therapy. *Antioxidants* **2023**, *12*, 1371. [CrossRef]
45. Ngo, V.T.-K.; Duennwald, M.L. Nrf2 and Oxidative Stress: A General Overview of Mechanisms and Implications in Human Disease. *Antioxidants* **2022**, *11*, 2345. [CrossRef]
46. Lian, D.; Chen, M.-M.; Wu, H.; Deng, S.; Hu, X. The Role of Oxidative Stress in Skeletal Muscle Myogenesis and Muscle Disease. *Antioxidants* **2022**, *11*, 755. [CrossRef] [PubMed]
47. Chen, J.; Simmen, T.; Makio, T. ER Stress as a Sentinel Mechanism for ER Ca^{2+} Homeostasis. *Cell Calcium* **2024**, *124*, 102961. [CrossRef]
48. Powers, S.K.; Hogan, M.C. Exercise and Oxidative Stress. *J. Physiol.* **2016**, *594*, 5079–5080. [CrossRef]
49. Romagnoli, M.; Sanchis-Gomar, F.; Alis, R.; Risso-Ballester, J.; Bosio, A.; Graziani, R.L.; Rampinini, E. Changes in Muscle Damage, Inflammation, and Fatigue-Related Parameters in Young Elite Soccer Players after a Match. *J. Sports Med. Phys. Fit.* **2016**, *56*, 1198–1205.
50. Nocella, C.; Cammisotto, V.; Pigozzi, F.; Borrione, P.; Fossati, C.; D'Amico, A.; Cangemi, R.; Peruzzi, M.; Gobbi, G.; Ettorre, E.; et al. Impairment between Oxidant and Antioxidant Systems: Short- and Long-Term Implications for Athletes' Health. *Nutrients* **2019**, *11*, 1353. [CrossRef]
51. Shalfawi, S.A.I.; Tjelta, L.I. A Critical Evaluation of the Aerobic Capacity Demands of Elite Male Soccer Players. *Int. J. Appl. Sports Sci.* **2016**, *28*, 200–212. [CrossRef]
52. Harper, D.; Harper, D.; Carling, C.; Kiely, J. High-Intensity Acceleration and Deceleration Demands in Elite Team Sports Competitive Match Play: A Systematic Review and Meta-Analysis of Observational Studies. *Sports Med.* **2019**, *49*, 1923–1947. [CrossRef] [PubMed]
53. Di Meo, S.; Napolitano, G.; Venditti, P. Mediators of Physical Activity Protection against ROS-Linked Skeletal Muscle Damage. *Int. J. Mol. Sci.* **2019**, *20*, 3024. [CrossRef]
54. Heyat, F. Cellular and Molecular Mechanisms of the Production of Free Radicals during Exercise and Their Function on Skeletal Muscles. *J. Fasa Univ. Med. Sci.* **2017**, *7*, 1–11.
55. Alghannam, A.F. Metabolic Limitations of Performance and Fatigue in Football. *Asian J. Sports Med.* **2012**, *3*, 65–73. [CrossRef] [PubMed]
56. Pérez-López, A.; Pérez-López, A.; Martin-Rincon, M.; Santana, A.; Perez-Suarez, I.; Dorado, C.; Calbet, J.A.L.; Morales-Alamo, D. Antioxidants Facilitate High-Intensity Exercise IL-15 Expression in Skeletal Muscle. *Int. J. Sports Med.* **2019**, *40*, 16–22. [CrossRef]
57. Petrick, H.L.; Dennis, K.M.J.H.; Miotto, P.M. The Importance of Exercise Intensity, Volume and Metabolic Signalling Events in the Induction of Mitochondrial Biogenesis. *J. Physiol.* **2018**, *596*, 4571–4572. [CrossRef]

58. Abbas, K.; Batool, U.; Memon, A.S.; Soomro, S. Oxidative Stress: A Double Edged Sword. *Biomed. Innov. Open Sci.* **2021**, *2*, 4–12. [CrossRef]
59. Meiliana, A.; Wijaya, A. Hormesis in Health and Disease: Molecular Mechanisms. *Indones. Biomed. J.* **2020**, *12*, 288–303. [CrossRef]
60. Demirci-Çekiç, S.; Özkan, G.; Avan, A.; Uzunboy, S.; Çapanoğlu, E.; Apak, R. Biomarkers of Oxidative Stress and Antioxidant Defense. *J. Pharm. Biomed. Anal.* **2021**, *209*, 114477. [CrossRef]
61. Li, J.; Pan, L.; Pan, W.; Li, N.; Tang, B. Recent Progress of Oxidative Stress Associated Biomarker Detection. *Chem. Commun.* **2023**, *59*, 7361–7374. [CrossRef] [PubMed]
62. Verstuyft, C.; Poinsignon, V.; Becquemont, L.; Chappell, K.; Tayeb, A.; Bouligand, J. Major Depressive Disorder and Oxidative Stress: A Review of Peripheral and Genetic Biomarkers According to Clinical Characteristics and Disease Stages. *Antioxidants* **2023**, *12*, 942. [CrossRef]
63. Pedlar, C.R.; Newell, J.; Lewis, N.A. Blood Biomarker Profiling and Monitoring for High-Performance Physiology and Nutrition: Current Perspectives, Limitations and Recommendations. *Sports Med.* **2019**, *49*, 185–198. [CrossRef]
64. Lewis, N.A.; Newell, J.; Burden, R.; Howatson, G.; Pedlar, C.R. Critical Difference and Biological Variation in Biomarkers of Oxidative Stress and Nutritional Status in Athletes. *PLoS ONE* **2016**, *11*, e0149927. [CrossRef] [PubMed]
65. Haller, N.; Behringer, M.; Reichel, T.; Wahl, P.; Simon, P.; Krüger, K.; Zimmer, P.; Stöggl, T. Blood-Based Biomarkers for Managing Workload in Athletes: Considerations and Recommendations for Evidence-Based Use of Established Biomarkers. *Sports Med.* **2023**, *53*, 1315–1333. [CrossRef] [PubMed]
66. Nowakowska, A.; Kostrzewa-Nowak, D.; Buryta, R.; Nowak, R. Blood Biomarkers of Recovery Efficiency in Soccer Players. *Int. J. Environ. Res. Public Health* **2019**, *16*, 3279. [CrossRef]
67. Schliep, E.M.; Schafer, T.L.J.; Hawkey, M. Distributed Lag Models to Identify the Cumulative Effects of Training and Recovery in Athletes Using Multivariate Ordinal Wellness Data. *J. Quant. Anal. Sports* **2021**, *17*, 241–254. [CrossRef]
68. Sun, Q.A.; Wang, B.; Miyagi, M.; Hess, D.T.; Stamler, J.S. Oxygen-Coupled Redox Regulation of the Skeletal Muscle Ryanodine Receptor/Ca^{2+} Release Channel (RyR1): Sites and Nature of Oxidative Modification. *J. Biol. Chem.* **2013**, *288*, 22961–22971. [CrossRef]
69. Sun, Q.A.; Hess, D.T.; Nogueira, L.; Yong, S.; Bowles, D.E.; Eu, J.; Laurita, K.R.; Meissner, G.; Stamler, J.S. Oxygen-Coupled Redox Regulation of the Skeletal Muscle Ryanodine Receptor-Ca^{2+} Release Channel by NADPH Oxidase 4. *Proc. Natl. Acad. Sci. USA* **2011**, *108*, 16098–16103. [CrossRef]
70. Kumar, R.; Coggan, A.R.; Ferreira, L.F. Nitric Oxide and Skeletal Muscle Contractile Function. *Nitric Oxide* **2022**, *122–123*, 54–61. [CrossRef]
71. Jackson, M.J.; Pollock, N.; Staunton, C.A.; Jones, S.W.; McArdle, A. Redox Control of Signalling Responses to Contractile Activity and Ageing in Skeletal Muscle. *Cells* **2022**, *11*, 1698. [CrossRef] [PubMed]
72. Pietrangelo, L.; Carraro, M.; Bernardi, P.; Raffaello, A.; Marcucci, L.; Nogara, L.; Reggiani, C.; Paolocci, N.; Canato, M.; Boncompagni, S.; et al. Mitochondria Can Substitute for Parvalbumin to Lower Cytosolic Calcium Levels in the Murine Fast Skeletal Muscle. *Acta Physiol.* **2024**, *240*, e14208. [CrossRef]
73. Cheng, A.J.; Yamada, T.; Rassier, D.E.; Andersson, D.C.; Westerblad, H.; Lanner, J.T. Reactive Oxygen/Nitrogen Species and Contractile Function in Skeletal Muscle during Fatigue and Recovery. *J. Physiol.* **2016**, *594*, 5149–5160. [CrossRef] [PubMed]
74. Martín-Prieto, E.; Márquez, C.; Fernández-Puente, E.; Palomero, J. Effect of RONS-Induced Intracellular Redox Homeostasis in 6-NBDG/Glucose Uptake in C2C12 Myotubes and Single Isolated Skeletal Muscle Fibres. *Int. J. Mol. Sci.* **2023**, *24*, 8082. [CrossRef]
75. Kjøbsted, R.; Hingst, J.R.; Fentz, J.; Foretz, M.; Sanz, M.N.; Pehmøller, C.; Shum, M.; Marette, A.; Mounier, R.; Treebak, J.T.; et al. AMPK in Skeletal Muscle Function and Metabolism. *FASEB J.* **2018**, *32*, 1741–1777. [CrossRef]
76. Popović, L.; Mitić, R.N. Influence of Free Radicals on Muscle Function and Adaptation in Exercise. *Praemeditatus* **2014**, *43*, 73–77. [CrossRef]
77. Kanda, K.; Sugama, K.; Hayashida, H.; Sakuma, J.; Kawakami, Y.; Miura, S.; Yoshioka, H.; Mori, Y.; Suzuki, K. Eccentric Exercise-Induced Delayed-Onset Muscle Soreness and Changes in Markers of Muscle Damage and Inflammation. *Exerc. Immunol. Rev.* **2013**, *19*, 72–85. [PubMed]
78. Devrnja, A.; Matković, B. The Effects of a Soccer Match on Muscle Damage Indicators. *Kinesiology* **2018**, *50*, 112–123. [CrossRef]
79. Schwiete, C.; Roth, C.; Skutschik, C.; Möck, S.; Rettenmaier, L.; Happ, K.; Broich, H.; Behringer, M. Effects of Muscle Fatigue on Exercise-Induced Hamstring Muscle Damage: A Three-Armed Randomized Controlled Trial. *Eur. J. Appl. Physiol.* **2023**, *123*, 2545–2561. [CrossRef]
80. Vargas-Mendoza, N.; Morales-González, Á.; Madrigal-Santillán, E.; Madrigal-Bujaidar, E.; Álvarez-González, I.; Garcia-Melo, L.F.; Anguiano-Robledo, L.; Fregoso-Aguilar, T.; Morales-González, J.A. Antioxidant and Adaptative Response Mediated by Nrf2 during Physical Exercise. *Antioxidants* **2019**, *8*, 196. [CrossRef]

81. Koyama, K. Exercise-Induced Oxidative Stress: A Tool for "Hormesis" and "Adaptive Response". *J. Phys. Fit. Sports Med.* **2014**, *3*, 115–120. [CrossRef]
82. Bartlett, M.F.; Miehm, J.D.; Fitzgerald, L.F.; Straight, C.R. Do Changes in Mitochondrial Quality Contribute to Increases in Skeletal Muscle Oxidative Capacity Following Endurance Training? *J. Physiol.* **2017**, *595*, 1861–1862. [CrossRef]
83. Mason, S.A.; Trewin, A.J.; Parker, L.; Wadley, G.D. Antioxidant Supplements and Endurance Exercise: Current Evidence and Mechanistic Insights. *Redox Biol.* **2020**, *35*, 101471. [CrossRef] [PubMed]
84. Rickards, L.; Lynn, A.; Harrop, D.; Baker, M.E.; Russell, M.; Ranchordas, M. Effect of Polyphenol-Rich Foods, Juices, and Concentrates on Recovery from Exercise-Induced Muscle Damage: A Systematic Review and Meta-Analysis. *Nutrients* **2021**, *13*, 2988. [CrossRef] [PubMed]
85. Devrim-Lanpir, A.; Hill, L.; Knechtle, B. How N-Acetylcysteine Supplementation Affects Redox Regulation, Especially at Mitohormesis and Sarcohormesis Level: Current Perspective. *Antioxidants* **2021**, *10*, 153. [CrossRef]
86. Peeri, M.; Shamsnia, E.; Azarbayjani, M.; Matinhomaee, H. The Effect of Aerobic Exercise on Oxidative Stress in Skeletal Muscle Tissue: A Narrative Review. *Gene Cell Tissue* **2023**, *10*, e131964. [CrossRef]
87. Ponce-González, J.; Corral-Pérez, J.; Villarreal, E.; Gutiérrez-Manzanedo, J.; De Castro-Maqueda, G.; Casals, C. Antioxidants Markers of Professional Soccer Players During the Season and Their Relationship with Competitive Performance. *J. Hum. Kinet.* **2021**, *80*, 113–123. [CrossRef]
88. Bell, P.; Stevenson, E.; Davison, G.; Howatson, G. The Effects of Montmorency Tart Cherry Concentrate Supplementation on Recovery Following Prolonged, Intermittent Exercise. *Nutrients* **2016**, *8*, 441. [CrossRef]
89. Carrera-Quintanar, L.; Funes, L.; Herranz-López, M.; Vicente-Salar, N.; Bonet-García, R.; Blasco-Peris, C.; Micol, V.; Pons, A.; Roche, E. Modulation of Oxidative Stress and Antioxidant Response by Different Polyphenol Supplements in Five-a-Side Football Players. *Nutrients* **2023**, *15*, 177. [CrossRef]
90. Higgins, M.R.; Izadi, A.; Kaviani, M. Antioxidants and Exercise Performance: With a Focus on Vitamin E and C Supplementation. *Int. J. Environ. Res. Public Health* **2020**, *17*, 8452. [CrossRef]
91. Bhol, N.; Dash, U.; Majhi, S.; Bhanjadeo, M.; Singh, A.; Duttaroy, A.; Ojha, R.; Jena, A. The Interplay Between Cytokines, Inflammation, and Antioxidants: Mechanistic Insights and Therapeutic Potentials of Various Antioxidants and Anti-Cytokine Compounds. *Biomed. Pharmacother.* **2024**, *178*, 117177. [CrossRef] [PubMed]
92. Nedelec, M.; McCall, A.; Carling, C.; Legall, F.; Berthoin, S.; Dupont, G. Recovery in Soccer: Part II—Recovery Strategies. *Sports Med.* **2013**, *43*, 9–22. [CrossRef]
93. White, G.E.; Caterini, J.E. Cold Water Immersion Mechanisms for Recovery Following Exercise: Cellular Stress and Inflammation Require Closer Examination. *J. Physiol.* **2017**, *595*, 631–632. [CrossRef] [PubMed]
94. Kim, T.-W.; Jeong, J.-H.; Hong, S.C. The Impact of Sleep and Circadian Disturbance on Hormones and Metabolism. *Int. J. Endocrinol.* **2015**, *2015*, 591729. [CrossRef]
95. Nedelec, M.; Halson, S.L.; Delecroix, B.; Abaidia, A.-E.; Ahmaidi, S.; Dupont, G. Sleep Hygiene and Recovery Strategies in Elite Soccer Players. *Sports Med.* **2015**, *45*, 1547–1559. [CrossRef]
96. Güllü, E.; Güllü, A. Recovery after High-Intensity Interval Training in Professional Soccer Players: Passive-Active Combination Recovery Technique. *Int. J. Relig.* **2024**, *5*, 221–229. [CrossRef]
97. Merry, T.L.; Ristow, M. Do Antioxidant Supplements Interfere with Skeletal Muscle Adaptation to Exercise Training? *J. Physiol.* **2016**, *594*, 5135–5147. [CrossRef]
98. Lamberti, I.; Buzzetti, E.; Cristani, A. Oxidative Stress Management in Elite Athletes. *Recent. Prog. Med.* **2013**, *104*, 490–492. [CrossRef]
99. Liu, C.; Lu, N.; Zhao, H.; Liu, J.; Song, G.; Teng, J.; Yun, H.; Zou, D.; Zhu, H. Assessment of Urolithin A Effects on Muscle Endurance, Strength, Inflammation, Oxidative Stress, and Protein Metabolism in Male Athletes with Resistance Training: An 8-Week Randomized, Double-Blind, Placebo-Controlled Study. *J. Int. Soc. Sports Nutr.* **2024**, *21*, 2419388. [CrossRef]
100. Lee, E.C.; Fragala, M.S.; Kavouras, S.A.; Queen, R.M.; Pryor, J.L.; Casa, D.J. Biomarkers in Sports and Exercise: Tracking Health, Performance, and Recovery in Athletes. *J. Strength Cond. Res.* **2017**, *31*, 2920–2937. [CrossRef]
101. Reid, M.B. Redox Interventions to Increase Exercise Performance. *J. Physiol.* **2016**, *594*, 5125–5133. [CrossRef]
102. Vassalle, C.; Pingitore, A.; De Giuseppe, R.; Vigna, L.; Bamonti, F. Biomarkers Part II: Biomarkers to Estimate Bioefficacy of Dietary/Supplemental Antioxidants in Sport. *Adv. Clin. Chem.* **2015**, *69*, 209–222.
103. Meckel, Y.; Doron, O.; Eliakim, E.; Eliakim, A. Seasonal Variations in Physical Fitness and Performance Indices of Elite Soccer Players. *Sports* **2018**, *6*, 14. [CrossRef] [PubMed]
104. Abate, M.; Pellegrino, R.; Di Iorio, A.; Salini, V. Oxidative Stress and Performance after Training in Professional Soccer (European Football) Players. *Antioxidants* **2023**, *12*, 1470. [CrossRef] [PubMed]

105. Flockhart, M.; Nilsson, L.; Tillqvist, E.; Vinge, F.; Millbert, F.; Lännerström, J.; Nilsson, P.; Samyn, D.; Apró, W.; Sundqvist, M.; et al. Glucosinolate-Rich Broccoli Sprouts Protect against Oxidative Stress and Improve Adaptations to Intense Exercise Training. *Redox Biol.* **2023**, *67*, 102873. [CrossRef]
106. Su, C.; Meng, Q. The Impact of Physical Exercise on Oxidative and Nitrosative Stress: Balancing the Benefits and Risks. *Antioxidants* **2024**, *13*, 573. [CrossRef]
107. Ranchordas, M.; Dawson, J.T.; Russell, M. Practical Nutritional Recovery Strategies for Elite Soccer Players When Limited Time Separates Repeated Matches. *J. Int. Soc. Sports Nutr.* **2017**, *14*, 35. [CrossRef] [PubMed]
108. Heaton, L.E.; Davis, J.K.; Rawson, E.S.; Nuccio, R.P.; Witard, O.C.; Stein, K.W.; Baar, K.; Carter, J.M.; Baker, L.B. Selected In-Season Nutritional Strategies to Enhance Recovery for Team Sport Athletes: A Practical Overview. *Sports Med.* **2017**, *47*, 2201–2218. [CrossRef]
109. Baldelli, S.; Ciccarone, F.; Limongi, D.; Checconi, P.; Palamara, A.T.; Ciriolo, M.R. Glutathione and Nitric Oxide: Key Team Players in Use and Disuse of Skeletal Muscle. *Nutrients* **2019**, *11*, 2318. [CrossRef]
110. Taherkhani, S.; Valaei, K.; Arazi, H.; Suzuki, K. An Overview of Physical Exercise and Antioxidant Supplementation Influences on Skeletal Muscle Oxidative Stress. *Antioxidants* **2021**, *10*, 1528. [CrossRef]
111. Poulios, A.; Papanikolaou, K.; Draganidis, D.; Tsimeas, P.; Chatzinikolaou, A.; Tsiokanos, A.; Jamurtas, A.Z.; Fatouros, I.G. The Effects of Antioxidant Supplementation on Soccer Performance and Recovery: A Critical Review of the Available Evidence. *Nutrients* **2024**, *16*, 3803. [CrossRef] [PubMed]
112. Besedovsky, L.; Lange, T.; Haack, M. The Sleep-Immune Crosstalk in Health and Disease. *Physiol. Rev.* **2019**, *99*, 1325–1380. [CrossRef] [PubMed]
113. Seifi-Skishahr, F.; Damirchi, A.; Farjaminezhad, M.; Babaei, P. Physical Training Status Determines Oxidative Stress and Redox Changes in Response to an Acute Aerobic Exercise. *Biochem. Res. Int.* **2016**, *2016*, 3757623. [CrossRef] [PubMed]
114. Devrim-Lanpir, A.; Bilgiç, P.; Kocahan, T.; Deliceoğlu, G.; Rosemann, T.; Knechtle, B. Total Dietary Antioxidant Intake Including Polyphenol Content: Is It Capable to Fight against Increased Oxidants within the Body of Ultra-Endurance Athletes? *Nutrients* **2020**, *12*, 1877. [CrossRef]
115. Ruderman, E.B.W.; White, G.E. Don't Mess with Stress: Vitamin Supplementation and the Role of Oxidative Stress for Aerobic Exercise Adaptation. *J. Physiol.* **2014**, *592*, 3951–3952. [CrossRef]
116. Canals-Garzón, C.; Guisado-Barrilao, R.; Martínez-García, D.; Chirosa-Ríos, I.; Jérez-Mayorga, D.; Guisado-Requena, I. Effect of Antioxidant Supplementation on Markers of Oxidative Stress and Muscle Damage after Strength Exercise: A Systematic Review. *Int. J. Environ. Res. Public Health* **2022**, *19*, 1803. [CrossRef]
117. Margaritelis, N.V.; Chatzinikolaou, P.N.; Theodorou, A.A.; Paschalis, V.; Vrabas, I.S.; Kyparos, A.; Fatouros, I.G.; Nikolaidis, M.G. Inter-Individual Variability in Redox and Performance Responses after Antioxidant Supplementation: A Randomized Double Blind Crossover Study. *Acta Physiol.* **2023**, *238*, e14017. [CrossRef]
118. Zare, M.; Shateri, Z.; Nouri, M.; Sarbakhsh, P.; Eftekhari, M.H.; Gargari, B.P. Association between Urinary Levels of 8-Hydroxy-2-Deoxyguanosine and F2α-Isoprostane in Male Football Players and Healthy Non-Athlete Controls with Dietary Inflammatory and Antioxidant Indices. *Front. Nutr.* **2023**, *9*, 1101532. [CrossRef]
119. Clemente, F.M.; González-Fernández, F.T.; Ceylan, H.İ.; Silva, R.; Younesi, S.; Chen, Y.-S.; Badicu, G.; Wolański, P.; Murawska-Ciałowicz, E. Blood Biomarkers Variations across the Pre-Season and Interactions with Training Load: A Study in Professional Soccer Players. *J. Clin. Med.* **2021**, *10*, 5576. [CrossRef]

antioxidants

MDPI

Article

Controlled Cultivation Confers *Rhodiola rosea* Synergistic Activity on Muscle Cell Homeostasis, Metabolism and Antioxidant Defense in Primary Human Myoblasts

Fortuna Iannuzzo [1,†], Elisabetta Schiano [2,†], Arianna Pastore [3], Fabrizia Guerra [2], Gian Carlo Tenore [3], Ettore Novellino [2] and Mariano Stornaiuolo [3,*]

1 Department of Pharmacy, University of Chieti-Pescara G. D'Annunzio, 66100 Chieti, Italy; fortuna.iannuzzo@unich.it

2 Inventia Biotech-Healthcare Food Research Center s.r.l., Strada Statale Sannitica KM 20.700, 81020 Caserta, Italy; elisabettaschiano@inventiabiotech.com (E.S.); fabrizia.guerra@inventiabiotech.com (F.G.); ettorenovellino@inventiabiotech.com (E.N.)

3 Department of Pharmacy, University of Naples Federico II, Via Domenico Montesano 59, 80131 Naples, Italy; arianna.pastore@unina.it (A.P.); giancarlo.tenore@unina.it (G.C.T.)

* Correspondence: mariano.stornaiuolo@unina.it

† These authors contributed equally to this work.

Abstract: *Rhodiola rosea* L. is recognized for its adaptogenic properties and ability to promote muscle health, function and recovery from exercise. The plethora of biological effects of this plant is ascribed to the synergism existing among the molecules composing its phytocomplex. In this manuscript, we analyze the activity of a bioactive fraction extracted from *Rhodiola rosea* L. controlled cultivation. Biological assays were performed on human skeletal myoblasts and revealed that the extract is able to modulate in vitro expression of transcription factors, namely Pax7 and myoD, involved in muscle differentiation and recovery. The extract also promotes ROS scavenging, ATP production and mitochondrial respiration. Untargeted metabolomics further reveals that the mechanism underpinning the plant involves the synergistic interconnection between antioxidant enzymes and the folic/acid polyamine pathway. Finally, by examining the phytochemical profiles of the extract, we identify the specific combination of secondary plant metabolites contributing to muscle repair, recovery from stress and regeneration.

Keywords: *Rhodiola rosea* L.; muscle health; antioxidant activity; nutraceuticals; controlled cultivation

1. Introduction

Rhodiola rosea L. (*genus Rhodiola* L., Crassulaceae family) is a plant with renowned phytopharmaceutical potential (some of them documented since ancient Greek times) [1]. Twenty of the approximately one hundred Rhodiola species are commonly used as adaptogens (substances that help the human body cope with stress), mostly to enhance physiological resilience and support normal bodily functions [2,3]. In virtue of their potential to promote endurance and improve physical performance during prolonged exercise, extracts of *Rhodiola rosea* L. are widely used as dietary supplements in the sports industry [4]. The extracts also exhibit antioxidant [5], anti-fatigue [6], anti-inflammatory [7], and antidepressant [8] activities, all properties that help athletes during training and competition. Tinsley et al. have recently reviewed pre-clinical and clinical studies where, in the context of exercise performance, *Rhodiola rosea* L. extracts have been tested as an adaptogen [2]. *Rhodiola rosea* L. supplementation supports prolonged exercise by increasing mitochondrial ATP production [9]. Furthermore, chronic supplementation promotes the accumulation of hepatic glycogen at rest and concomitant attenuation of muscle glycogen depletion during exercise, implying that changes in glycogen turnover may potentially contribute to the ergogenic effects of the extracts [10]. Additionally, supplementation with *Rhodiola rosea* L.

reduces post-exercise fatigue, improves tissue oxygenation and promotes the expression of proteins involved in lipid metabolism [11,12]. Finally, the antioxidant properties of *Rhodiola rosea* L. increase the expression of antioxidant enzymes, counteracting oxygen free radicals and lipid peroxidation that occurs after exercise [13].

Pharmacological studies on *Rhodiola rosea* L. are numerous. The medicinal activities of *Rhodiola rosea* L. are attributed to its phytochemicals profile, especially to phenylethanoids and phenylpropanoids present in its roots and their glycosidic derivatives [14]. Salidroside and Rosavins (a collective name for Rosavin, Rosarin, and Rosin) are among the active components of *Rhodiola rosea* L. bioactive extracts [15]. Salidroside and Rosavins contributed to the adaptogenic properties of *Rhodiola rosea* L. [14], while antioxidant activity is primarily due to organic acids and flavonoids [16]. *Rhodiola rosea* L. phytochemicals allow the plant to adapt and grow at high altitudes and in cold regions of the globe (Arctic regions of Europe, Asia, and North America). However, growing market demand has led to the development of *Rhodiola rosea* L. controlled cultivations in other regions of the globe. The phytochemical profile of controlled cultivations presents similar Salidroside and Rosavins content compared to wild harvests [17,18]. The United States Pharmacopeia standards are products titrated to 0.3% in rosavins and 0.08% in salidroside [19]. According to the Russian State Pharmacopeia, the content of rosavins should be not less than 1%, and salidroside should be not less than 0.8% [20]. The Australian standard for the extract is not less than 1.8% phenylpropanoids, 1.2% rosavins, and 0.6% salidroside [21].

The molecular mechanism underpinning *Rhodiola rosea* L. is complex, and not all of its biological activities can be attributed to salidroside or rosavins. Recently, Liu et al. (2024) demonstrated that pure salidroside might, in part, contribute to the adaptogenic effect of the extracts by effectively attenuating cardiac dysfunction, myocardial hypertrophy, myocardial fibrosis, and cardiac inflammation in animal models [22]. In other scientific settings, pure Salidroside and Rosavins failed to recapitulate *Rhodiola rosea* L. effects. This is probably to be ascribed to a synergistic effect of the phytocomplex of the plant. About 140 organic compounds have been isolated in *Rhodiola rosea* L., including polyphenols, organic acids, sugars, tannins, terpenes, and essential oils, with the root containing over 100 chemical compounds [23]. These bioactive constituents synergistically contribute to the adaptogenic properties. This synergy makes each different *Rhodiola rosea* L. cultivation of particular interest since specific climatic and cultivation conditions could generate a phytochemical profile endowed with specific properties and physiological effects [24].

In this manuscript, we analyze the biological effect of a *Rhodiola rosea* L. extract from plants harvested from controlled cultivation (RRcc) plots situated at altitudes ranging from 20 m to 600 m above sea level in Central Europe from a cultivation line called Rhodiofarm (Pharmaplant Germany a Martin Bauer company). The bioactive effects were investigated on human skeletal myoblasts and included the analysis of pathways involved in muscle differentiation and renewal, ROS scavenging, ATP production and mitochondrial respiration. The bioactive effects of RRcc were compared with those of *Rhodiola rosea* L. wild harvest (RRwh).

2. Materials and Methods

2.1. *Rhodiola rosea* L. Wild Harvest (RRwh) and Controlled Cultivation (RRcc)

For *Rhodiola rosea* L. wild harvest (RRwh), plants were collected in a region that straddles the borders of Kazakhstan, China and Mongolia at altitudes ranging from 2000 to 3000 m. Once collected, the roots were washed, cut into thin pieces and artificially dried to preserve medicinal properties. For *Rhodiola rosea* L. controlled cultivation (RRcc), *Rhodiola rosea* L. seeds were cultivated in specialized, isolated plots (known as Rhodiofarm, Pharmaplant Germany a Martin Bauer company (Martin Bauer, Alveslohe, Germany) to ensure pure lineage. The process was conducted under strict conditions to maintain the genetic quality and vitality of the seeds, selecting mother plants based on criteria such as disease resistance, growth rate, and active compound content. From the seeds, seedlings were raised under controlled conditions within greenhouses. This stage was crucial for ensuring a high germination rate and fostering the healthy growth of young plants. Once the seedlings had reached a suitable size, these were transplanted into cultivation plots situated at altitudes ranging from 20 m to 600 m above sea level in Central Europe. Cultivation sites were selected considering soil quality, sunlight exposure and irrigation needs to create optimal growing conditions. During the whole process, conditions such as lighting, temperature and humidity were controlled to promote the development of robust and vigorous plants. The harvest occurred three years after planting, timed to coincide with the peak in the plants' maturity and active compound concentration. Post-harvest, the roots of Rhodiola were washed, coarsely cut and subjected to artificial drying.

2.2. *Rhodiola rosea L. Extracts*

All extracts were prepared by Finzelberg GmbH & Co. KG, Andernach/Germany, under full GMP conditions. *Rhodiola rosea* L. roots, from both controlled cultivation and wild collection, were initially prepared by cutting them to a size suitable for equipment. An exhaustive extraction was then conducted with ethanol 70% v/v using multi-stage extraction, specifically employing percolation. This method ensured thorough extraction of the bioactive compounds present in the Rhodiola root. The solvent-to-solid ratio utilized in the extraction process was 1:12. Following extraction, the eluate obtained from the process was subjected to evaporation and homogenization to yield the soft extract. This soft extract was then further processed by adding excipients (maltodextrin + silicium dioxide). These excipients were carefully chosen to enhance the stability and handling properties of the final extract. After the addition of excipients, the extract was dried by using a vacuum belt dryer and subsequently adjusted with excipients to achieve the desired composition. Lastly, the dried extract underwent additional processing steps, including grinding, mixing, sieving, and packing.

2.3. *Cell Culture Experiments*

Human skeletal muscle myoblasts (CC-2580; Lonza (Basel, Switzerland) were cultured in Dulbecco's modified Eagle Medium (DMEM) (Gibco, (Calrsbad, CA, USA)), supplemented with 10% Fetal Bovine Serum (FBS) (Gibco, (Calrsbad, CA, USA)), 2 mM Glutamine, 100 U mL^{-1} Penicillin and 100 μg mL^{-1} Streptomycin (Gibco) and incubated at 37 °C and 5% CO_2. These conditions were consistent throughout the experiments to ensure reproducibility and maintain the physiological relevance of the in vitro model. Cell confluence was maintained at 60–80%, and subculturing was performed at a 1:10 ratio twice a week using 25 cm^2 flasks (Nunc, (Koln, Germany)). To induce differentiation, cells were cultivated in DMEM supplemented with 2% (v/v) Horse Serum (HS) for 7 days.

2.3.1. In Vitro Treatment of Cell Cultures

Primary cultures were seeded at a density of 2×10^5 cells/mL in 25 cm^2 flasks. For cell culture experiments, the dried extract, obtained as described in Section 2.2, was solubilized using dimethyl sulfoxide (DMSO). Subsequently, the solution was diluted with water to achieve a final concentration of 30 μg/mL, maintaining a final concentration of 1% DMSO. Upon 16 h from plating, cells were treated with either RRwh or RRcc extracts. The treatments were maintained for 48 h. Untreated cells and cells treated with equal volumes of vehicles (DMSO) were used as negative controls. Upon treatment, the medium was removed, and cells were rinsed several times in phosphate-buffered saline (PBS, composed of 137 mM sodium chloride, 2.7 mM potassium chloride, 10 mM phosphate buffer, pH 7.4) to be then processed for metabolomic analysis. When indicated, cells were supplemented with Folic acid 5 μM, Cystine (Cys2) 1 mM, Quercetin (Q, 1 μM), Quercetin 3-glycoside (QG, 1 μM), Rutin (Q2G, 1 μM), and Quercetin 3-rhamnoglucoside 7-glucoside (Q3G, 1 μM), all from Merck (Darmstadt, Germany).

2.3.2. Measurement of Intracellular ATP Content

Primary myoblasts were seeded into a 96-well plate at a density of 2×10^4 cells/well in 100 µL of growth medium. Upon incubation with test substances for 48 h, intracellular ATP content measurement was performed using an ATP assay kit (Toyo Ink (Tokyo, Japan)), following the manufacturer's instructions. First, cells were washed twice with PBS. Next, 50 µL of serum-free DMEM and 50 µL of ATP assay reagent were added to each well. After 15 min of incubation, cells were washed twice with PBS to remove any residual media and unmetabolized substances. Subsequently, a lysis buffer from the ATP assay kit was added to the cells. The total cellular ATP content was measured by light emission with the Envision 2105 spectrofluorometer, Perkin Elmer, Waltham, MA, USA.

2.3.3. Measurement of Mitochondrial Potential

Mitochondria staining of primary myocytes was achieved by incubation with MitoTracker®Red CMXRos (Thermo Fisher Scientific). A dye working solution was prepared by diluting a stock solution (10 µM in DMSO) in DMEM to yield a final concentration of 100 nM. For staining of in vitro samples, cells were rinsed twice in PBS before adding the dye. Thus, cells were incubated in the presence of the probe for 40 min in a cell incubator at 37 °C and 5% CO_2. At the end of the incubation, cells and tissues were rinsed three times in DMEM and once in PBS, then fixed in 4% formaldehyde. The quantitative measurement of MitoTracker®Red CMXRos fluorescence was performed with the spectrofluorometer Envision 2105 from Perkin Elmer.

2.3.4. Metabolomic Profiling of In Vitro Cultured Human Myoblasts

Metabolites were extracted as follows: cell pellets were thawed on ice and 100 µL of ice-cold MeOH/H_2O (80:20 *v*/*v*) was added. The samples were extracted in an ultrasonic bath for 6 min, vortexed for 30 s, and finally centrifuged at 14,000 rpm for 10 min at 4 °C. The supernatants were dried under nitrogen and were reconstituted in 100 µL of ACN/H_2O (70:30) (*v*/*v*) before HRMS analysis. A pooled quality control (QC) sample was prepared by pooling an aliquot of cellular extract from each sample. Unless otherwise described, all solvents and additives were LCMS grade and purchased by Merck (Darmstadt, Germany). HRMS analyses were performed on a Thermo Ultimate RS 3000 coupled online to a Exploris 120 hybrid quadrupole Orbitrap mass spectrometer (Thermo Fisher Scientific, Bremen, Germany) equipped with a heated electrospray ionization probe (HESI II). The MS was calibrated by Thermo calmix PierceFlexmix™ calibration solutions in both polarities. The analysis of cellular metabolites was performed in HILIC mode, with an Acquity BEH Amide (100 × 2.1 mm; 1.7 µm) protected with a VanGuard amide precolumn (5 × 2.1 mm; 1.7 µm) (Waters, Milford, MA, USA). The column temperature was set at 45 °C, and the flow rate was 0.400 mL/min. The mobile phase was (A) 10 mM CH_3COONH_4 in H_2O/ACN (95:5 *v*/*v*) and (B) 10 mM CH_3COONH_4 in H_2O/ACN (5:95 *v*/*v*). Full MS (80–800 *m*/*z*) and data-dependent MS/MS were performed at a resolution of 60,000 and 15,000 FWHM, respectively, and normalized collision energy (NCE) values of 10, 30 and 50 were used. Source parameters were as follows: sheath gas pressure, 40 arbitrary units; auxiliary gas flow, 15 arbitrary units; spray voltage, +3.5 kV, −2.8 kV; capillary temperature, 300 °C; auxiliary gas heater temperature, 280 °C. Data were analyzed with Compound Discoverer 3.3 (Thermo) and MetaboAnalyst 5.0.

2.3.5. Isolation of RNA, cDNA Synthesis and qPCR Analysis Myoblast mRNAs

Total RNA was extracted according to the manufacturing protocol of RNeasy Plus Mini Kit (# 74136, Qiagen (Hilden, Germany)). First, 2 µL of total RNA was quantified by NANODROP2000 spectrophotometer (Thermo Fisher). Then, 2 µg of total RNA was converted into cDNA using SuperScript VILO Master Mix (# 11755050, Thermo Fisher) as indicated by the manufacturer, and qPCR analysis was performed using QuantiNova SYBR PCR Master Mix (# 208252, Qiagen (Hilden, Germany)). The amplification was performed using the Real-Time PCR System StepOne-Plus, Applied Biosystems (Thermo

Fisher), with the following cycling conditions: 95 °C for 3 min followed by 40 cycles of 30 s of denaturation at 95 °C, 30 s of annealing at 52 °C, and 30 s product amplification at 60 °C. The relative gene expression analysis of target genes was conducted in comparison with the β-actin housekeeping control gene following the comparative $2^{-\Delta Ct}$ method. Primers used for qPCR analysis were all from Biorad (Hercules, CA, USA). Primers for myogenic differentiation 1 (myoD) (qHsaCED0023842), paired box 7 (Pax7) (qHsaCID0016103) and myogenin (qHsaCED0043933), as well as primers for gapdh; superoxide dismutase (SOD1); catalase (CAT); glutathione peroxidase (GPX1) and glutathione reductase (GR); peroxiredoxins (PRDX1); thioredoxin (TXN) and thioredoxin reductase (TXNRD1); and heme oxygenase-1 (HO-1) all belonged to the Oxidative Stress and Antioxidant Defense preassembled primer plate H96 code #10034219 from Biorad.

2.4. Phytochemical Profiling of *Rhodiola rosea* L. by UHPLC Analysis

10 mg of *Rhodiola rosea* L. extracts were solubilized in 1 mL of ethanol/H_2O (70:30 *v*/*v*). Each sample was filtered through a 0.45 µm membrane filter and analyzed by either Reverse Phase Ultra-High Performance Liquid Chromatography associated with Photodiode Array Detection (RP-UHPLC-PDA) or High-Resolution Mass Spectrometry (RP-UHPLC-HRMS). RP-UHPLC-PDA analyses were performed on a Shimadzu Nexera UHPLC system consisting of a CBM-20A controller, two LC-30AD dual-plunger parallel-flow pumps, a DGU-20 AR5 degasser, an SPD-M20A photodiode array detector, a CTO-20A column oven and an SIL-30AC autosampler. PDA detection parameters were a sampling rate of 12 Hz and a time constant of 0.160 s, and chromatograms were extracted at 220, 280 and 330 nm. RP-UHPLC-HRMS analysis was performed on a Thermo Ultimate RS 3000 coupled online to a Q-Exactive hybrid quadrupole Orbitrap mass spectrometer (Thermo Fisher Scientific, Bremen, Germany) equipped with a heated electrospray ionization probe (HESI II). The MS was calibrated by Thermo calmix PierceTM calibration solutions in both polarities. Separation was performed with a Luna Omega C18 100 mm × 2.1 mm, 1.6 µm (L × I.D, particle size, Phenomenex®, Bologna, Italy) column at a flow rate of 0.4 mL/min. The mobile phases consisted of (A) 0.1% CH_3COOH in H_2O and (B) 0.1% CH_3COOH in ACN. Analysis was performed in gradient as follows: 0.01 min, 2% B; 18 min, 35% B; 20.0 min, 70% B; 22.0 min, 95% B; isocratic for 2.0 min; and returning to initial conditions in 0.1 min. The column oven was set to 40 °C, and 3 µL was injected. The same chromatographic conditions (i.e., column, flow rate and mobile phases) were used for both RP-UHPLC-PDA and RP-UHPLC-HRMS analysis. MS analyses were performed in negative ionization mode (ESI-). Full MS (100–1500 *m*/*z*) and data-dependent MS/MS were performed at resolutions of 70,000 and 15,000 FWHM, respectively, and normalized collision energy (NCE) values of 15, 25 and 30 were used. Source parameters were as follows: sheath gas pressure, 50 arbitrary units; auxiliary gas flow, 13 arbitrary units; spray voltage, +3.5 kV, −2.8 kV; capillary temperature, 310 °C; and auxiliary gas heater temperature, 300 °C. Two replicates of each sample were performed. A node-based processing workflow was custom-built in Compound Discoverer™ software v.3.3 (Thermo Fisher Scientific) to search and identify metabolites. Specifically, metabolite annotation was based on accurate mass measurement, MS/MS fragmentation pattern, and comparison within silico spectra with MS database searching. Salidroside was selected as an external standard for quantitative analysis. Stock solution (1 mg/mL) was prepared in 1 mL of ethanol/H_2O (70:30). A calibration curve was obtained in the concentration range of 1–100 µg/mL (R^2 = 0.999). For sensitivity evaluation, limits of detection (LODs) and quantification (LOQs) were calculated by the ratio between the standard deviation (SD) and analytical curve slope multiplied by 3 and 10, respectively. The compound salidroside LOD was 0.031 µg/mL and LOQ was 0.103 µg/mL.

3. Results

All biological experiments were conducted on primary human myoblasts supplemented with RRcc (30 µg/mL). As controls, treatment with RRwh (30 µg/mL), untreated cells (untr) and cells treated with the same amount of vehicle DMSO (veh) were analyzed in parallel. Primary human myoblasts were very sensitive to DMSO. However, treatments of up to 30 μg/mL of RRcc and RRwh did not change the overall growth rate of the cell cultures, with myoblasts presenting an equal degree of confluency at the end of every experiment (Figure 1A).

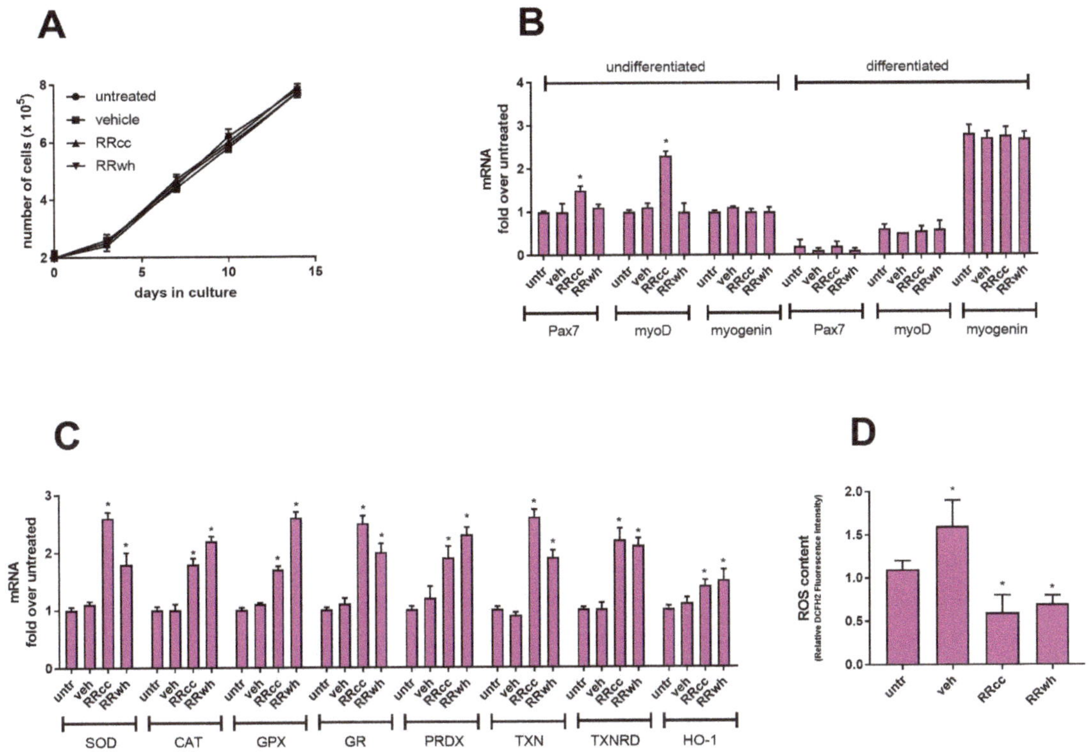

Figure 1. RRcc modulates undifferentiated muscle cells and promotes defense from oxidative stress. Growth (**A**), qPCR analysis of differentiation-related muscle cell transcription factors (**B**), antioxidant enzymes (**C**) and total ROS content (**D**) in primary muscle myoblasts treated for 48 h in the presence of RRcc (30 μg/mL), RRwh (30 μg/mL), an equal amount of vehicle (veh) or left untreated (untr). Data are shown as mean ± SD of five independent experiments. Statistical analysis was performed by ANOVA test comparing each mean with that of untreated cells. p value = * <0.05, otherwise differences were not statistically significant.

3.1. RRcc Promotes the Transcription of myoD and Pax7

First, we analyzed the effect of RRcc on the expression of differentiation markers of skeletal muscle cells. Pure salidroside has been shown to maintain murine skeletal muscle cells in an undifferentiated state by promoting the expression of transcription factors involved in muscle recovery and regeneration [25]. The quantitative measurement of salidroside by UHPLC-LRMS revealed that RRcc and RRwh contain 5.724 ± 0.170 μg/mg (~0.8% *w*/*w*) and 9.136 ± 0.192 μg/mg (~1.4% *w*/*w*) salidroside, respectively. In terms of the effect on the differentiation of skeletal muscle cells, the two extracts should have had similar effects. However, analyzed by qPCR, treatment with RRcc but not with RRwh promoted the upregulation of two mRNA markers of muscle cell undifferentiation: myogenic differentiation 1 *myoD* (a marker of proliferating undifferentiated myoblasts) and paired box 7 *Pax7* (a marker of undifferentiated reserve muscle satellite cells) (Figure 1B).

In vivo, Pax7-positive cells are placed underneath the myofiber basal lamina, where they act as a muscle stem cell reservoir and allow the repair and maintenance of myofibers. When cultivated for 6 days in the appropriate differentiation medium, primary human myoblasts spontaneously differentiate into myotubes, as shown by the appearance of the myotube marker myogenin (Figure 1B). During differentiation, supplementation with RRcc or RRwh did not alter the expression of myogenin, suggesting that the two extracts do not retard the differentiation of myoblasts into myotubes. Similarly, when cultivated in the absence of differentiation medium, RRcc and RRwh do not promote the expression of myogenin, suggesting no effect on the subpopulation of myogenin-positive cells. RRwh is thus inactive in intracellular pathways regulating muscle homeostasis. On the contrary, RRcc is active on a subpopulation of Pax7- and myoD-positive myoblast cells involved in muscle renewal. Since both Pax7 and myoD are upregulated by RRcc, and myoD is expressed at later time points during myoblast differentiation, RRcc very likely allows the bypass of early differentiation control points, promoting the proliferation of Pax7/myoD-positive myoblasts.

3.2. RRcc and RRwh Regulate the Redox State of Skeletal Muscle Cells

Considering that the differentiation state of muscle cells depends on their redox state [26], we wondered if the effect of RRcc on *myoD* and *Pax7* expression could be the result of a different antioxidant activity of RRcc compared to RRwh. Despite the different effect on the transcription of undifferentiation-related genes, both extracts promoted the transcription of mRNAs coding for antioxidant enzymes. As shown in Figure 1C, mRNA for superoxide dismutase (SOD), catalase (CAT), glutathione peroxidase (GPX) and glutathione reductase (GR), peroxiredoxins (PRDX), thioredoxin (TXN), thioredoxin reductase (TXNRD) and heme oxygenase-1 (HO-1) were all upregulated by both RRcc and RRwh. As a consequence, myoblasts treated with RRcc and RRwh showed decreased ROS content when compared to cells treated with vehicle, confirming that both RRcc and RRwh protect muscle cells from endogenous or culture-derived ROS.

3.3. RRcc and RRwh Both Promote ATP Production and Mitochondrial Activity

We thus wondered if the different effects of the two extracts might have been the result of a different metabolic consequence of their treatment.

Along with the support of muscle regeneration, RR is known to promote mitochondrial activity and ATP production in muscle cells. We thus measured intracellular ATP concentration in human muscle cells treated with RRcc or RRwh by luminescence assay. As shown in Figure 2A and compared to untreated or vehicle-treated cells, 48 h of incubation with RRcc or RRwh promoted a statistically significant increase in ATP production (untr: 6.9 ± 0.1 μM; veh: 6.5 ± 0.2 μM; RRcc: 8.2 ± 0.2 μM; RRwh: 8.1 ± 0.3 μM; *** = ANOVA p value < 0.001; $n = 3$ independent experiments). However, no statistically significant difference was measured between RRcc and RRwh in terms of ATP production.

In order to confirm the mitochondrial boosting potential of the two extracts, the transcriptional upregulation of *Pgc1α* (a master regulator of mitochondrial activity) and the intermembrane mitochondrial potential were measured in RRcc- and RRwh-treated cells. Both the extracts strongly upregulated *Pgc1α*. The intermembrane mitochondrial potential was measured with the mitochondrial probe Mito Tracker CMX-ROS. The dye accumulates in mitochondria depending on their membrane potential and, thus, on their activity. As shown in Figure 3B, when analyzed by fluorimetry, human myoblast cells treated with RRwh or RRcc showed increased mitochondrial activity compared to vehicle-treated cells (untr: 1.1 ± 0.1-fold increase in fluorescence; veh: 0.9 ± 0.3; RRcc: 1.7 ± 0.1; RRwh: 1.6 ± 0.2; * = ANOVA p value < 0.05; $n = 3$ independent experiments), confirming that both RRcc and RRwh stimulate mitochondrial activity. No difference between RRcc and RRwh in terms of PGC1α mRNA levels or intermembrane mitochondrial potential could be detected.

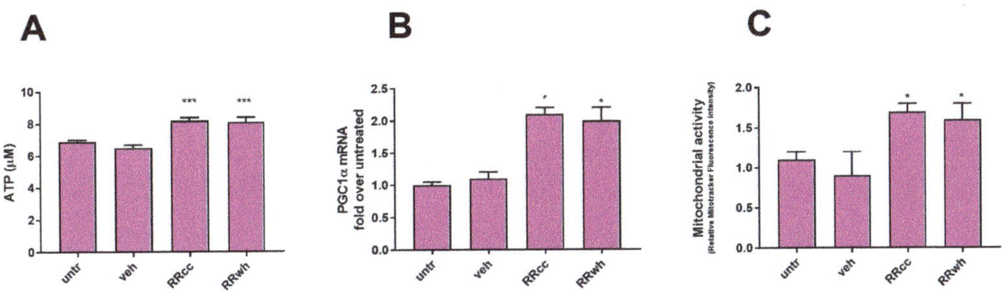

Figure 2. RRcc promotes ATP production and mitochondrial activity in cultured human myoblasts. Intracellular ATP content (**A**), *Pgc1α* mRNA levels (**B**) and mitochondrial activity(**C**) in primary muscle myoblasts treated for 48 h in the presence of RRcc (30 μg/mL), RRwh (30 μg/mL), an equal amount of vehicle (veh) or left untreated (untr). Data are shown as mean ± SD of five independent experiments. Statistical analysis was performed by ANOVA test comparing each mean with that of untreated cells. p value = * <0.05, *** < 0.001, otherwise differences were not statistically significant.

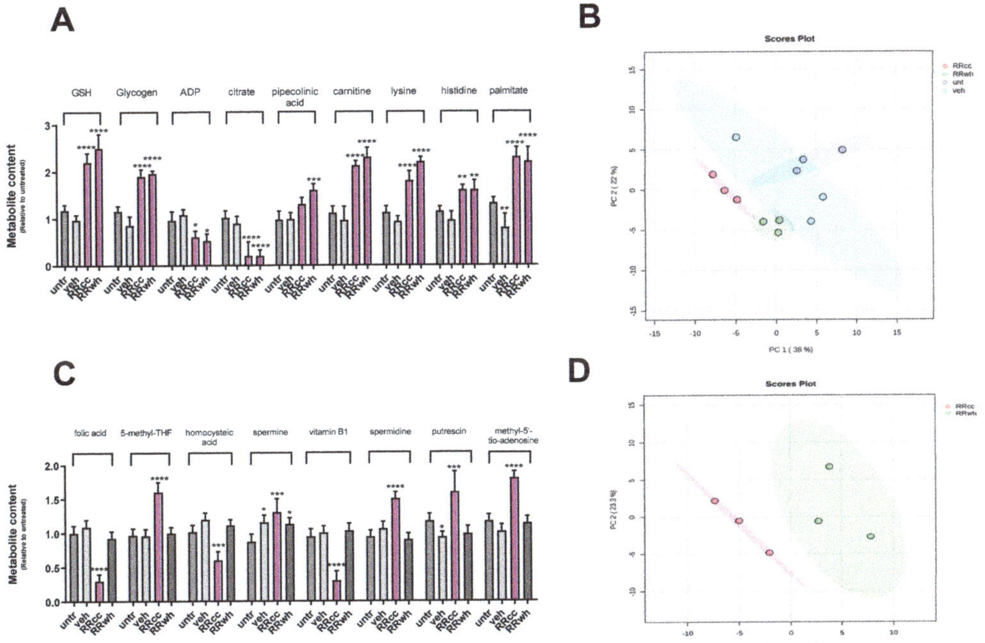

Figure 3. RRcc presents folic acid-cycle and polyamine pathway modulatory activity in human myoblasts. Metabolic profile of primary muscle myoblasts treated for 48 h with RRcc (30 μg/mL), RRwh (30 μg/mL), an equal amount of vehicle (veh) or left untreated (untr). (**A**) Content (fold variation vs. untr) of the indicated metabolites as measured in the four different experimental conditions; (**B**) primary component analysis (PCA) score plot showing that the metabolic profiles of RRcc and RRwh are different from veh and untr cells; (**C**) content (fold variation vs. untr) of the folic acid-cycle and polyamine pathway metabolites as measured in the four different experimental conditions (n = 3 independent experiments); (**D**) PCA score plot showing the metabolic profile of RRcc being different from RRwh. In A and C, values are represented as mean ± SD. Two-way ANOVA and Bonferroni post-test analysis were performed; * = $p < 0.05$; ** = $p < 0.01$; *** = $p < 0.001$; **** = $p < 0.0001$; otherwise differences were not statistically significant.

So far, differently from RRwh, RRcc presents an unprecedented regulatory activity on transcription factors involved in the maintenance of undifferentiated myoblasts Pax7 and myoD. However, this peculiar effect of RRcc does not depend on a different antioxidant, ATP-stimulatory, mitochondriogenic activity of the extract.

3.4. Metabolomic Profiling of RRcc- and RRwh-Treated Human Myoblasts

To investigate the metabolic pathways selectively modulated by RRcc in muscle cells, we performed HR mass-spectrometry metabolic profiling of cells undergoing treatment for 48 h with 30 μg/mL of RRcc. For comparison, metabolic profiling was also performed in RRwh (30 μg/mL), vehicle and untreated cells as controls. One hundred and ninety metabolites were unambiguously identified by comparison with spectral data from an in-house library of standards from HMDB. As shown in Figure 3, principal component analysis (PCA) showed that RRcc- and RRwh-supplemented cells were clearly clustered from untr and veh cells. The metabolite fluctuation (Figure 3A) indicates specific metabolites (reduced GSH, glycogen, ADP, citrate, DL-Pipecolinic acid, L-Carnitine, L-Lysine, L-Histidine and palmitic acid) that are strongly associated with the separation of both RRcc and RRwt from untreated cells. The increase in reduced GSH (untr: 1.18 ± 0.12; veh: 0.97 ± 0.11; RRcc: 2.2 ± 0.2; RRwh: 2.5 ± 0.3; fold difference compared to untr) strongly confirms the antioxidant potential of both RRcc and RRwh extracts. The decrease in intracellular glycogen (untr: 1.16 ± 0.11-fold increase in fluorescence; veh: 0.86 ± 0.19; RRcc: 1.90 ± 0.15; RRwh: 1.96 ± 0.06) proves that RRcc and RRwh promote glycogen production in muscle cells, at least in vitro. The decrease in ADP (untr: 0.95 ± 0.20-fold increase; veh: 1.08 ± 0.12; RRcc: 0.60 ± 0.13; RRwh: 0.52 ± 0.15) induced by the RR extracts confirms the increase in ATP measured by luminescence, confirming the energy-stimulating activity of the extract in muscle cells.

Moreover, RRcc and RRwh treatment both cause an increase in intracellular levels of both L-Lysin (untr: 1.12 ± 0.13; veh: 0.92 ± 0.12; RRcc: 1.80 ± 0.20; RRwh: 2.2 ± 0.1) and L-Histidine (untr: 1.14 ± 0.10; veh: 0.95 ± 0.19; RRcc: 1.60 ± 0.10; RRwh: 1.60 ± 0.20) that reveals the sparing of intracellular substrates for energy production. Lysin sparing may also be supported by increased levels of its non-enzymatic by-product DL-Pipecolinic acid (untr: 0.97 ± 0.19; veh: 0.98 ± 0.14; RRcc: 1.30 ± 0.14; RRwh: 1.60 ± 0.13). Substrates of acylcarnitines production—L-Carnitine (untr: 1.12 ± 0.14; veh: 0.95 ± 0.30; RRcc: 2.12 ± 0.10; RRwh: 2.30 ± 0.2) and palmitic acid (untr: 1.32 ± 0.11; veh: 0.78 ± 0.30; RRcc: 2.3 ± 0.2; RRwh: 2.2 ± 0.3)—both increase, suggesting the sparing of fatty acids for mitochondrial metabolism. Finally, citrate (untr: 1.02 ± 0.13; veh: 0.89 ± 0.15; RRcc: 0.20 ± 0.3; RRwh: 0.20 ± 0.13) decreases in both RRcc and RRwh cells, suggesting the promotion of the Krebs cycle and, most likely, the inhibition of the mevalonate pathway.

3.5. RRcc Presents Folic Acid Cycle Modulatory Activity

As shown in Figure 3D, PCA demonstrated that RRcc-supplemented cells grouped distinctly from RRwh cells. Metabolite fluctuation (Figure 3C) revealed that specific metabolites (folic acid, 5-methyl-THF, homocysteic acid, spermidine, vitamin B1, putrescein, methyl-5′-tio-adenosine) were strongly associated with the separation of RRcc from RRwt. In RRcc cells, folic acid (untr: 1.02 ± 0.10; veh: 1.09 ± 0.10; RRcc: 0.30 ± 0.10; RRwh: 0.92 ± 0.09), 5-methyl-THF (untr: 0.96 ± 0.09; veh: 0.95 ± 0.09; RRcc: 1.60 ± 0.15; RRwh: 0.99 ± 0.2) and vitamin B1 (untr: 0.94 ± 0.11-fold increase; veh: 1.00 ± 0.09; RRcc: 0.30 ± 0.14; RRwh: 1.02 ± 0.11) are decreased, suggesting that RRcc stimulates the folate pathway. Specifically, the significant reduction in vitamin B1 observed in RRcc compared to RRwh could be attributed to the stimulatory effect of RRcc on the folate-mediated one-carbon metabolism, which is linked to the pentose phosphate pathway, where thiamine serves as a critical cofactor. This seems to be further confirmed by the relative levels of homocisteic acid (untr: 1.00 ± 0.09-fold increase; veh: 1.19 ± 0.09; RRcc: 0.60 ± 0.13; RRwh: 1.10 ± 0.08). Similarly, RRcc (but not RRwh) stimulates four metabolites of the polyamine pathway: spermine (untr: 0.87 ± 0.10; veh: 1.15 ± 0.10; RRcc: 1.30 ± 0.20; RRwh:

1.12 ± 0.09), spermidine (untr: 0.93 ± 0.09-fold; veh: 1.05 ± 0.10; RRcc: 1.50 ± 0.10; RRwh: 0.89 ± 0.09), putrescine (untr: 1.17 ± 0.10; veh: 0.93 ± 0.07; RRcc: 1.60 ± 0.30; RRwh: 0.98 ± 0.11) and methyl-5-tioadenosine (untr: 1.16 ± 0.09; veh: 1.01 ± 0.10; RRcc: 1.80 ± 0.10; RRwh: 1.13 ± 0.09).

Together with metabolically promoting redox maintenance, ATP and protein production, RRcc presents an unprecedented stimulatory activity on folate-mediated one-carbon metabolism, absent in RRwh and resulting in an intracellular increase in polyamines.

3.6. Both Folic Acid Stimulatory and Antioxidant Activity Contribute to RRcc Activity

In order to verify whether folic acid pathway stimulatory activity was indeed the key element responsible for the difference between RRcc and RRwh, we treated skeletal muscle cells with RRwh in the presence of folic acid. As shown in Figure 4, the presence of 5 μM folic acid indeed confers on RRwh the ability to upregulate the mRNA of myoD and Pax7 in skeletal muscle cells. Surprisingly, 5 μM folic acid alone failed in promoting the transcription of the two genes, suggesting that the other pathways stimulated by RRcc synergize for the overall activity of the extract. Since antioxidant activity was the other pathway mainly stimulated by the RRcc, we attempted inhibition of the latter by co-supplementation with pro-oxidant molecule cystine (Cys2), given at the concentration of 1 mM. Indeed, both RRcc and RRwh supplemented with folic acid lost their activity in upregulating myoD and Pax7 in the presence of the pro-oxidant cystine, confirming that RRcc relies on the synergistic effect of folic acid and antioxidant pathways to exert its effect on the differentiation state of muscle cells (Figure 4).

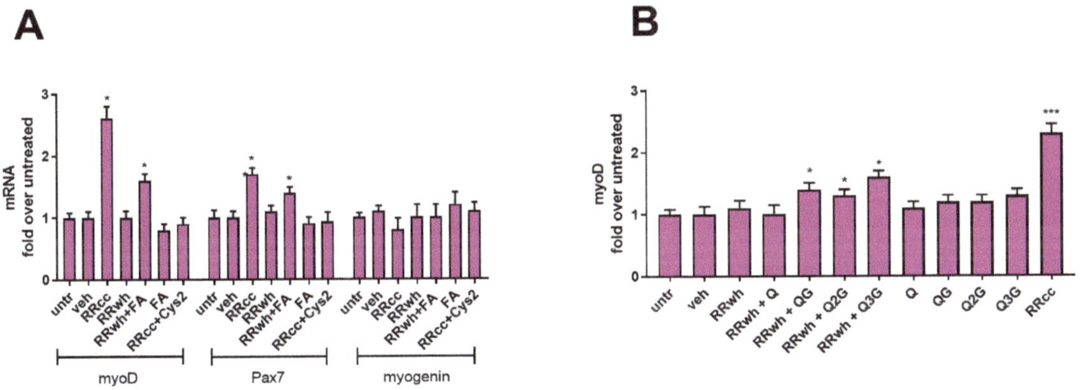

Figure 4. RRcc activity relies on the synergism between the folic acid pathway and antioxidant activity and on the presence of quercetin-glycosides. (**A**,**B**) mRNA levels of muscle cell differentiation markers in primary muscle myoblasts treated for 48 h in the presence of RRcc (30 μg/mL), RRwh (30 μg/mL), an equal amount of vehicle (veh) or left untreated (untr). When indicated, folic acid (FA, 5μM), cystine (Cys2, 1 mM), quercetin (Q, 1 μM), quercetin 3-glycoside (QG, 1 μM), rutin (Q2G, 1 μM) and quercetin 3-rhamnoglucoside 7-glucoside (Q3G, 1 μM) were included in the treatment. Data are shown as mean ± SD of five independent experiments. Statistical analysis was performed by ANOVA test comparing each mean with that of untreated cells. p value = * <0.05, *** <0.001, otherwise differences not statistically significant.

3.7. Phytochemical Profile of RRcc vs. RRwh

In order to identify the phytochemical profile responsible for the RRcc activity, we analyzed its bioactive fraction by high-resolution mass spectrometry. Untargeted metabolomics (Figure 5) allowed the identification of a further 42 metabolites (Table 1). Compared to RRwh, RRcc is enriched in various chemical classes of metabolites (Table 2), including the non-cyanogenic hydroxynitriles (e.g., rhodiocyanoside A (**2**)) and the cyanogenic glycosides (e.g., heterodendrin (**4**)). RRcc also demonstrates to be enriched in various classes of polyphenols, including flavanols (e.g., gallocatechin (**8**)), hydroxycinnamic acids (caffeic acid (**13**) and its hexoside (**10**), as well as ferulic acid (**23**)), flavonol (rhodiolin (**43**), and flavonol glycosides (e.g., quercetin hexoside (**22**), quercetin 3-(2G-glucosylrutinoside (**20**) and rutin (**21**)). Additional glycosides in RRcc include sacranoside A (**35**) and its isomer (**37**), along with rhodiolatuntoside (**36**). The flavanone eriodictyol (**38**), the flavone C-glycoside vicenin 2 (**39**) and rhodiooctanoside (**41**), belonging to the class of acyclic alcohol glycosides, are also found in higher amounts compared to RRwh (Figure 5). On the contrary, RRcc contains—in lower quantities—cyanogenic glycoside and rosiridin (**28**), as well as glycosides such as rosin (**27**), rhodioloside C (**40**), rhodioloside D (**9**), salidroside hexoside (**7**) and rhodioloside A (**12**). The hydroxycinnamic acid, coumaric acid (**5**), and the flavanol catechin (**11**) are present in reduced quantities as well. Additionally, a rosiridin isomer (**29**) and the gallic acid derivative sachalinoside A (**33**) are also lower in RRcc.

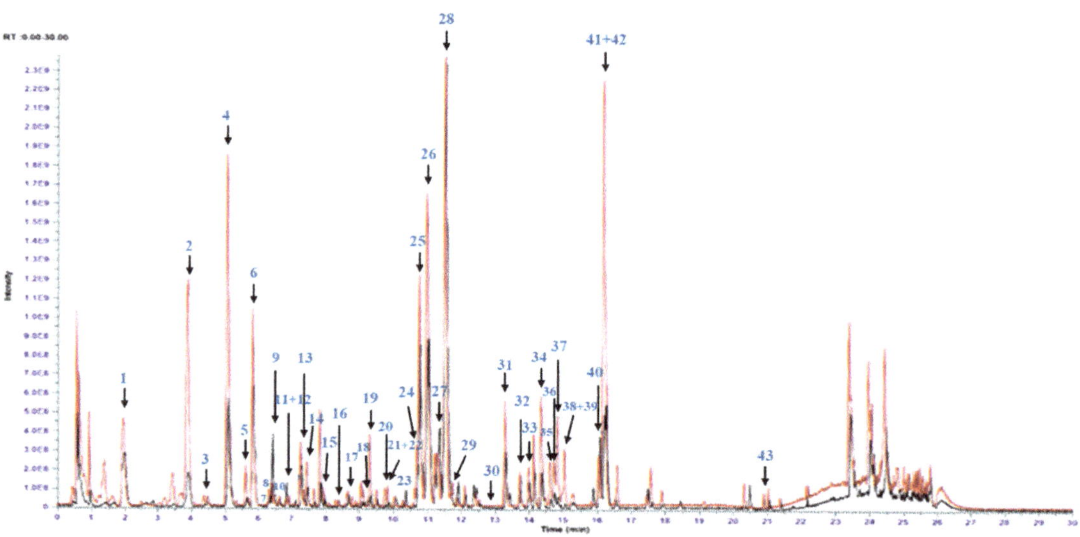

Figure 5. Overlapped UHPL-HRMS base peak chromatogram of *Rhodiola rosea* L. wild harvest (RRwh, black chromatogram) and *Rhodiola rosea* L. controlled cultivation (RRcc, red chromatogram) with peak annotation.

Table 1. UHPLC-HRMS analysis of phytochemical compounds identified in RRcc and RRwh extracts. Retention time, identity, m/z of molecular ions, fragments, and mass accuracy are indicated.

Peak	Rt (min)	Compound	Formula	m/z [M + H]$^-$ or [M + CH_3COO]$^-$	MS/MS	Mass Accuracy (Error ppm)
1	1.97	Gallic acid	$C_7H_5O_5$	169.0132	125.0232	0.47
2	3.91	Rhodiocyanoside A	$C_{11}H_{16}NO_6$	318.1195 [M + CH_3COO]$^-$	258.9975; 161.0445; 59.0127	1.97
3	4.67	*p*-hydroxyphenacyl-β-D-glucopyranoside	$C_{14}H_{17}O_8$	313.0922	151.039	3.71
4	5.2	Heterodendrin	$C_{11}H_{18}NO_6$	320.1349 [M + CH_3COO]$^-$	261.1170; 161.0445; 59.0127	−0.07
5	5.6	Coumaric acid	$C_9H_7O_3$	163.0391	119.049	0.66

Table 1. *Cont.*

Peak	Rt (min)	Compound	Formula	m/z [M + H]⁻ or [M + CH$_3$COO]⁻	MS/MS	Mass Accuracy (Error ppm)
6	5.84	Salidroside	$C_{14}H_{19}O_7$	299.1136	179.0552; 161.0446; 119.0339	3.64
7	6.06	Salidroside hexoside	$C_{20}H_{29}O_{12}$	461.1667	299.1137; 179.0552	2.96
8	6.36	Gallocatechin	$C_{14}H_{13}O_7$	305.0667	125.0232	3.51
9	6.42	Rhodioloside D	$C_{16}H_{29}O_8$	349.1855	179.0552; 119.0338	−0.51
10	6.66	Caffeic acid hexoside	$C_{15}H_{17}O_9$	341.0888	179.0341	2.66
11	6.81	Catechin	$C_{15}H_{13}O_6$	289.0722	169.0134	4.8
12	6.83	Rhodioloside A	$C_{16}H_{27}O_8$	347.1713	179.0552; 119.0338	−4.04
13	7.27	Caffeic acid	$C_9H_7O_4$	179.0342	135.044	1.39
14	7.47	Viridoside	$C_{15}H_{21}O_7$	313.1249	151.0754	3.98
15	7.96	Unknown	-	755.2054	609.1464; 300.0274	−4.19
16	8.37	Epicatechin	$C_{15}H_{13}O_6$	289.0722	169.0134	4.8
17	8.68	Epigallocatechine 3-gallate	$C_{22}H_{17}O_{11}$	457.0781	169.0133	2.81
18	9.02	Quercetin hexoside isomer 1	$C_{21}H_{19}O_{12}$	463.0889	300.0276	1.4
19	9.35	Epigallocatechin 3-gallate	$C_{22}H_{17}O_{11}$	457.0781	169.0133	2.81
20	9.79	Quercetin 3-(2G-glucosylrutinoside)	$C_{33}H_{39}O_{21}$	771.2002	463.0879; 301.0344	1.48
21	9.87	Rutin	$C_{27}H_{29}O_{16}$	609.1461	463.0884; 300.0276	0.88
22	9.9	Quercetin hexoside isomer 2	$C_{21}H_{19}O_{12}$	463.0887	300.0276	1.4
23	10.37	Ferulic acid	$C_{10}H_9O_4$	193.05	178.0263	−3.16
24	10.7	Rhodioloside E	$C_{21}H_{37}O_{11}$	465.2344	191.0554; 149.0445	−0.67
25	10.77	Rosavin	$C_{20}H_{27}O_{10}$	487.1823 [M + CH$_3$COO]⁻	427.1610; 293.0882	2.59
26	11.01	Rosarin	$C_{20}H_{27}O_{10}$	487.1823 [M + CH$_3$COO]⁻	427.1610; 293.0882	2.59
27	11.34	Rosin	$C_{15}H_{19}O_6$	355.1398 [M + CH$_3$COO]⁻	161.0446	−0.01
28	11.51	Rosiridin isomer 1	$C_{16}H_{27}O_7$	391.1975 [M + CH$_3$COO]⁻	331.1752; 179.0555; 161.0447	3.2
29	11.9	Rosiridin isomer 2	$C_{16}H_{27}O_7$	391.1975 [M + CH$_3$COO]⁻	331.1752; 179.0555; 161.0447	3.2
30	13.06	Rhodiolgin	$C_{21}H_{19}O_{12}$	463.0868	317.0302	−0.67
31	13.31	Unknown	-	737.5194	677.4975	1.69

Table 1. *Cont.*

Peak	Rt (min)	Compound	Formula	*m/z* [M + H]$^-$ or [M + CH_3COO]$^-$	MS/MS	Mass Accuracy (Error ppm)
32	13.75	Sachaloside II	$C_{21}H_{33}O_{10}$	505.2295 [M + CH_3COO]$^-$	445.2080;	3.17
33	14	Sachalinoside A	$C_{23}H_{31}O_{11}$	483.1875	271.0461	1.97
34	14.51	Rhodiosin	$C_{27}H_{29}O_{16}$	609.1466	301.0354	2.62
35	14.64	Sacranoside A isomer 1	$C_{21}H_{33}O_{10}$	505.2295 [M + CH_3COO]$^-$	445.2080;	3.17
36	14.75	Rhodiolatuntoside	$C_{21}H_{19}O_{11}$	447.0927	301.0352	1.017
37	14.86	Sacranoside A isomer 2	$C_{21}H_{33}O_{10}$	505.2295 [M + CH_3COO]$^-$	445.2080;	3.17
38	15.1	Eriodictyol	$C_{15}H_{12}O_6$	287.056	151.0026; 135.0404	0.97
39	15.18	Vicenin 2	$C_{27}H_{29}O_{15}$	593.1519	285.0404	1.3
40	16.06	Rhodioloside C	$C_{22}H_{37}O_{12}$	493.2296	447.2239	4.61
41	16.23	Rhodiooctanoside	$C_{19}H_{35}O_{10}$	423.2239	291.1815	3.46
42	16.24	Rhodioloside B	$C_{22}H_{37}O_{12}$	493.2296	447.2239	4.61
43	20.9	Rhodiolin	$C_{25}H_{19}O_{10}$	479.0989	299.0199	−4.3

Table 2. Phytochemicals enriched in RRcc compared to RRwh.

Peak	Compound	% AUC Ration (RRcc/RRwh)
2	Rhodiocyanoside A	341%
4	Heterodendrin	185%
8	Gallocatechin	156%
10	Caffeic acid hexoside	117%
13	Caffeic acid	165%
18	Quercetin hexoside	208%
20	Quercetin 3-(2G-glucosylrutinoside)	362%
21	Rutin	186%
22	Quercetin hexoside	208%
23	Ferulic acid	227%
34	Rhodiosin	205%
35	Sacranoside A	239%
36	Rhodiolatuntoside	165%
37	Sacranoside A isomer	395%
38	Eriodictyol	168%
39	Vicenin 2	282%
41	Rhodiooctanoside	314%
43	Rhodiolin	475%

%AUC ratio (RRcc/RRwh) represents the percentage ratio of the Area Under the Curve for compounds measured in *Rhodiola rosea* L. controlled cultivation compared to wild harvest. It quantifies the relative abundance of each compound in the controlled cultivation extracts versus those in the wild, thereby indicating variances in phytochemical profiles due to cultivation conditions.

Considering the enrichment of quercetin glycosides found in RRcc compared to RRwh, we wondered if these glycosylated quercetins could be contributing to the RRcc effect on the myoD-positive population of myoblasts. As shown in Figure 4B, RRwh supplemented with 1 μM quercetin glycosides (quercetin 3-glucosides, rutin, quercetin 3-O-rutinoside-7-O-glucoside) but not with 1 μM quercetin aglycone presented myoD-promoting activity in myoblasts. Interestingly, when supplemented alone, the three quercetin glycosides presented a weaker potency compared to RRcc, indicating that they cannot fully recapitulate the extract and pointing instead toward a synergistic effect among different compounds in the RR extract being responsible for it.

4. Discussion

Rhodiola rosea L., recognized for its adaptogenic properties, has shown promising results in muscle health due to its influence on key biological pathways that are crucial for muscle function and recovery [3]. Scientific evidence indicated that *Rhodiola rosea* L. may enhance physical endurance and aid in muscle repair, possibly due to its effects on stress-response systems and increase in antioxidant capacity [4]. Moreover, its bioactive components are thought to mitigate fatigue and improve recovery times, making it of interest to athletes and those undergoing physical rehabilitation [27]. Interestingly, the evaluation of specific results obtained from the two plant matrices herein tested, i.e., RRcc and RRwh, adds depth to the discussion on novel mechanisms through which *Rhodiola rosea* L. may exert its beneficial effects on muscle health. In this regard, the results showed that RRcc led to the activation of folate-mediated one-carbon metabolism and polyamine biosynthesis, especially stimulating metabolites such as spermine, spermidine and putrescine. The effect of RRcc on the polyamine pathway might also be related to the increased consumption of folic acid and vitamin B1. As shown by others [28] and by us in our manuscript, RRcc extracts transcriptionally regulate *Pax7* and *MyoD*, contributing to the maintenance of an undifferentiated pool of *MyoD*-positive myoblasts, which are important for muscle regeneration. Notwithstanding, RRcc does not impair myoblast differentiation into myotubes. The involvement of the cellular antioxidant machinery and the polyamine pathway has already been demonstrated in muscle cells [29,30]. Moreover, the polyamine pathway has been involved in many other aspects of muscle biology, including protein turnover and metabolic adaptation to anoxia [31]. In this regard, the modulation of the transcription and expression of specific genes involved in these processes, potentially induced by RRcc, warrants further investigation.

Our results also indicate that controlled cultivation conditions for RRcc may result in an improved bioactive compound profile, leading to more effective stimulation of these critical metabolic pathways. Our data point toward quercetin-glycosides as contributing to the effect of RRcc on muscle cells, at least in vitro. However, the effect of a phytocomplex is rarely recapitulated by one or few of its components and, instead, is most of the time attributable to the synergistic activity of many of its secondary metabolites. We could thus not exclude that other molecules and their synergism might contribute to RRcc's overall activity.

Our results regarding the metabolic profile of RRcc, together with the analysis of its mechanism of action, are particularly important in the context of the modern economy and globalization. As a consequence of growing market demand, controlled cultivations of *Rhodiola rosea* L. are starting to appear in several regions of the globe. The different growing conditions (altitude, temperature, water availability) alter the phytochemical profiles of the plants and their bioactive fractions. Notwithstanding, the phytocomponents of these *Rhodiola rosea* L. controlled cultivations are rarely analyzed in detail. In most cases, this limitation results from the use of chromatographic techniques such as HPLC-DAD and from not using high-resolution instrumentation [32,33]. The study conducted by Vouillamoz et al. in 2012 that focused on the first synthetic variety of *Rhodiola rosea* L. from the Swiss Alps, called "Mattmark", showed limitations in metabolite analysis with the quantification of only six major compounds using HPLC-DAD analysis [34]. It is noteworthy that only one previous study by Alperth et al. combined UHPLC-DAD analysis with mass spectrometry, allowing the identification of only 18 metabolites [35]. In our research, we performed a UHPLC-HRMS analysis of *Rhodiola rosea* L. RRcc and compared it to RRwh, with the unequivocal identification of 43 metabolites, including phenylpropanoids, phenylethanoids, non-cyanogenic hydroxynitriles, glycosides and polyphenols (flavanols, hydroxycinnamic acids, flavonols and flavonol glycosides). This analysis stands out from previous studies in the literature, which have mainly focused on the identification and quantification of a few key bioactive compounds, particularly salidroside and total rosavins (rosavin, rosin, rosarin) [17,36–38]. In this context, our study represents an innovative and significant contribution, as it is the first to provide a comprehensive analysis of the compounds present in *Rhodiola rosea* L. by combining conventional methods such as HPLC-DAD with high-resolution mass spectrometry analysis, thus providing a more detailed understanding of the metabolic profile of this medicinal plant.

Our study demonstrates that the transition from wild to controlled cultivation does not compromise the bioactive potential and the antioxidant profile of *Rhodiola rosea* extract. Moreover, while the quantities of rosavin and salidroside in RRcc remained within the limits specified by The United States Pharmacopeia [19], the controlled cultivation allowed an increase in other metabolites. This effect could be related to factors such as consistent environmental conditions, nutrient availability, and selective breeding practices in controlled cultivation settings that could potentially enhance the production of secondary metabolites with bioactive properties [17]. In this regard, the unique phytochemical profile of RRcc, enriched in various metabolites such as flavonoids, phenolic acids, and specifically polyphenolic compounds, strongly suggests its improved biological activities compared to wild harvests. These compounds are renowned for their antioxidant, anti-inflammatory, and anti-fatigue properties, which are beneficial in managing the oxidative stress and inflammatory responses implicated in the pathophysiology of sarcopenia [39]. Furthermore, based on Table 2, a comparative dose of the extracts of *Rhodiola rosea* L. can be calculated. Specifically, based on the calculated comparative doses of various compounds (e.g., rhodiocyanoside A, heterodendrin, gallocatechin and others) between RRcc and RRwh, it could be speculated that to induce an effect on folic acid and polyamine pathways, approximately a 2.5-fold higher dose of RRwh extract would be required compared to RRcc extract.

Although our study focused on metabolomic analysis and in vitro observations, which are inherently limited in confirming direct in vivo effects, this approach was crucial for identifying potential metabolic pathways involved and preliminary mechanisms of action of RRcc. Nonetheless, previous studies provided solid evidence of the beneficial effects of *Rhodiola rosea* supplementation in vivo in terms of response to muscle damage, increased antioxidant capacity and enhanced physical performance [4,40]. Therefore, the findings of this study perfectly align with this evidence, supporting the hypothesis that the metabolic changes observed in vitro could reflect similar biological effects in vivo. Moreover, compared to existing research, the obtained results not only confirm previous hypotheses about the beneficial role of *Rhodiola rosea* L. but also unveil new research directions. Based on the observed stimulatory effects of RRcc on folate and polyamine metabolic pathways, which are essential for the maintenance, growth and repair of muscle cells, it is indeed conceivable to speculate on its potential effect in the treatment of sarcopenia. Sarcopenia, characterized by progressive and generalized loss of skeletal muscle mass and strength, poses significant challenges, particularly in the aging population [41]. In this regard, the present study demonstrated the efficacy of RRcc in stimulating ATP production and improving mitochondrial respiration. This is particularly relevant for the management of sarcopenia, as improved mitochondrial function is associated with increased energy availability and efficiency in muscle cells, potentially counteracting the energy deficit reported in sarcopenic muscle [42].

5. Limitations of the Study

While this study offers valuable insights into the potential benefits of controlled cultivation of *Rhodiola rosea* L., various limitations must be acknowledged. First, the experimental design primarily focused on in vitro assessments of bioactivity, which may not fully capture the complex interplay of phytochemicals and physiological responses in vivo. However, muscle biology is subjected to great in vivo variability (e.g., influence of food, feeding scheme, training), and preliminary in vitro experiments serve as valuable initial steps as they help with designing the proper experimental platforms. Additionally, the scope of this study did not encompass comprehensive clinical trials to evaluate the efficacy and safety of *Rhodiola rosea* L. extracts in human subjects. Moreover, while the study identified salidroside, rosavins and quercetin glycosides as promising bioactive compounds in *Rhodiola rosea* L., their specific mechanisms of action and potential synergistic effects remain to be fully elucidated. Future research endeavors should aim to address these limitations by conducting robust clinical trials, employing advanced analytical techniques to elucidate molecular mechanisms, and optimizing cultivation protocols to ensure the consistency and reproducibility of results. Despite these limitations, the findings of this study provide a solid foundation for the further exploration of the controlled cultivation of *Rhodiola rosea* L. as a potential therapeutic agent for various health applications.

6. Conclusions

This study has demonstrated that controlled cultivation of *Rhodiola rosea* L. significantly enhances its bioactive potential. Our comprehensive phytochemical analysis revealed different bioactive compounds in extracts from RRcc compared to RRwh. Moreover, the properties of the RRcc extract measured on primary human myocytes revealed that, along with the expected ATP stimulatory activity, the extract derived from this specific controlled cultivation shows an unprecedented transcriptional regulation of two markers of muscle regeneration, *Pax7* and *myoD*, and a stimulatory effect on the folate and polyamine pathway, metabolic processes that play a significant role in resilience muscle regeneration during injuries or, more physiologically, as a consequence of physical exercise. Overall, the implications of our findings extend beyond the field of sports nutrition and physical rehabilitation, opening new avenues for research into the role of natural supplements in combating age-related muscle loss. Future studies could focus on clinical trials to validate these effects in human subjects and explore the molecular mechanisms underlying the observed bioactivities, potentially guiding the optimization of cultivation practices to yield even more potent health benefits.

Author Contributions: Conceptualization, M.S.; investigation, F.I., E.S. and F.G.; writing—original draft preparation, M.S.; writing—review and editing, M.S., A.P., E.N. and G.C.T.; funding acquisition, E.N. All authors have read and agreed to the published version of the manuscript.

Funding: This research received no external funding.

Institutional Review Board Statement: Not applicable.

Informed Consent Statement: Not applicable.

Data Availability Statement: The data used to support the findings of this study are included within the article.

Conflicts of Interest: Authors Elisabetta Schiano, Fabrizia Guerra and Ettore Novellino are from Inventia Biotech-Healthcare Food Research Center s.r.l. The company had no role in the design of the study; in the collection, analysis, or interpretation of data; in the writing of the manuscript; or in the decision to publish the results. The remaining authors declare that the research was conducted in the absence of any commercial or financial relationships that could be construed as a potential conflict of interest.

References

1. Li, X.; Chen, W.; Simal-Gandara, J.; Georgiev, M.I.; Li, H.; Hu, H.; Wu, X.; Efferth, T.; Wang, S. West meets east: Open up a dialogue on phytomedicine. *Chin. Med.* **2021**, *16*, 57. [CrossRef] [PubMed]
2. Tinsley, G.M.; Jagim, A.R.; Potter, G.D.M.; Garner, D.; Galpin, A.J. Rhodiola Rosea as an Adaptogen to Enhance Exercise Performance: A Review of the Literature. *Br. J. Nutr.* **2023**, *131*, 461–473. [CrossRef] [PubMed]
3. Stojcheva, E.I.; Quintela, J.C. The Effectiveness of Rhodiola rosea L. Preparations in Alleviating Various Aspects of Life-Stress Symptoms and Stress-Induced Conditions—Encouraging Clinical Evidence. *Molecules* **2022**, *27*, 3902. [CrossRef] [PubMed]
4. Lu, Y.; Deng, B.; Xu, L.; Liu, H.; Song, Y.; Lin, F. Effects of Rhodiola Rosea Supplementation on Exercise and Sport: A Systematic Review. *Front. Nutr.* **2022**, *9*, 856287. [CrossRef]
5. Kosakowska, O.; Bączek, K.; Przybył, J.L.; Pióro-Jabrucka, E.; Czupa, W.; Synowiec, A.; Gniewosz, M.; Costa, R.; Mondello, L.; Węglarz, Z. Antioxidant and antibacterial activity of roseroot (*Rhodiola rosea* L.) dry extracts. *Molecules* **2018**, *23*, 1767. [CrossRef]

6. Kang, D.Z.; Hong, H.D.; Kim, K.I.; Choi, S.Y. Anti-fatigue effects of fermented Rhodiola rosea extract in mice. *Prev. Nutr. Food Sci.* **2015**, *20*, 38–42. [CrossRef] [PubMed]
7. Pu, W.L.; Zhang, M.Y.; Bai, R.Y.; Sun, L.K.; Li, W.H.; Yu, Y.L.; Zhang, Y.; Song, L.; Wang, Z.X.; Peng, Y.F.; et al. Anti-inflammatory effects of *Rhodiola rosea* L.: A review. *Biomed. Pharmacother.* **2020**, *121*, 109552. [CrossRef] [PubMed]
8. Mao, J.J. Rhodiola Rosea Therapy for Major Depressive Disorder: A Study Protocol for a Randomized, Double-Blind, Placebo-Controlled Trial. *J. Clin. Trials* **2014**, *4*, 170. [CrossRef]
9. Abidov, M.; Crendal, F.; Grachev, S.; Seifulla, R.; Ziegenfuss, T. Effect of extracts from *Rhodiola rosea* and *Rhodiola crenulata* (*Crassulaceae*) roots on ATP content in mitochondria of skeletal muscles. *Bull. Exp. Biol. Med.* **2003**, *136*, 585–587. [CrossRef] [PubMed]
10. Parisi, A.; Tranchita, E.; Duranti, G.; Ciminelli, E.; Quaranta, F.; Ceci, R.; Cerulli, C.; Borrione, P.; Sabatini, S. Effects of chronic Rhodiola Rosea supplementation on sport performance and antioxidant capacity in trained male: Preliminary results. *J. Sports Med. Phys. Fitness* **2010**, *50*, 57–63.
11. Hezhang, Y.; Bin, L.; Wenbo, S.; Junjie, W.; Jing, Z.; Zhaolong, W.; Yi, L.; Yaowei, S.; Chang, L. Combined effects of Rhodiola rosea and caffeine supplementation on aerobic endurance and muscle explosiveness: A synergistic approach. *Front. Nutr.* **2024**, *11*, 1335950. [CrossRef]
12. Pomari, E.; Stefanon, B.; Colitti, M. Effects of two different Rhodiola rosea extracts on primary human visceral adipocytes. *Molecules* **2015**, *20*, 8409–8428. [CrossRef] [PubMed]
13. Sist, P.; Tramer, F.; Lorenzon, P.; Urbani, R.; Vrhovsek, U.; Bernareggi, A.; Sciancalepore, M. Rhodiola rosea, a protective antioxidant for intense physical exercise: An in vitro study. *J. Funct. Foods* **2018**, *48*, 27–36. [CrossRef]
14. Perfumi, M.; Mattioli, L. Adaptogenic and central nervous system effects of single doses of 3% rosavin and 1% salidroside *Rhodiola rosea* L. extract in mice. *Phyther. Res.* **2007**, *21*, 37–43. [CrossRef] [PubMed]
15. Zhang, S.; Jiang, S.; Deng, N.; Zheng, B.; Li, T.; Liu, R.H. Phytochemical Profiles, Antioxidant Activity and Antiproliferative Mechanism of *Rhodiola rosea* L. Phenolic Extract. *Nutrients* **2022**, *14*, 3602. [CrossRef]
16. Polumackanycz, M.; Konieczynski, P.; Orhan, I.E.; Abaci, N.; Viapiana, A. Chemical Composition, Antioxidant and Anti-Enzymatic Activity of Golden Root (*Rhodiola rosea* L.) Commercial Samples. *Antioxidants* **2022**, *11*, 919. [CrossRef]
17. Galambosi, B.; Galambosi, Z.; Slacanin, I. Comparison of natural and cultivated roseroot (*Rhodiola rosea* L.) roots in Finland. *J. Med. Spice Plants* **2007**, *12*, 141–147.
18. Marchev, A.S.; Aneva, I.Y.; Koycheva, I.K.; Georgiev, M.I. Phytochemical variations of *Rhodiola rosea* L. wild-grown in Bulgaria. *Phytochem. Lett.* **2017**, *20*, 386–390. [CrossRef]
19. Committee on Herbal Medicinal Products (HMPC). Community herbal monograph on *Rhodiola rosea* L., rhizoma et radix. *Eur. Med. Agency* **2012**, *44*, 1–5.
20. Shikov, A.N.; Pozharitskaya, O.N.; Makarov, V.G.; Wagner, H.; Verpoorte, R.; Heinrich, M. Medicinal Plants of the Russian Pharmacopoeia; Their history and applications. *J. Ethnopharmacol.* **2014**, *154*, 481–536. [CrossRef]
21. Therapeutic Goods Administration (TGA). Compositional Guideline: *Rhodiola rosea* Dried Root (Powdered) Extract. Available online: https://www.tga.gov.au/resources/resource/compositional-guidelines/rhodiola-rosea-dried-root-powdered-extract (accessed on 1 March 2024).
22. Liu, Q.; Luo, Q.; Zhong, B.; Tang, K.; Chen, X.; Yang, S.; Li, X. Salidroside attenuates myocardial remodeling in DOCA-salt-induced mice by inhibiting the endothelin 1 and PI3K/AKT/NFκB signaling pathways. *Eur. J. Pharmacol.* **2023**, *962*, 176236. [CrossRef] [PubMed]
23. Erst, A.A.; Kotsupiy, O.V.; Erst, A.S.; Kuznetsov, A.A. Individual Differences in Growth and in Accumulation of Secondary Metabolites in *Rhodiola rosea* Cultivated in Western Siberia. *Int. J. Mol. Sci.* **2023**, *24*, 11244. [CrossRef] [PubMed]
24. Peschel, W.; Kump, A.; Horváth, A.; Csupor, D. Age and harvest season affect the phenylpropenoid content in cultivated European *Rhodiola rosea* L. *Ind. Crops Prod.* **2016**, *83*, 787–802. [CrossRef]
25. Zhang, P.; Li, W.; Wang, L.; Liu, H.; Gong, J.; Wang, F.; Chen, X. Salidroside Inhibits Myogenesis by Modulating p-Smad3-Induced Myf5 Transcription. *Front. Pharmacol.* **2018**, *9*, 209. [CrossRef] [PubMed]
26. Hansen, J.M.; Klass, M.; Harris, C.; Csete, M. A reducing redox environment promotes C2C12 myogenesis: Implications for regeneration in aged muscle. *Cell Biol. Int.* **2007**, *31*, 546–553. [CrossRef] [PubMed]
27. Bernatoniene, J.; Jakstas, V.; Kopustinskiene, D.M. Phenolic Compounds of Rhodiola rosea L. as the Potential Alternative Therapy in the Treatment of Chronic Diseases. *Int. J. Mol. Sci.* **2023**, *24*, 12293. [CrossRef] [PubMed]
28. Shulpekova, Y.; Nechaev, V.; Kardasheva, S.; Sedova, A.; Kurbatova, A.; Bueverova, E.; Kopylov, A.; Malsagova, K.; Dlamini, J.C.; Ivashkin, V. The concept of folic acid in health and disease. *Molecules* **2021**, *26*, 3731. [CrossRef] [PubMed]
29. Galasso, L.; Cappella, A.; Mulè, A.; Castelli, L.; Ciorciari, A.; Stacchiotti, A.; Montaruli, A. Polyamines and Physical Activity in Musculoskeletal Diseases: A Potential Therapeutic Challenge. *Int. J. Mol. Sci.* **2023**, *24*, 9798. [CrossRef]
30. Soda, K. Overview of Polyamines as Nutrients for Human Healthy Long Life and Effect of Increased Polyamine Intake on DNA Methylation. *Cells* **2022**, *11*, 164. [CrossRef]
31. Tabbaa, M.; Ruz Gomez, T.; Campelj, D.G.; Gregorevic, P.; Hayes, A.; Goodman, C.A. The regulation of polyamine pathway proteins in models of skeletal muscle hypertrophy and atrophy: A potential role for mTORC1. *Am. J. Physiol. Cell Physiol.* **2021**, *320*, C987–C999. [CrossRef]

32. Magsar, J.; Sharkhuu, A.; Bączek, K.; Przybył, J.L.; Węglarz, Z. Intraspecific variability of roseroot (*Rhodiola rosea*) naturally occurring in Mongolian Altai. *Acta Hortic.* **2012**, *955*, 51–58. [CrossRef]
33. Adamczak, A.; Buchwald, W.; Gryszczyńska, A. Biometric features and content of phenolic compounds of roseroot (*Rhodiola rosea* L.). *Acta Soc. Bot. Pol.* **2016**, *85*, 11–14. [CrossRef]
34. Vouillamoz, J.F.; Carron, C.A.; Malnoë, P.; Baroffio, C.A.; Carlen, C. Rhodiola rosea "Mattmark", the first synthetic cultivar is launched in Switzerland. *Acta Hortic.* **2012**, *955*, 185–189. [CrossRef]
35. Alperth, F.; Turek, I.; Weiss, S.; Vogt, D.; Bucar, F. Qualitative and quantitative analysis of different rhodiola rosea rhizome extracts by UHPLC-DAD-ESI-MSn. *Sci. Pharm.* **2019**, *87*, 8. [CrossRef]
36. Altantsetseg, K.; Przybył, J.L.; Węglarz, Z.; Geszprych, A. Content of biologically active compounds in roseroot (*Rhodiola* sp.) raw material of different derivation. *Herba Pol.* **2007**, *53*, 20–26.
37. Elameen, A.; Kosman, V.M.; Thomsen, M.; Pozharitskaya, O.N.; Shikov, A.N. Variability of Major Phenyletanes and Phenylpropanoids in 16-Year-Old *Rhodiola rosea* L. Clones in Norway. *Molecules* **2020**, *25*, 3463. [CrossRef] [PubMed]
38. Węglarz, Z.; Przybył, J.L.; Geszprych, A. Roseroot (*Rhodiola rosea* L.): Effect of Internal and External Factors on Accumulation of Biologically Active Compounds. *Bioact. Mol. Med. Plants* **2008**, *16*, 297–315. [CrossRef]
39. Gomes, M.J.; Martinez, P.F.; Pagan, L.U.; Damatto, R.L.; Cezar, M.D.M.; Lima, A.R.R.; Okoshi, K.; Okoshi, M.P. Skeletal muscle aging: Influence of oxidative stress and physical exercise. *Oncotarget* **2017**, *8*, 20428–20440. [CrossRef] [PubMed]
40. Sanz-Barrio, P.M.; Noreen, E.E.; Gilsanz-Estebaranz, L.; Lorenzo-Calvo, J.; Martínez-Ferrán, M.; Pareja-Galeano, H. Rhodiola rosea supplementation on sports performance: A systematic review of randomized controlled trials. *Phytother. Res.* **2023**, *37*, 4414–4428. [CrossRef]
41. Santilli, V.; Bernetti, A.; Mangone, M.; Paoloni, M. Clinical definition of sarcopenia. *Clin. Cases Miner. Bone Metab.* **2014**, *11*, 177–180. [CrossRef]
42. Wu, J.; Ding, P.; Wu, H.; Yang, P.; Guo, H.; Tian, Y.; Meng, L.; Zhao, Q. Sarcopenia: Molecular regulatory network for loss of muscle mass and function. *Front. Nutr.* **2023**, *10*, 1037200. [CrossRef] [PubMed]

antioxidants

MDPI

Article

Irisin, Brain-Derived Neurotrophic Factor (BDNF), and Redox Balance in Geriatric Dynapenia

Aleksandra Wojszel [1,2,*], Jakub Śliwowski [1], Justyna Rentflejsz [2], Joanna Rogalska [3], Małgorzata Michalina Brzóska [3] and Zyta Beata Wojszel [4,*]

1 Interdisciplinary Scientific Group at the Department of Geriatrics, Medical University of Bialystok, 15-089 Bialystok, Poland; 42216@student.umb.edu.pl
2 Doctoral School, Medical University of Bialystok, 15-089 Bialystok, Poland; justyna.rentflejsz@sd.umb.edu.pl
3 Department of Toxicology, Medical University of Bialystok, 15-089 Bialystok, Poland; joanna.rogalska@umb.edu.pl (J.R.); malgorzata.brzoska@umb.edu.pl (M.M.B.)
4 Department of Geriatrics, Medical University of Bialystok, 15-089 Bialystok, Poland
* Correspondence: 42230@student.umb.edu.pl (A.W.); zyta.wojszel@umb.edu.pl (Z.B.W.); Tel.: +48-7104242 (Z.B.W.)

Abstract: Irisin and brain-derived neurotrophic factor (BDNF) are considered potential biomarkers for sarcopenia; however, their interplay and relationship with oxidative stress remain unclear. Therefore, the aim of this study was to assess the serum concentration of irisin and BDNF in patients over 60 years of age, as well as their relationship with dynapenia and redox homeostasis. Dynapenia was diagnosed using the Five Times Sit-to-Stand Test (5TSST). Serum levels of irisin, BDNF, total oxidative status (TOS), and total antioxidative status (TAS) were measured, and the oxidative stress index (OSI) was calculated. A total of 110 patients from a geriatric ward (72.7% women, mean age 78.2 ± 7.1 years) participated in the study. BDNF concentration was negatively associated with dynapenia, irisin, and the irisin/BDNF ratio. TOS, TAS, and OSI were negatively associated with BDNF and positively associated with irisin and dynapenia. No significant association was found between irisin and sarcopenia parameters. In regression analysis, significantly higher odds of dynapenia were observed for older age, female sex, a greater number of chronic diseases, and higher OSI values, after adjusting for TOS, BDNF, and the irisin-to-BDNF ratio. These results confirm redox imbalance as an independent predictor of sarcopenia. A lower BDNF concentration and a higher irisin-to-BDNF ratio may indicate a protective role of BDNF in the development of sarcopenia in geriatric patients; however, this finding requires further confirmation.

Keywords: sarcopenia; myokine; irisin; neurotrophin; brain-derived neurotrophic factor; total oxidative status; total antioxidative status; oxidative stress index

1. Introduction

Sarcopenia, initially considered a geriatric syndrome and a complication of the aging process, was included in the International Classification of Diseases (ICD-10; M62.84) in 2019 and is now recognized as a medical condition that requires diagnosis and treatment. It is primarily characterized by a progressive decrease in skeletal muscle mass and function. Sarcopenia diagnosis is based on anatomical and functional criteria [1]. The latest consensus on sarcopenia prioritized muscle strength as a key prognostic factor and proposed dynapenia as its primary screening criterion. For this purpose, the measurement of grip strength can be used, or the chair stand test result can be employed [2].

Several complex pathogenetic mechanisms are thought to be involved in sarcopenia development, including disturbances in protein homeostasis with a predominance of catabolic over anabolic processes, chronic inflammation, and hormonal imbalances [3]. Therefore, apart from the anatomical biomarkers of sarcopenia proposed in its clinical diagnostic algorithms [1,4,5], several specific biochemical biomarkers have been identified for use in research and clinical trials. They can be specified as markers of musculoskeletal status (myokines, markers of the neuromuscular junction, those specific to muscle mass and muscle turnover) and markers of processes underlying the pathogenesis of sarcopenia (adipokines, hormones, and inflammatory parameters) [6]. Among them, irisin and brain-derived neurotrophic factor (BDNF) are two substances regarded as potential biomarkers for sarcopenia. Irisin is a myokine, a polypeptide compound secreted by muscles in response to physical exercise. It can prevent muscle degradation associated with aging through its antioxidant properties and influence the regeneration and growth of muscle tissue by activating the appropriate signaling pathways [7]. BDNF is a neurotrophin that may play a crucial role in regulating neuromuscular junction function, which is essential not only for nerve–muscle coordination but also for maintaining muscle strength and function [8]. Studies conducted in recent years indicate a link between the secretion of irisin during physical exercise and the expression of BDNF (named the muscle–brain axis) [9]. However, the mutual interaction of these substances in the development of sarcopenia is unclear. Especially in the context of oxidative stress, an essential player in the complex pathomechanisms of sarcopenia development [10].

Therefore, the study aimed to measure the serum concentration of irisin and BDNF in older geriatric patients and to assess their association with markers of dynapenia and redox homeostasis parameters.

2. Materials and Methods

The study was conducted with patients over 60 years of age in the sub-acute geriatric ward who were able to participate in the functional assessment. The study was approved by the Bioethics Committee at the Medical University of Bialystok (no.APK.002.513.2022). All procedures performed in the study were conducted following the ethical standards of the Medical University of Bialystok's research committee and in compliance with the Helsinki Declaration and its subsequent amendments. All study participants provided their informed consent to participate. Patients participated in the study voluntarily, knowing they could withdraw at any time, and the study was not burdensome for them. Functional assessment was part of a routine comprehensive geriatric assessment at the department, and blood samples for additional biochemical tests were collected during scheduled blood draws. Serum samples were obtained from fasting venous blood and then stored at −80 °C until assayed.

The inclusion criteria were: age 60 years or older, provision of informed consent, stable clinical condition, and health and fitness status sufficient to allow participation in a functional assessment.

Exclusion criteria included: neuromuscular disorders such as Parkinson's disease and amyotrophic lateral sclerosis; moderate to severe dementia (defined as a Mini-Mental State Examination score of <18 points); advanced or terminal stages of chronic diseases (e.g., severe chronic obstructive pulmonary disease requiring oxygen therapy, severe cardiovascular disease, decompensated heart failure, and severe osteoarthritis); active phases of acute illnesses; the presence of a pacemaker; ongoing cancer treatment; and lack of informed consent.

Information on patient's age, gender, and number of chronic diseases frequently occurring in geriatric patients, associated with the risk of sarcopenia, and in the pathogenesis of which oxidative stress plays an important role (out of 19: atrial fibrillation, chronic arthritis, chronic cardiac failure, chronic obstructive pulmonary disease, asthma, chronic renal disease, dementia, depression, urinary incontinence, diabetes/prediabetes, hypertension, orthostatic hypotension, ischemic heart disease, myocardial infarction, neoplasm, osteoporosis, peripheral arterial disease, and stroke) was collected; two or more diseases was classified as "multimorbidity".

Screening for sarcopenia included:

- The SARC-F questionnaire (an acronym that stands for Strength, Assistance in walking, Rise from a chair, Climb stairs, and Falls), a self-assessment tool for identifying deficiencies suggestive of sarcopenia. A score of 0 to 2 points may be assigned for each item (with a maximum total score of 10). A result of ≥4 points indicates probable sarcopenia and a potential risk of poor functional outcomes [11,12].
- Functional test assessing the strength of lower extremities—5 Times Sit-to-Stand Test, 5TSST. The patient sits in a chair with a straight back, feet flat on the floor, and arms folded across their chest. The time it takes the person to stand up and sit down five times as quickly as possible is measured. A result greater than 15 seconds indicates lower limb dynapenia, which is recognized as probable sarcopenia.

The serum concentration of irisin was determined with the use of a Human Irisin ELISA Kit (No. EH4702) and that of BDNF with the use of a Human BDNF ELISA Kit (No. EH0043) by FineTest (Wuhan, China). The kits were based on sandwich enzyme-linked immunosorbent assay technology (ELISA). The precision of the irisin assay, expressed as intra- and inter-assay coefficients of variation (CV), was <2% (for both used kits, intra-assay CV <2%) and 2%, respectively. The intra-assay CV of the BDNF measurement was <2% for the first kit and <4% for the second kit, while the inter-assay CV was <3%. The ratio of irisin to BDNF (irisin/BDNF) was mathematically calculated.

Total oxidative status (TOS) and total antioxidative status (TAS) were measured in the serum using spectrophotometric PerOx (TOS/TOC) Kit (No. KC5100) and ImAnOx (TAS/TAC) Kit (No. KC5200), respectively, provided by Immundiagnostik AG (Bensheim, Germany). To assay TOS, the total lipid peroxides present in the serum were quantified through the reaction of peroxidase and peroxides in the tested sample, as well as the transformation of 3,3′,5,5′-tetramethylbenzidine added to the sample into a colored product. TAS was determined based on the reaction of antioxidants present in the serum and the added hydrogen peroxide. TOS values measured in control samples included with the first kit (271.0 ± 11.31 μmol/L and 602.5 ± 13.44 μmol/L; mean ± standard error for two repetitions) and the second kit (141.8 ± 3.111 μmol/L and 352.0 ± 6.364 μmol/L) fell within the ranges specified by the manufacturer (first kit 204–338 μmol/L and 450–750 μmol/L, respectively; second kit 122–203 μmol/L and 260–433 μmol/L, respectively). The intra-assay CV was <2% for the first kit and <4% for the second kit. TAS values determined in control samples included with the first kit (221.58 ± 6.611 μmol/L and 265.0 ± 18.38 μmol/L; mean ± standard error for two repetitions) and the second kit (214.66 ± 5.037 μmol/L and 252,1 ± 5374 μmol/L) fell within the ranges specified by the manufacturer (180–242 μmol/L and 212–286 μmol/L in both kits). The intra-assay CV was <1% for the first kit and <2% for the second kit, while the inter-assay CV was <2%. Serum assays were performed in duplicate to ensure the reliability, precision, and validity of the results, and to enhance the diagnostic and analytical value of the obtained data, using an Epoch spectrophotometer (BioTek Instruments, Inc., Winooski, VT, USA).The oxidative stress index (OSI) was mathematically calculated as the ratio of TOS and TAS (OSI = TOS/TAS) [13].

Statistical analysis was performed using IBM SPSS Version 18 software (SPSS, Chicago, IL, USA). The distribution of variables was checked using the Shapiro–Wilk test and presented as frequencies and percentages (categorical variables), as means and standard deviations (normally distributed quantitative variables), or as medians and interquartile ranges (non-normally distributed quantitative variables). Proportions were compared using χ^2 tests, while Student's *t*-test for independent samples and the Mann–Whitney U test were used to compare means and medians. Spearman's correlation was used to assess the relationship between variables. Multivariable logistic regression models of dynapenia, as the dependent variable, were built, including all predictors with a *p*-value less than 0.1, excluding those with a high multicollinearity effect. In all analyses, a two-tailed *p*-value of less than 0.05 was regarded as significant.

3. Results

3.1. Study Group Characteristics

One hundred ten patients from the geriatric ward (72.7% women, mean age 78.2 ± 7.1 years) participated in the study. Table 1 presents the characteristics of the study group. Dynapenia of the lower extremities was more frequently observed in older patients and in females.

Table 1. Study group characteristics.

Characteristic	Total	Dynapenia "−"	Dynapenia "+"	p [1]
N (%)	110 (100)	37 (33.6)	73 (66.4)	
Age (y), M (SD)	78.2 (7.1)	74.3 (7.8)	80.1 (5.9)	<0.001
Sex				0.003
female, n (%)	80 (72.7)	20 (25.0)	60 (75.0)	
male, n (%)	30 (27.3)	17 (56.7)	13 (43.3)	
SARC-F, points, Me (IQR)	2.0 (1.0, 4.0)	1 (0, 2)	3 (1.5, 5)	<0.001
Sarcopenia by SARC-F				<0.001
no (<4 points), n (%)	74 (67.3)	36 (48.6)	38 (51.4)	
yes (≥4 points), n (%)	36 (32.7)	1 (2.8)	35 (97.2)	
Number of chronic diseases, Me (IQR)	4 (3, 5)	3 (2, 4.5)	4 (3, 6)	0.003
Multimorbidity				0.006
no, n (%)	7 (6.4)	6 (85.7)	1 (14.3)	
yes, n (%)	103 (93.6)	31 (30.1)	72 (69.9)	
5TSST, s, Me (IQR)	17.4 (13.5, 24.0)	12.9 (11.7, 13.6)	20.4 (17.4, 28.0)	<0.001

IQR—Interquartile Range; M—Mean; Me—Median; N—number; SD—Standard Deviation; 5TSST—5 Times Sit-to-Stand Test. [1] Student's *t*-test or the Mann–Whitney U test. Multimorbidity—2 or more diseases of 19 assessed. Dynapenia "−"—5TSST ≤ 15 s; Dynapenia "+"—5TSST > 15 s.

Dynapenia, assessed with the 5-TSST, occurred in as many as 66.4% of the study group, significantly more often in those with SARC-F scores of 4 or higher. Individuals with reduced lower limb muscle strength had a considerably higher number of chronic diseases and were substantially more frequently diagnosed with multimorbidity.

3.2. Association Between Dynapenia, Irisin, BDNF and Redox Parameters

The analysis revealed that BDNF concentration was significantly lower in patients with dynapenia, while there was no significant association between dynapenia and irisin. The irisin/BDNF ratio was considerably higher in dynapenic patients. TAS did not differ significantly between the two analyzed groups. In contrast, TOS was higher in the dynapenia group (p on the verge of significance, 0.05), and OSI was significantly higher in the dynapenia group (Table 2).

Table 2. Dynapenia and irisin, BDNF, irisin/BDNF ratio and redox balance parameters.

Characteristic	All	Dynapenia "−"	Dynapenia "+"	p [1]
N (%)	110 (100)	37 (33.6)	73 (66.4)	
BDNF, ng/mL, Me (IQR)	4.02 (2.4, 5.9)	5.45 (3.1, 6.9)	3.7 (2.9, 5.6)	0.03
Irisin, µg/mL, Me (IQR)	2.1 (1.7, 2.4)	1.9 (1.6, 2.3)	2.2 (1.7, 2.4)	0.11
Irisin/BDNF ratio, Me(IQR)	584 (272, 760)	379 (230, 705)	648 (297, 773)	0.02
TOS, µmol/L, Me (IQR)	656.5 (401.1, 1041.4)	495.4 (399.7, 818.1)	799.6 (401.9, 1111.1)	0.05
TAS, µmol/L, Me (IQR)	359.4 (277.4, 386.1)	375.8 (265.4, 386.6)	353.1 (291.6, 385.1)	0.94
OSI, M (SD)	2.1 (1.4, 2.8)	1.9 (0.7)	2.3 (1.1)	0.03

BDNF—brain-derived neurotrophic factor; IQR—Interquartile Range; M—Mean; Me—Median; N—number; OSI—oxidative stress index; SD—Standard Deviation; TAS—total antioxidative status; TOS—total oxidative status; [1] Student's t-test or the Mann–Whitney U test. Dynapenia "−"—5TSST ≤ 15 s; Dynapenia "+"—5TSST > 15 s.

5TSST results were positively associated with age, the number of chronic diseases, SARC-F scores, the irisin/BDNF ratio, TAS, and OSI and negatively associated with BDNF concentration (Table 3). BDNF concentration was negatively associated with irisin concentration, the irisin/BDNF ratio, TOS, TAS, and OSI. In contrast, irisin concentration was positively associated with the irisin/BDNF ratio, TOS, TAS, and OSI.

Table 3. Spearman's rank correlation coefficients between the analyzed parameters.

Variable		SARCF	5TSST	BDNF	Irisin	Irisin/BDNF	TOS	TAS	OSI	Age
SARC-F			0.62 **	−0.11	−0.01	0.06	0.03	−0.09	0.09	0.36 **
	p		<0.001	0.24	0.94	0.51	0.78	0.34	0.34	<0.001
5TSST		0.62 **		−0.27 **	0.17	0.27 **	0.27 **	0.06	0.30 **	0.40 **
		<0.001		0.004	0.08	0.005	0.005	0.52	0.001	<0.001
BDNF		−0.11	−0.27 **		−0.53 **	−0.95 **	−0.95 **	−0.95 **	−0.12	−0.12
	p	0.24	0.004		<0.001	<0.001	<0.001	<0.001	0.22	0.22
Irisin		−0.01	0.17	−0.53 **		0.71 **	0.55 **	0.69 **	0.33 **	0.04
	p	0.94	0.08	<0.001		<0.001	<0.001	<0.001	<0.001	0.71
Irisin/BDNF		0.064	0.27 **	−0.95 **	0.71 **		0.50 **	0.59 **	0.32 **	0.12
	p	0.51	0.005	<0.001	<0.001		<0.001	<0.001	<0.001	0.22
TOS		0.03	0.27 **	−0.95 **	0.55 **	0.50 **		0.48 **	0.92 **	−0.07
	p	0.78	0.005	<0.001	<0.001	<0.001		<0.001	<0.001	0.47
TAS		−0.09	0.06	−0.95 **	0.69 **	0.59 **	0.48 **		0.18	0.001
	p	0.34	0.52	<0.001	<0.001	<0.001	<0.001		0.05	0.99
OSI		0.09	0.30 **	−0.12	0.33 **	0.32 **	0.92 **	0.18		−0.07
	p	0.34	0.001	0.22	<0.001	<0.001	<0.001	0.05		0.50
Age (years)		0.36 **	0.40 **	−0.12	0.04	0.12	−0.07	0.001	−0.07	
	p	<0.001	<0.001	0.22	0.71	0.22	0.47	0.99	0.50	
Number of diseases		0.38 **	0.33 **	−0.02	−0.03	0.03	−0.10	−0.260 *	−0.04	0.33 **
	p	<0.001	<0.001	0.87	0.75	0.76	0.28	0.006	0.69	<0.001

BDNF—Brain Derived Neurotrophic Factor; OSI—Oxidative Stress Index; TOS—Total Oxidant Status; TAS—Total Antioxidant Status; 5TSST—time achieved in 5 Times Sit-to-Stand Test.

The concentrations of irisin and BDNF, as well as their ratio, were not associated with age or the number of chronic diseases. TAS showed a negative association with the number of chronic diseases.

3.3. Predictors of Dynapenia in Regression Analysis

A direct multivariable logistic regression analysis was conducted to investigate dynapenia as the outcome and seven predictors: age, sex, number of chronic diseases, OSI, TOS, BDNF concentration, and the irisin/BDNF ratio (Table 4). Significantly higher odds for dynapenia were observed only for female sex (odds ratio, 5.45; 95% CI, 1.75–16.96; p = 0.003), age (odds ratio, 1.13; 95% CI, 1.04–1.23; p = 0.005), and number of chronic diseases (odds ratio, 1.48; 95% CI, 1.01–2.16; p = 0.04). For the model, an overall prediction success rate of 81.8% was observed, with 93.2% of the dynapenia "+" status (sensitivity) and 59.5% of dynapenia "−" status (specificity) correctly predicted.

Table 4. Determinants of dynapenia "+" status—direct multivariable logistic regression model.

	OR	95% CI	p
		MODEL 1	
Age (years)	1.13	1.04–1.23	0.005
Sex (female)	5.45	1.75–16.96	0.003
Number of chronic diseases	1.48	1.01–2.16	0.04
OSI	2.23	0.52–9.54	0.28
TOS	1.00	0.996–1.004	0.99
BDNF	0.81	0.61–1.06	0.13
Irisin/BDNF	0.999	0.997–1.001	0.52

BDNF—Brain Derived Neurotrophic Factor; CI—confidence interval; OSI—Oxidative Stress Index; OR—Odds ratio; TOS—Total Oxidative Status.

A stepwise backward logistic regression analysis was performed on dynapenia, using seven predictors to construct Model 1 (Table 5). The results of the study suggested the model with four variables only: age (odds ratio, 1.13; 95% CI, 1.04–1.22; p = 0.003), female sex (odds ratio, 5.67; 95% CI, 1.86–317.27; p = 0.002), number of chronic diseases (odds ratio, 1.44; 95% CI, 1.01–2.06; p = 0.04), and OSI (odds ratio, 2.20; 95% CI, 1.23–3.94; p = 0.008). Prediction success was slightly worse in the case of Model 2, with 90.4% of dynapenia "+" cases and 79.1% of dynapenia "−" cases correctly predicted, resulting in an overall success rate of 79.1%.

Table 5. Determinants of dynapenia "+" status—stepwise backward multivariable logistic regression analysis.

	OR	95% CI	p
		MODEL 2	
Age (years)	1.13	1.04–1.22	0.003
Sex (female)	5.67	1.86–17.27	0.002
Number of chronic diseases	1.44	1.01–2.06	0.04
OSI	2.20	1.23–3.94	0.008

Variables included in the analysis: age, sex, number of chronic diseases, OSI, BDNF, irisin/BDNF ratio, and TOS. CI—confidence interval; OSI—Oxidative Stress Index; OR—Odds ratio.

4. Discussion

The study has confirmed a high prevalence (66.9%) of dynapenia, as assessed with the 5TSST, among geriatric patients. It was positively associated with older age, female sex, and the number of chronic diseases. Additionally, the time required to perform the test increased with age and the number of diseases diagnosed. This finding is consistent with previous research results, which confirm that muscle strength declines with age and is also associated with several diseases [14,15]. All expert groups involved in the development of existing algorithms for sarcopenia diagnosis emphasize the crucial role of muscle strength in assessment. In older adults, it is a predictor of adverse health-related outcomes (such as mobility limitation, falls, activities of daily living disability, and mortality in community-dwelling older adults) [5].

In our study, the 5TSST results and the diagnosis of dynapenia were not associated with irisin concentration; however, they showed a positive relationship with the irisin-to-BDNF ratio and a negative relationship with BDNF concentration. According to the latest systematic review on the relationship between irisin and sarcopenia, cross-sectional studies have demonstrated lower serum irisin concentration in individuals with sarcopenia compared to healthy controls [16,17]. The majority of the available 12 human studies results suggested that irisin might serve as a potential diagnostic marker for sarcopenia in the older population and postmenopausal women. Moreover, animal and cellular experiments indicated that increased concentrations of irisin helped improve muscle mass [7]. However, the authors of this review noted that the majority of the studies analyzed had a high or unclear risk of selection bias, performance bias, and detection bias. They also pointed out that conflicting evidence exists, as some researchers failed to establish a definitive connection between irisin and sarcopenia [7,18]. Our study also did not confirm that the concentration of irisin itself differed significantly between patients with and without dynapenia, suggesting that irisin may not accurately predict sarcopenia in older, multimorbid individuals. In a recent study conducted among sedentary adults, a decrease in irisin concentrations was observed after an 8-week intensive lifestyle intervention program [19]. This finding was contrary to its supposed mechanisms of action and dynamics, suggesting that there may be yet undiscovered impacts on the secretion of irisin [19]. A recently published analyses of research on irisin in the context of its association with age-related problems, including sarcopenia, showed that the evidence is not consistent regarding the association between irisin concentration and health or disease in older adults and indicated the need for further research in this area, taking into account the influence of various factors, including not only age, lifestyle, diet, and ethnicity, but also comorbidities and concomitant medications [20].

BDNF is a neurotrophin produced primarily by the brain, involved in synaptic modulation and neurogenesis, influencing neuronal survival and brain angiogenesis [21–23]. It has also been shown to be a myokine, a molecule synthesized in skeletal muscle that controls biological processes occurring within it. BDNF is believed to play a role in cellular mechanisms that regulate muscle function, maintenance, and plasticity [8]. When released during exercise, BDNF directly contributes to the strengthening of neuromuscular junctions [24] as well as insulin-regulated glucose uptake and beta-oxidation processes in muscle tissue. It is primarily produced in contracting skeletal muscle and plays a vital role in metabolic regeneration and exercise-induced skeletal muscle remodeling [8,25]. Some studies have confirmed that its concentration in the blood significantly increases during physical activity [26,27]. In contrast, a meta-analysis of studies on the relationship between BDNF in the blood and physical activity in patients with diabetes showed that the results obtained by different researchers were inconsistent—some confirmed that BDNF level increases, others that it decreases, and in some studies, no significant effect of exercise on BDNF concentration was observed [28]. Perhaps the type of physical activity undertaken by the respondents may also have an impact; moderate-intensity exercises seem to be more effective in promoting an increase in the peripheral levels of BDNF in older adults [29]. Nevertheless, extraocular eye muscles are those that are less susceptible to age-related loss of function and mass. It has been confirmed that the satellite cells of these muscles express higher levels of neurotrophins, including BDNF, and their receptors compared to limb muscles [30]. Some researchers suggested that neurotrophins may be hidden behind their highly regenerative properties and play a role in protecting craniofacial muscles against inflammatory and wasting diseases [31].

Some studies suggest that irisin, secreted during physical exercise, can cross the blood–brain barrier and enhance BDNF expression [32–34]. However, in our study, irisin concentration showed the inverse relationship with BDNF, further highlighting the complexity of the interaction between these two molecules. Our study demonstrated that the time required to perform the 5TSST and the occurrence of dynapenia were inversely associated with BDNF concentration and positively associated with the irisin-to-BDNF ratio. These findings may support the protective role of this neurotrophin in the development of sarcopenia in geriatric, multimorbid patients, and highlight the importance of the irisin-to-BDNF ratio in the context of sarcopenia.

Oxidative stress and mitochondrial dysfunction have been considered as critical pathophysiologic mechanisms involved in the development of sarcopenia [10,35,36]. Skeletal muscle is particularly susceptible to oxidative stress-related damage and dysfunction because it is an organ characterized by a high metabolic rate [37,38]. TOS, TAS, and OSI are parameters of redox homeostasis. TOS evaluates the overall oxidative state of the body. At the same time, TAS measures the overall antioxidative status, while OSI reflects the imbalance between the oxidative-reductive processes, considered a most precise indicator of oxidative stress in the body [13,39]. The results of our study indicate that TOS and OSI were strongly associated with the occurrence of dynapenia in patients in the geriatric ward. A positive association was found between these redox homeostasis markers and both the longer time needed to perform the 5TSST and the frequency of dynapenia diagnosis based on this test. No association was found with TAS. It suggests that antioxidative mechanisms are not sufficiently effective, and the redox balance has shifted towards oxidation in the studied group of people with dynapenia. The regression analysis has confirmed that OSI was the only significant biochemical marker of sarcopenia in the studied group of geriatric patients, controlling for age, sex, and the number of chronic diseases. Several studies have confirmed that oxidative stress may contribute to sarcopenia by affecting the neuromuscular junction, leading to loss of innervation, a negative impact on muscle mitochondrial function, and promoting atrophy pathways, ultimately disrupting muscle contractile function [10].

This study has several limitations that warrant emphasis. First, as noted above, the term "dynapenia" used in the manuscript refers specifically to reduced lower extremity muscle strength, as assessed by the Five Times Sit-to-Stand Test (5TSTS). Some studies suggest that lower limb muscle weakness, in particular, is a strong predictor of adverse outcomes in older adults [40]. However, the evidence is not entirely consistent, and handgrip strength (HGS) is often cited as a more reliable indicator of overall muscle strength and a key diagnostic criterion for sarcopenia [41,42]. That said, HGS measurement also has its limitations, as highlighted by experts. These include patient-related factors common in older populations—such as pain due to osteoarthritis, fatigue, or low motivation (e.g., in patients with depression)—as well as measurement-related factors, including variability in testing protocols and differences in dynamometer design and calibration [43–45]. Nevertheless, at the time this study was conducted, the prevailing consensus on sarcopenia diagnosis recognized the Five Times Sit-to-Stand Test as a valid alternative for assessing muscle strength when HGS measurement was not feasible [2].

The 5TSST test is also a fitness test, and we cannot rule out the possibility that it may not accurately reflect general muscle strength, mass, and quality, providing a reasonable basis for research on biochemical markers of dynapenia/sarcopenia. Additionally, the first global consensus on the conceptual definition of sarcopenia, published in 2024, recommended that muscle mass, muscle strength, and muscle-specific force be included in the diagnostic algorithm, while physical performance is viewed as an outcome [1].

The results of this study can be generalized to populations with similar characteristics—namely, individuals of more advanced age and with a high burden of multimorbidity—but not to the broader population of older adults. It is possible that the specific characteristics of this group, being relatively older and with a higher prevalence of disease compared to cohorts previously studied for BDNF or irisin levels, also influenced the results obtained, including the finding that redox imbalance was a significant, independent predictor of dynapenia.

The results of our study suggest a potential protective role of BDNF; however, longitudinal studies and more specific measures of neuromuscular function are needed to confirm this directly. Additionally, although the irisin/BDNF ratio appears to be a promising indicator of sarcopenia, its interpretation remains uncertain—it is unclear whether it represents a reliable biomarker or merely reflects a state of physiological imbalance. As this was an observational study, we cannot definitively determine the direction of causality between OSI and reduced muscle strength. Moreover, several confounding factors that influence oxidative stress—such as diet, physical activity, inflammatory status, or medication use—were not taken into account. These limitations should be addressed in future research.

Furthermore, our study was static and lacked any intervention involving physical activity. In the available literature, these myokines have often been assessed using dynamic models, particularly in the context of the impact of physical exercise on their concentrations [9]. Additionally, several studies have confirmed that irisin levels are influenced not only by physical activity but also by metabolic health conditions such as obesity and insulin resistance. Its concentration in the body is associated with the amount of intramuscular fat tissue, which is actively utilized during exercise [46–50]. It is possible that different mechanisms are responsible for baseline levels and exercise-induced changes in these myokines.

5. Conclusions

The study results have confirmed a lower status of BDNF and a higher irisin/BDNF ratio in patients with dynapenia, suggesting that lower BDNF expression may occur in aging muscles despite adequate irisin concentration. Higher levels of oxidative stress parameters can also play a role in the development of sarcopenia.

A lower concentration of BDNF and a higher irisin/BDNF ratio in patients with dynapenia may indicate a protective role of BDNF in the development of sarcopenia in geriatric patients. Higher levels of oxidative stress parameters point to redox imbalance involvement in sarcopenia development. Older age, female gender, and the higher number of chronic diseases and OSI were the only independent predictors connected with significantly higher odds for dynapenia status, controlling for age, sex, and number of chronic diseases. The findings may suggest the need to enhance the body's antioxidant potential to reduce the severity and frequency of dynapenia in older patients.

Author Contributions: Conceptualization, Z.B.W., A.W. and J.R. (Justyna Rentflejsz); methodology, J.R. (Justyna Rentflejsz), M.M.B. and Z.B.W.; software, Z.B.W.; validation, Z.B.W.; formal analysis, A.W., J.Ś. and Z.B.W.; investigation, J.R. (Justyna Rentflejsz), J.R. (Joanna Rogalska), M.M.B. and Z.B.W.; resources, J.R. (Justyna Rentflejsz), M.M.B. and Z.B.W.; data curation, J.R. (Justyna Rentflejsz), J.R. (Joanna Rogalska) and Z.B.W.; writing—original draft preparation, A.W., J.Ś. and Z.B.W.; writing—review and editing, Z.B.W., A.W., J.Ś., J.R. (Justyna Rentflejsz), J.R. (Joanna Rogalska) and M.M.B.; visualization, A.W., J.Ś. and Z.B.W.; supervision, Z.B.W.; project administration, Z.B.W.; funding acquisition, Z.B.W. All authors have read and agreed to the published version of the manuscript.

Funding: This research was funded by the Medical University of Bialystok, grant number B.SUB.25.268.

Institutional Review Board Statement: The study was conducted in accordance with the Declaration of Helsinki and is part of a larger research project approved by the Bioethics Committee at the Medical University of Bialystok (protocol no.APK.002.513.2022).

Informed Consent Statement: Informed consent was obtained from all subjects involved in the study.

Data Availability Statement: The data that support the findings of this study are available from the corresponding author upon reasonable request. The raw data supporting the findings of this study are available from the corresponding authors upon reasonable request.

Acknowledgments: We are grateful for the help of Elżbieta Tryskuć, who was involved in laboratory work, and of the nurses working in the Department of Geriatrics of the Medical University of Bialystok, who were involved in blood collection.

Conflicts of Interest: The authors declare no conflicts of interest. The funders had no role in the design of the study; in the collection, analysis, or interpretation of data; in the writing of the manuscript; or in the decision to publish the results. The preliminary data from the study were presented as a poster at the Free Radicals in Biology, Medicine, Sport, and Nutrition Conference held in Gdansk (Poland) on 27–28 June 2025.

Abbreviations

The following abbreviations are used in this manuscript:

BDNF	Brain-derived neurotrophic factor
CI	Confidence interval
CV	Coefficient of variation
EWGSOP2	The Second European Working Group on Sarcopenia in Older People
ICD-10	The 10th International Classification of Diseases
IQR	Interquartile range
M	Mean
Me	Median
N	Number
OR	Odds ratio
OSI	Oxidative stress index
SARC-F	S(trength), A(ssistance with walking), R(ise from a chair), C(limbing stairs), and F(alls) questionnaire
SD	Standard deviation
TAS	Total antioxidative status
TOS	Total oxidative status
5TSST	5 Times Sit-to-Stand Test

References

1. Kirk, B.; Cawthon, P.M.; Arai, H.; Ávila-Funes, J.A.; Barazzoni, R.; Bhasin, S.; Binder, E.F.; Bruyere, O.; Cederholm, T.; Chen, L.-K.; et al. The Conceptual Definition of Sarcopenia: Delphi Consensus from the Global Leadership Initiative in Sarcopenia (GLIS). *Age Ageing* **2024**, *53*, afae052. [CrossRef]
2. Cruz-Jentoft, A.J.; Bahat, G.; Bauer, J.; Boirie, Y.; Bruyère, O.; Cederholm, T.; Cooper, C.; Landi, F.; Rolland, Y.; Sayer, A.A.; et al. Sarcopenia: Revised European consensus on definition and diagnosis. *Age Ageing* **2018**, *48*, 16–31. [CrossRef]
3. Tu, S.; Hao, X.; Xu, S.; Jin, X.; Liao, W.; Xia, H.; Wang, S.; Sun, G. Sarcopenia: Current Insights into Molecular Mechanisms, Diagnostics, and Emerging Interventional Approaches. *Int. J. Mol. Sci.* **2025**, *26*, 6740. [CrossRef]
4. Chen, L.-K.; Woo, J.; Assantachai, P.; Auyeung, T.-W.; Chou, M.-Y.; Iijima, K.; Jang, H.C.; Kang, L.; Kim, M.; Kim, S. Asian Working Group for Sarcopenia: 2019 consensus update on sarcopenia diagnosis and treatment. *J. Am. Med. Dir. Assoc.* **2020**, *21*, 300–307.e302. [CrossRef]
5. Bhasin, S.; Travison, T.G.; Manini, T.M.; Patel, S.; Pencina, K.M.; Fielding, R.A.; Magaziner, J.M.; Newman, A.B.; Kiel, D.P.; Cooper, C. Sarcopenia definition: The position statements of the sarcopenia definition and outcomes consortium. *J. Am. Geriatr. Soc.* **2020**, *68*, 1410–1418. [CrossRef] [PubMed]
6. Ladang, A.; Beaudart, C.; Reginster, J.-Y.; Al-Daghri, N.; Bruyère, O.; Burlet, N.; Cesari, M.; Cherubini, A.; da Silva, M.C.; Cooper, C.; et al. Biochemical Markers of Musculoskeletal Health and Aging to be Assessed in Clinical Trials of Drugs Aiming at the Treatment of Sarcopenia: Consensus Paper from an Expert Group Meeting Organized by the European Society for Clinical and Economic Aspects of Osteoporosis, Osteoarthritis and Musculoskeletal Diseases (ESCEO) and the Centre Académique de Recherche et d'Expérimentation en Santé (CARES SPRL), Under the Auspices of the World Health Organization Collaborating Center for the Epidemiology of Musculoskeletal Conditions and Aging. *Calcif. Tissue Int.* **2023**, *112*, 197–217. [CrossRef] [PubMed]

7. Chen, Y.; Qian, M.; Gao, F.; Li, G.; Peng, K.; Sun, Q.; Sun, Y.; Liu, G.; Ge, Y.; Yang, M.; et al. Potential effect of Irisin on sarcopenia: A systematic review. *BMC Musculoskelet. Disord.* **2025**, *26*, 520. [CrossRef]
8. Rentería, I.; García-Suárez, P.C.; Fry, A.C.; Moncada-Jiménez, J.; Machado-Parra, J.P.; Antunes, B.M.; Jiménez-Maldonado, A. The Molecular Effects of BDNF Synthesis on Skeletal Muscle: A Mini-Review. *Front. Physiol.* **2022**, *13*, 934714. [CrossRef] [PubMed]
9. Villamil-Parra, W.; Moscoso-Loaiza, L. Effects of physical exercise on Irisin and BDNF concentrations, and their relationship with cardiometabolic and mental health of individuals with Metabolic Syndrome: A Systematic Review. *Exp. Gerontol.* **2024**, *198*, 112640. [CrossRef]
10. Xu, H.; Brown, J.L.; Bhaskaran, S.; Van Remmen, H. Reactive oxygen species in the pathogenesis of sarcopenia. *Free Radic. Biol. Med.* **2025**, *227*, 446–458. [CrossRef]
11. Malmstrom, T.; Morley, J. SARC-F: A Simple Questionnaire to Rapidly Diagnose Sarcopenia. *J. Am. Med. Dir. Assoc.* **2013**, *14*, 531–532. [CrossRef]
12. Malmstrom, T.K.; Miller, D.K.; Simonsick, E.M.; Ferrucci, L.; Morley, J.E. SARC-F: A symptom score to predict persons with sarcopenia at risk for poor functional outcomes. *J. Cachexia Sarcopenia Muscle* **2016**, *7*, 28–36. [CrossRef] [PubMed]
13. Erel, O. A novel automated direct measurement method for total antioxidant capacity using a new generation, more stable ABTS radical cation. *Clin. Biochem.* **2004**, *37*, 277–285. [CrossRef] [PubMed]
14. Larsson, L.; Degens, H.; Li, M.; Salviati, L.; Lee, Y.i.; Thompson, W.; Kirkland, J.L.; Sandri, M. Sarcopenia: Aging-Related Loss of Muscle Mass and Function. *Physiol. Rev.* **2019**, *99*, 427–511. [CrossRef] [PubMed]
15. Sivritepe, R.; Siyer, O.K.; Tiril, S.M.; Basat, S.U. Do we know about dynapenia? *North. Clin. Istanb.* **2024**, *11*, 593–599. [CrossRef] [PubMed]
16. Park, H.S.; Kim, H.C.; Zhang, D.; Yeom, H.; Lim, S.K. The novel myokine irisin: Clinical implications and potential role as a biomarker for sarcopenia in postmenopausal women. *Endocrine* **2019**, *64*, 341–348. [CrossRef]
17. Chang, J.S.; Kim, T.H.; Nguyen, T.T.; Park, K.S.; Kim, N.; Kong, I.D. Circulating irisin levels as a predictive biomarker for sarcopenia: A cross-sectional community-based study. *Geriatr. Gerontol. Int.* **2017**, *17*, 2266–2273. [CrossRef]
18. Baek, J.Y.; Jang, I.Y.; Jung, H.W.; Park, S.J.; Lee, J.Y.; Choi, E.; Lee, Y.S.; Lee, E.; Kim, B.J. Serum irisin level is independent of sarcopenia and related muscle parameters in older adults. *Exp. Gerontol.* **2022**, *162*, 111744. [CrossRef]
19. Radikova, Z.; Mosna, L.; Eckerstorfer, C.; Bajer, B.; Havranova, A.; Imrich, R.; Vlcek, M.; Penesova, A. Plasma irisin and the brain-derived neurotrophic factor levels in sedentary subjects: Effect of 8-weeks lifestyle intervention. *Endocr. Regul.* **2024**, *58*, 115–128. [CrossRef]
20. Plácido, A.I.; Azevedo, D.; Herdeiro, M.T.; Morgado, M.; Roque, F. Understanding the Role of Irisin in Longevity and Aging: A Narrative Review. *Epidemiologia* **2025**, *6*, 1. [CrossRef]
21. Woźniak, W. Brain-derived neurotrophic factor (BDNF): Role in neuronal development and survival. *Folia Morphol.* **1993**, *52*, 173–181. [PubMed]
22. Miranda, M.; Morici, J.F.; Zanoni, M.B.; Bekinschtein, P. Brain-Derived Neurotrophic Factor: A Key Molecule for Memory in the Healthy and the Pathological Brain. *Front. Cell. Neurosci.* **2019**, *13*, 363. [CrossRef]
23. Numakawa, T.; Kajihara, R. The Role of Brain-Derived Neurotrophic Factor as an Essential Mediator in Neuronal Functions and the Therapeutic Potential of Its Mimetics for Neuroprotection in Neurologic and Psychiatric Disorders. *Molecules* **2025**, *30*, 848. [CrossRef]
24. Zhang, J.; Kwan, H.-L.R.; Chan, C.B.; Lee, C.W. Localized release of muscle-generated BDNF regulates the initial formation of postsynaptic apparatus at neuromuscular synapses. *Cell Death Differ.* **2025**, *32*, 546–560. [CrossRef]
25. Chan, W.S.; Ng, C.F.; Pang, B.P.S.; Hang, M.; Tse, M.C.L.; Iu, E.C.Y.; Ooi, X.C.; Yang, X.; Kim, J.K.; Lee, C.W.; et al. Exercise-induced BDNF promotes PPARδ-dependent reprogramming of lipid metabolism in skeletal muscle during exercise recovery. *Sci. Signal.* **2024**, *17*, eadh2783. [CrossRef] [PubMed]
26. Shobeiri, P.; Karimi, A.; Momtazmanesh, S.; Teixeira, A.L.; Teunissen, C.E.; van Wegen, E.E.H.; Hirsch, M.A.; Yekaninejad, M.S.; Rezaei, N. Exercise-induced increase in blood-based brain-derived neurotrophic factor (BDNF) in people with multiple sclerosis: A systematic review and meta-analysis of exercise intervention trials. *PLoS ONE* **2022**, *17*, e0264557. [CrossRef] [PubMed]
27. Garcia-Suarez, P.C.; Renteria, I.; Plaisance, E.P.; Moncada-Jimenez, J.; Jimenez-Maldonado, A. The effects of interval training on peripheral brain derived neurotrophic factor (BDNF) in young adults: A systematic review and meta-analysis. *Sci. Rep.* **2021**, *11*, 8937. [CrossRef]
28. Jamali, A.; Shahrbanian, S.; Morteza Tayebi, S. The Effects of Exercise Training on the Brain-Derived Neurotrophic Factor (BDNF) in the Patients with Type 2 Diabetes: A Systematic Review of the Randomized Controlled Trials. *J. Diabetes Metab. Disord.* **2020**, *19*, 633–643. [CrossRef]
29. Coelho, F.G.; Gobbi, S.; Andreatto, C.A.; Corazza, D.I.; Pedroso, R.V.; Santos-Galduroz, R.F. Physical exercise modulates peripheral levels of brain-derived neurotrophic factor (BDNF): A systematic review of experimental studies in the elderly. *Arch. Gerontol. Geriatr.* **2013**, *56*, 10–15. [CrossRef]

30. Carrero-Rojas, G.; Benítez-Temiño, B.; Pastor, A.M.; Davis López de Carrizosa, M.A. Muscle Progenitors Derived from Extraocular Muscles Express Higher Levels of Neurotrophins and their Receptors than other Cranial and Limb Muscles. *Cells* **2020**, *9*, 747. [CrossRef] [PubMed]
31. Gustafsson, T.; Ulfhake, B. Aging Skeletal Muscles: What Are the Mechanisms of Age-Related Loss of Strength and Muscle Mass, and Can We Impede Its Development and Progression? *Int. J. Mol. Sci.* **2024**, *25*, 10932. [CrossRef]
32. Wrann, C.D.; White, J.P.; Salogiannnis, J.; Laznik-Bogoslavski, D.; Wu, J.; Ma, D.; Lin, J.D.; Greenberg, M.E.; Spiegelman, B.M. Exercise induces hippocampal BDNF through a PGC-1α/FNDC5 pathway. *Cell Metab.* **2013**, *18*, 649–659. [CrossRef] [PubMed]
33. Pignataro, P.; Dicarlo, M.; Zerlotin, R.; Zecca, C.; Dell'Abate, M.T.; Buccoliero, C.; Logroscino, G.; Colucci, S.; Grano, M. FNDC5/Irisin System in Neuroinflammation and Neurodegenerative Diseases: Update and Novel Perspective. *Int. J. Mol. Sci.* **2021**, *22*, 1605. [CrossRef]
34. Waseem, R.; Shamsi, A.; Mohammad, T.; Hassan, M.I.; Kazim, S.N.; Chaudhary, A.A.; Rudayni, H.A.; Al-Zharani, M.; Ahmad, F.; Islam, A. FNDC5/Irisin: Physiology and Pathophysiology. *Molecules* **2022**, *27*, 1118. [CrossRef]
35. Kubat, G.B.; Bouhamida, E.; Ulger, O.; Turkel, I.; Pedriali, G.; Ramaccini, D.; Ekinci, O.; Ozerklig, B.; Atalay, O.; Patergnani, S.; et al. Mitochondrial dysfunction and skeletal muscle atrophy: Causes, mechanisms, and treatment strategies. *Mitochondrion* **2023**, *72*, 33–58. [CrossRef]
36. Hyatt, H.W.; Powers, S.K. Mitochondrial dysfunction is a common denominator linking skeletal muscle wasting due to disease, aging, and prolonged inactivity. *Antioxidants* **2021**, *10*, 588. [CrossRef]
37. Pascual-Fernández, J.; Fernández-Montero, A.; Córdova-Martínez, A.; Pastor, D.; Martínez-Rodríguez, A.; Roche, E. Sarcopenia: Molecular Pathways and Potential Targets for Intervention. *Int. J. Mol. Sci.* **2020**, *21*, 8844. [CrossRef] [PubMed]
38. Musaro, A.; Fulle, S.; Fano, G. Oxidative stress and muscle homeostasis. *Curr. Opin. Clin. Nutr. Metab. Care* **2010**, *13*, 236–242. [CrossRef]
39. Harma, M.; Erel, O. Increased oxidative stress in patients with hydatidiform mole. *Swiss Med. Wkly.* **2003**, *133*, 563–566. [CrossRef]
40. Kawamura, K.; Iwase, T.; Ishino, S.; Nakao, Y.; Kagaya, H.; Akatsu, H.; Arai, H. Low five-repetition chair stand test and usual gait speed scores predict falls within one year in an outpatient clinic for frailty. *Eur. Geriatr. Med.* **2025**, *16*, 1295–1302. [CrossRef] [PubMed]
41. Verstraeten, L.M.G.; de Haan, N.J.; Verbeet, E.; van Wijngaarden, J.P.; Meskers, C.G.M.; Maier, A.B. Handgrip strength rather than chair stand test should be used to diagnose sarcopenia in geriatric rehabilitation inpatients: REStORing health of acutely unwell adulTs (RESORT). *Age Ageing* **2022**, *51*, afac242. [CrossRef]
42. Li, Y.H.; Wang, X.H.; Ya, S. Prevalence and Diagnostic Agreement of Sarcopenia Based on Handgrip Strength and 5-Time Chair-Stand Test Among Chinese Community-Dwelling Older Adults. *Int. J. Older People Nurs.* **2024**, *19*, e12635. [CrossRef] [PubMed]
43. Núñez-Cortés, R.; Cruz, B.d.P.; Gallardo-Gómez, D.; Calatayud, J.; Cruz-Montecinos, C.; López-Gil, J.F.; López-Bueno, R. Handgrip strength measurement protocols for all-cause and cause-specific mortality outcomes in more than 3 million participants: A systematic review and meta-regression analysis. *Clin. Nutr.* **2022**, *41*, 2473–2489. [CrossRef]
44. Gränicher, P.; Maurer, Y.; Spörri, J.; Haller, B.; Swanenburg, J.; de Bie, R.A.; Lenssen, T.A.F.; Scherr, J. Accuracy and Reliability of Grip Strength Measurements: A Comparative Device Analysis. *J. Funct. Morphol. Kinesiol.* **2024**, *9*, 274. [CrossRef]
45. Quattrocchi, A.; Garufi, G.; Gugliandolo, G.; De Marchis, C.; Collufio, D.; Cardali, S.M.; Donato, N. Handgrip Strength in Health Applications: A Review of the Measurement Methodologies and Influencing Factors. *Sensors* **2024**, *24*, 5100. [CrossRef] [PubMed]
46. Mohammad Rahimi, G.R.; Hejazi, K.; Hofmeister, M. The effect of exercise interventions on Irisin level: A systematic review and meta-analysis of randomized controlled trials. *EXCLI J.* **2022**, *21*, 524–539. [CrossRef] [PubMed]
47. Shi, X.; Lin, M.; Liu, C.; Xiao, F.; Liu, Y.; Huang, P.; Zeng, X.; Yan, B.; Liu, S.; Li, X.; et al. Elevated circulating irisin is associated with lower risk of insulin resistance: Association and path analyses of obese Chinese adults. *BMC Endocr. Disord.* **2016**, *16*, 44. [CrossRef]
48. Yuksel Ozgor, B.; Demiral, I.; Zeybek, U.; Celik, F.; Buyru, F.; Yeh, J.; Bastu, E. Effects of Irisin Compared with Exercise on Specific Metabolic and Obesity Parameters in Female Mice with Obesity. *Metab. Syndr. Relat. Disord.* **2020**, *18*, 141–145. [CrossRef]
49. Dianatinasab, A.; Koroni, R.; Bahramian, M.; Bagheri-Hosseinabadi, Z.; Vaismoradi, M.; Fararouei, M.; Amanat, S. The effects of aerobic, resistance, and combined exercises on the plasma irisin levels, HOMA-IR, and lipid profiles in women with metabolic syndrome: A randomized controlled trial. *J. Exerc. Sci. Fit.* **2020**, *18*, 168–176. [CrossRef]
50. Li, J.; Yi, X.; Li, T.; Yao, T.; Li, D.; Hu, G.; Ma, Y.; Chang, B.; Cao, S. Effects of exercise and dietary intervention on muscle, adipose tissue, and blood IRISIN levels in obese male mice and their relationship with the beigeization of white adipose tissue. *Endocr. Connect.* **2022**, *11*, e210625. [CrossRef]

MDPI AG
Grosspeteranlage 5
4052 Basel
Switzerland
Tel.: +41 61 683 77 34

Antioxidants Editorial Office
E-mail: antioxidants@mdpi.com
www.mdpi.com/journal/antioxidants

www.ingramcontent.com/pod-product-compliance
Lightning Source LLC
Chambersburg PA
CBHW061603210926
51850CB00009B/2200
* 9 7 8 3 7 2 5 8 7 2 0 0 8 *